Handbook of
Condition Monitoring

Published by Chapman & Hall, an imprint of Thomson Science,
2–6 Boundary Row, London SE1 8HN, UK

Thomson Science, 2–6 Boundary Row, London SE1 8HN, UK

Thomson Science, 115 Fifth Avenue, New York, NY 10003, USA

Thomson Science, Suite 750, 400 Market Street, Philadelphia, PA 19106, USA

Thomson Science, Pappelallee 3, 69469 Weinheim, Germany

First edition 1998

© 1998 Chapman & Hall
Thomson Science is a division of International Thomson Publishing I(T)P®

Typeset in 10/12pt Times by Doyle Graphics, Tullamore, Ireland
Printed in Great Britain by T.J. International Ltd, Padstow, Cornwall

ISBN   0 412 61320 4

A catalogue record for this book is available from the British Library

Handbook of Condition Monitoring

Techniques and Methodology

by

A. Davies (ed.)

*Systems Division, School of Engineering,
University of Wales–Cardiff (UWC), P.O Box 688,
Queens Buildings, Newport Road,
Cardiff, CF2 3TE, UK.*

CHAPMAN & HALL

London · Weinheim · New York · Tokyo · Melbourne · Madras

Dedication
For my wife – Sylvia

Contents

Contents

Contributors

Armitt, T. (Chapter 5) Lavender International NDT Ltd, Unit 7, Penistone Station, Sheffield, S30 6HJ, UK

Cempel, Dr C. (Chapter 13) Head of Applied Mechanics Institute, Technical University of Poznan, 3 Piotrowo Street, PL-60-965, Poznan, Warsaw, Poland; tel: 48 061 782238; fax: 48 061 782307

Davies, A. (editor) (Preface, Chapters 3, 6, 8, 10, 16, 18, 23) ENGIN, UWC, PO Box 688, Newport Road, Cardiff CF2 3TE; tel: 01222 874000 x5912; fax: 01222 874003

Dibley, D.A.G. (Chapter 4) Agema Infrared, Arden House, West Street, Leighton Buzzard, Bedfordshire LU7 7DD; tel: 01525 375660; fax: 01525 379271

Drake, Dr P.R. (Chapter 20) ENGIN, UWC, PO Box 688, Newport Road, Cardiff CF2 3TE; tel: 01222 874000 x5919; fax: 01222 874003

Eade, G. (Chapter 19) Managing Director, Asset Management Centre Ltd, Ludworth Trout Farm, Marple Bridge, Cheshire SK6 5NS; tel/fax: 0161 449 9520

Gardiner, D. (Chapter 11) Entek IRD International, Bumpers Lane, Sealand Industrial Estate, Chester CH1 4LT; tel: 01244 374914; fax: 01244 379870

Hands, G. (Chapter 5) Godfrey Hands Ltd, Unit 14, Hammond Business Centre, Hammond Close, Attleborough, Nuneaton, CV11 6RY, UK; tel: 01203 320812; fax: 01203 320813

Heron, R.A. (Chapter 7) Engineering Consultancy, 30 Church Lane, Stagsden, Bedfordshire MK43 8SH; tel: 01234 823124; fax: 01234 826035

Jones, R. (Chapter 21) MCP Manufacturing & Maintenance Consultants, Business & Innovation Centre, Aston Science Park, Love Lane, Birmingham B7 4BJ; tel: 0121 693 9313; fax: 0121 693 9315

Martin, Dr K.F. (Chapter 10) ENGIN, UWC, PO Box 688, Newport Road, Cardiff, CF2 3TE; tel: 01222 874000 x4429; fax: 01222 874003

Mathew, Dr J. (Chapter 12) Centre for Machine Condition Monitoring, Monash University, Department of Mechanical Engineering, Wellington Road, Clayton, Victoria 3168, Australia; tel: 03 9905 5699; fax: 03 9905 5726

Milne, Dr R. (Chapter 22) Managing Director, Intelligent Applications Ltd, 1 Michaelson Square, Livingston, West Lothian, Scotland EH54 7DP; tel: 01506 472047; fax: 01506 472282

Mobley, R.K. (Chapter 2) President, Integrated Systems Inc., 215 South Rutgers Avenue, Oak Ridge, Tennessee 37830, USA; tel: (615) 482 1999; fax: (615) 481 0921

Price, Dr A.L. (Chapter 15) Tribology Centre, University of Wales, College of Swansea, Singlton Park, Swansea, SA2 8PP; tel: 01792 295222; fax: 01792 295674

Rao, Professor Raj B.K.N. (Chapter 1) Director COMADEM International, 307 Tiverton Road, Selly Oak, Birmingham B29 6DA; tel/fax: 0121 472 2338

Riley, N.H. (Chapter 17) Fuchs Lubricants (UK) Plc, Hanley Plant, New Century Street, Hanley, Stoke-on-Trent, ST1 5HU; tel: 01782 202521; fax: 01782 202072

Roylance, Dr B.J. (Chapter 15) Tribology Centre, University of Wales, College of Swansea, Singlton Park, Swansea SA2 8PP; tel: 01792 295222; fax: 01792 295674

Shrieve, P.F. (Chapter 14) ATL Consulting Services Ltd, 36–38 The Avenue, Southampton SO17 1XN; tel: 01703 325000; fax: 01703 335251

Thorpe, P. (Chapter 10) ENGIN, UWC, PO Box 688, Newport Road, Cardiff, CF2 3TE; tel: 01222 874000 x5912; fax: 01222 874003

Williams, Dr J.H. (Chapter 6, 9) ENGIN, UWC, PO Box 688, Newport Road, Cardiff CF2 3TE; tel: 01222 874000 x4429; fax: 01222 874003

Preface

Although the technology and application of condition monitoring is continually evolving, its conceptual basis can be traced back to the earliest development of machinery, and the use of human senses to monitor the state of industrial equipment. This approach of looking, listening, and using our senses of touch, smell and taste is still valid, although nowadays heavily augmented by scientific and sophisticated instrumentation. The use of such instrumentation allows us to quantify the health or condition of industrial equipment, so that problems can be diagnosed early in their development, and corrected by suitable maintenance, before they become serious enough to cause plant breakdown. Condition monitoring therefore involves the design and use of sensing arrangements on industrial plant, together with data acquisition and analysis systems, plus predictive and diagnostic methods, with the objective of implementing equipment maintenance in a planned way using actual condition knowledge.

Accordingly, the methods and techniques involved in condition monitoring have over the years, become widely adopted in many industries. These include power generation, petrochemicals, manufacturing, coal mining and steelmaking. In general, there is still a great deal of scope for further research, development, and application in respect of condition monitoring methods and techniques, especially where the benefits are known to be quite substantial. In today's competitive climate, for example, the economics of production have become a critical factor, and therefore achieving cost-effective plant maintenance is highly important. It is in this context that condition monitoring is now playing a vital role, by sustaining the reliable operation of industrial plant and machinery in the pursuit of economic whole life operation, and where high availability is one of the main criteria for satisfactory performance.

Thus in future, condition monitoring is likely to advance both in terms of the range of techniques available and in their application. It should be noted, therefore, that industry has already entered the machine management age, where the monitoring of industrial systems can be remotely performed, via the use of intelligent knowledge-based software to aid fault prognosis, and so avoid costly downtime. However, successful maintenance is based upon good management, the correct organizational structure and upon the employment of the right methods or techniques in any given situation. Accordingly, the *Handbook of Condition Monitoring* has been compiled to focus attention on the methods and techniques used in this area of maintenance activity, with the objective of increasing management awareness as to what is available, and thereby improving industrial performance and profitability. It is also hoped, that the book will enable both the undergraduate and postgraduate student to appreciate the wide diversity of techniques available in condition monitoring, and to stimulate research interest in this area further.

A. Davies,
Cardiff,
July 1996

Acknowledgement

Grateful acknowledgement is made to Professor D.V. Morgan, Head of the School of Engineering (ENGIN), UWC, Cardiff, UK, for his encouragement and permission to use School facilities in the preparation of this handbook. Acknowledgement is also made to all the individual authors and companies who generously contributed important material for the chapters contained in the handbook, related to their areas of expertise. In addition, special thanks are made to the editor's colleagues in the machine monitoring group at Cardiff for their valuable support in the preparation of this text, and also to Mr D. Peterson and Mrs J. Jones for their skill in the production of the illustrations contained in the handbook.

PART ONE

Introduction to Condition Monitoring

CHAPTER 1

Condition monitoring and the integrity of industrial systems

B.K.N. Rao

Director COMADEM International

307 Tiverton Road, Selly Oak, Birmingham, B29 6A

1.1 INTRODUCTION

Although to engineer and manufacture is human, in today's marketplace all manufactured and processed goods are subjected to severe international competition. Accordingly, consumers' perceptions towards order-winning criteria such as total quality, reliability, health and safety, environmental issues, energy conservation and the cost of ownership are changing day by day. To match the dynamics of the marketplace, and to provide the goods required, industrial machinery and systems are therefore also changing. For example, the speed of rotating machinery, together with the capabilities of their control and processing systems, are now improving at an ever-increasing rate. Many of these machines are currently being designed to operate almost supercritically, and this trend is likely to continue under the competitive demands of the marketplace.

Thus, the monitoring and control of excessive vibration, noise, thermal variation, dust and dangerous emissions from plant and machinery of various types are now required by law in many countries. In addition, a large number of countries are also applying strict liability legislation in respect of defective products. This has been done under various Acts and Directives, which have been formulated to protect a nation's society from the unscrupulous application of technology. Accordingly, anyone who suffers harm as a result of a defective product or process can now recover compensation from those liable for the injury. This is a significant change and makes manufacturers, importers and suppliers of any product directly responsible for their actions.

Handbook of Condition Monitoring
Edited by A. Davies
Published in 1998 by Chapman & Hall, London. ISBN 0 412 61320 4.

Today's engineer, therefore, shares with management the legal and moral responsibility to minimize risks of all kinds associated with industrial processing. All accidents, near misses and dangerous occurrences are now regarded as symptoms of a failure of management. Also, and in addition to the global economic and environmental pressures that companies are now facing, the energy debate is being transformed into an unprecedented mission for industrial management, to improve efficiency and minimize wastage in production operations. One further aspect, which should also be mentioned in relation to the current industrial climate, is the long-awaited change now taking place in public attitudes towards industry. This is the growing realization by government and society that manufacturing is not a marginal add-on to a nation's economy but that it really does matter.

In this context, inadequate company attention to maintenance management, in respect of many industrial processes, manufacturing and control activities, has been the death knell of several firms. Poor and dangerous maintenance practices, in both the process and batch manufacturing industries, have cost the UK alone millions of pounds each year! As prevention is better than cure, and remembering the stimulus of legislation, the use of Condition Monitoring to predict machinery failure does make a lot of sense. Accordingly, the use of Condition Monitoring and Diagnostic Engineering Management (COMADEM) will be advocated in the sections which follow, in an attempt to convince practising industrial managers of the veracity of this view.

1.2   CONDITION MONITORING AND DIAGNOSTIC ENGINEERING MANAGEMENT (COMADEM)

Until a few decades ago, we were living in a conformist society where the pace of technological progress was relatively slow and the marketplace was closely organized and protected, often by cartels. The general attitudes of those who were managing the economy and industry were very rigid, formal and restrictive. Monitoring instruments were very basic and bulky, while signal processing and control techniques were at best very rudimentary. Education, training and opportunities for technician staff were also extremely limited, and the field of human factors unheard of! Yet it was not uncommon to observe many technicians employing their highly trained senses to monitor effectively the performance of the plant and machinery in their charge. Top management, meanwhile, viewed the maintenance function as a vast overhead on their operation, and the maintenance budget was regarded by most company accountants as simply another overhead on the operating cost.

In this environment a maintenance department was seen to have fulfilled its task if it remained within its budget, yet its impact on the

Table 1.1 Major causes of 100 large accidents

Causes	Frequency
Mechanical failure	38
Operational error	26
Unknown/miscellaneous	12
Process upset	10
Natural hazards	7
Design error	4
Arson/sabotage	3

cost of production was several orders of magnitude greater than its direct cost. For example, poor and dangerous maintenance practices were followed by the petrochemical industry during this period – a fact which was revealed in a survey, that reviewed 100 worldwide petrochemical plant accidents. These accidents took place between 1958 and 1987, and resulted in extensive property damage (Garrison, 1988). The survey report contains clear evidence which indicates that poor maintenance practice and catastrophic accidents are not restricted by company or national boundaries. Table 1.1 indicates the major causes of these accidents, while Figure 1.1 outlines the considerable increase in the frequency of major incidents in the chemical industry worldwide (Carson and Mumford, 1979). Other evidence is also available to indicate poor maintenance practice, such as that in Table 1.2, which lists all the major industrial accidents recorded from 1975 to 1977 (Goverts-Lepicard, 1990).

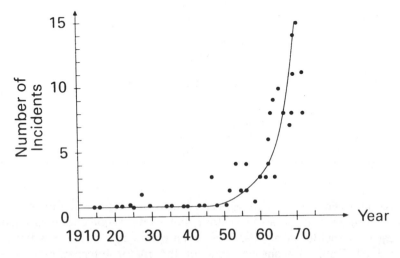

Fig. 1.1 Frequency of major accidents in the chemical industry worldwide.

Table 1.2   Major industrial accidents (1975–7)

Year	Location	Description	Killed/ injured
1975	Belgium	Ethylene from polyethylene plant explosion	6/15
	Beek, Netherlands	Propylene explosion	14/104
	Germany (FDR)	Naphtha plus hydrogen exploded	0/4
	California	Hydrogen explosion	0/2
	Louisiana	Butadiene escaped without ignition	0
	Louisiana	270 tons propane escaped but failed to ignite	0
	Czechoslovakia	Explosion of light hydrocarbons	14/
	Netherlands	Ethylene explosion	4/35
	France	Large confined vapour explosion	1/35
	South Africa	Methane explosion	7/7
	Philadelphia	Crude oil explosion	8/2
	United Kingdom	Electrolytor plant explosion	1/3
1976	Texas	Ethylene explosion at alcohol plant	1/15
	Texas	Natural gas leakage ignited	1/4
	Puerto Rico	C_5 hydrocarbons ignited	1/2
	New Jersey	Propylene explosion	2/?
	Lake Charles	Isobutane explosion	7/?
	Baton Rouge	Chlorine release: 10 000 evacuated	0
	Norway	Flammable liquid escaping from ruptured pipe explosion	6/?
	Seveso, Italy	Escape of TCDD resulting in evacuation of entire area	0
1977	United Kingdom	Fire and explosion involving sodium chloride plant	?
	Mexico	Ammonia escaped and leaked into sewer system	2/102
	Quatar	LPG explosion damaging villages distant from source and closing airport	7/many
	Mexico	Vinylchloride release	0/90
	Cassino, Italy	Propane/butane explosion	1/9
	Jacksonville	LPG incident resulting in evacuation of 2000	?
	Gela, Italy	Ethylene oxide explosion	1/2
	India	Hydrogen explosion	0/20
	Italy	Ethylene explosion	3/22
	Columbia	Ammonia escape	30/22

Aside from the human aspect of any accident, the impact on the environment may be so important that some major chemical incidents must be considered as disasters, even when no human casualties are involved. Table 1.3 shows some of the major international disasters which occurred from 1980 to 1989 with their attendant effects. Thus, in

Table 1.3 Some major international disasters

Year	Location	Causes	Effects
1980	North Sea (UK)	Oil rig capsize	123 killed
1984	Bhopal (India)	Toxic release	2700 killed and about 10 times as many injured
1986	Chernobyl (USSR)	Nuclear reactor fire	31 killed and 135 000 residents were evacuated
1988	North Sea (UK)	Piper Alpha explosion	167 killed
1984	Ixhuatepec (Mexico)	LPG explosion	500 killed
1989	USSR	Gas pipe explosion	500 killed

order that safety may be improved, nothing is more important to the process industry than up-to-date maintenance information. The information requirement is also true when we are trying to improve process performance, productivity, overall profitability and product time to market. Hence, there is no conflict between these objectives, and the results achieved in any sphere can only be as good as our information on the process concerned. Information is to industry what air and water are to us as human beings. That is why information technology has taken such a central role in making industry more competitive.

However, for various reasons, studies have shown that many of the process control industries are not successful in taking full advantage of this new technology. Figure 1.2, for example, shows the current trend in the speed and power of rotating machinery. The increase in speed to what is now available has resulted in the design of machines which are more and more susceptible to vibratory phenomena. At present, there is insufficient knowledge to understand and predict the behaviour of such machinery in real-life situations. Thus, the monitoring and control of vibration and noise in high-speed machinery, hydraulic systems and fans etc. is receiving close and increased attention. With the introduction of the provisions in the Health & Safety at Work Act in the UK, and the growing customer demand for quiet systems, this trend is set to continue worldwide. Such high-speed machines do impose a limit on the operator's capabilities, and accordingly, a lack of education or training to cope with this environment may lead to unpredictive systems failure and an expensive shutdown.

In addition, environmental pressure groups are now claiming that the manufacturing sector across the world is depleting the earth's limited resources of raw materials, and churning out waste products, which further undermine the ecological balance of the world around us. Process industries, on the whole, are also becoming increasingly aware

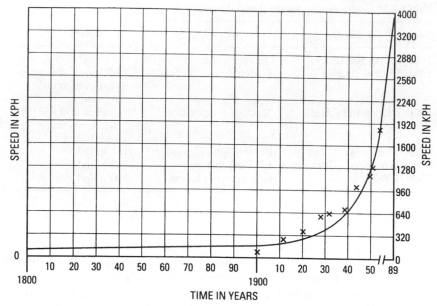

Fig. 1.2 Speed of travel from the year 1800 to 1989.

of the risks now facing the environment as a result of their activities. They are, accordingly, seriously attempting to address the issues which have arisen, with many responsible companies facing up to the fact that care for the environment is a corporate concern and that the challenge is an ongoing one. On the issue of energy conservation, for example, the current world population is around 5.3 billion, and it is predicted that in 25 years' time it will be over 8 billion. According to the World Energy Council's report this population will consume energy to the tune of 88% more than in 1990. No wonder our existing limited resources are becoming stretched, and other environmental issues may well arise.

Thus, the global economic and environmental pressures outlined above are pushing industry to improve their energy efficiency and to minimize waste. Process industries in particular are major consumers of energy. Hence, the adoption of an effective and efficient energy plan will not only reduce the national energy bill, but improve company competitiveness and profitability, and make a substantial contribution to the global environment. In today's highly competitive marketplace, survival is a major factor underlying all business activities. Only through profitable projects and optimization schemes can this be achieved. Thus, a key factor in any survival strategy must be the effective, cost-efficient control of maintenance activity, using the most appropriate condition monitoring and diagnostic techniques that are currently available. It is therefore important to note that the maintenance challenge is central to the issues and opportunities that face any industrial organization.

Unfortunately, in many companies, maintenance is still viewed as essentially a technical activity, rather than a legitimate management discipline. This prejudice is costing industry dearly. For research carried out in 1988 by the Department of Trade and Industry in the UK (DTI, 1989) showed that Britain's factories could save up to £1.3 billion per year if 'best practice' maintenance organization and techniques were used. Several factors tend to reinforce this prejudice. The first is that the maintenance department is highly visible only when it is administering emergency first aid to some malfunctioning machine. Secondly, maintenance managers are themselves often seduced by their image as 'fire-fighters' and technical experts. This is the 'you break it and we mend it syndrome'. Finally, maintenance departments are often accused of being a law unto themselves, with unpredictable response times and erratic priorities. Although understandable, none of the above factors contributes anything to the purpose of maintenance which is to provide:

The management, control, execution and achievement of quality in those activities which ensure, optimum levels of availability and overall performance of plant, in order to meet business objectives.

Thus it may be seen that the maintenance function has a major impact on the efficiency of the business. That impact increases as our manufacturing systems become more sophisticated and their elements become more interdependent. This requires an interdisciplinary approach based on monitoring, prediction, prevention and control. It should be noted that, when plant or machinery fails or functions inefficiently, the true costs are not just the cost of the repairs. Lurking beneath the surface are hidden costs due to loss of production, machine outage, damaged or poor-quality product, poor service to customers, lost sales opportunities, damage to other plant and buildings, hire of replacement equipment and increased energy usage. These costs eat into profit, and if someone is injured, who pays for the harm to corporate image, loss of customer goodwill or employee alienation?

The DTI's study which revealed the potential saving of £1.3 billion per year was based on a modest 5% increase in machine availability. This figure could also secure a 30% profitability improvement for some companies. Hence, the lesson must be that it is no good saying that failures and accidents will happen. They need not and should not happen, for accidents, breakdowns and stoppages can cost a great deal, both in compensation and in lost production. According to British Petroleum (BP) and the Atomic Energy Authority (AEA) Technology, the visible cost of injuries represents only the tip of a large but less obvious 'iceberg' of financial loss which may include many of the factors mentioned above. A review of recent disasters showed that outright technical failure is rarely the cause. They usually result from a combination of events, and the prime contributory features are a lack of communication, failure of supervision and shortcomings within the organization for not imposing the correct corporate culture, plus a structured approach to risk management.

Not preparing to face a crisis is a way of inviting a crisis to occur. Accordingly, steps should be taken to deal effectively with a crisis via the appropriate management techniques. Determining early on that a crisis has started is an obvious point, while formulating a strategy for an effective response, and developing techniques for crisis management via training, is frequently overlooked. It is a proverbial remark that, our attitudes are such that we never begin to make a reform or improvement in particular conditions until disaster takes its toll. This 'sacrificial-lamb' approach to industrial process technology is to be regretted, and frequently is, in many firms. Very few people, for instance, recognize or really understand our total dependence on the wealth creating jobs of those who work in industry. Accordingly, we should be aware of the necessity and value of 'goodness' of the things which industry produces. These are enhancing both the quality and standard of living in the world today. It should be noted that it is only from the value added, or profitability of manufacturing industry that we can pay for all the services a modern society requires, such as government, defence, health and education.

Although a lot has been achieved in the field of education within the last 20 years, it is not really related to the way in which a nation earns its living. For the majority of young people, their education stops at the age of 16 or, at the very latest at, say, 21. Only recently is it being seen as a life-long process via the introduction of Continuing Professional Development (CPD) while industry, be it noted, is dependent upon a steady supply of well-educated and trained personnel. This fact is now stressed by all political parties, trade unions, management federations and leaders in the field of education. For as we all know, good engineers and managers are trained over a lifetime and do not just appear. Manufacturing industry, however, still suffers from inadequate human resource planning, i.e. arranging to have the right people in the right place at the right time with the right skills, knowledge and attitudes.

In particular, there is an inadequate supply of properly educated and trained personnel in condition monitoring and diagnostic technology. The areas of maintenance technology management, accident and risks management, clean technology and waste disposal management, health and safety management, environmental technology management plus energy technology and conservation management also suffer from a lack of trained personnel and this is costing industry a fortune. Condition monitoring and diagnostic engineering management is now recognized by many as multi-disciplinary in nature and overlapping some of the areas mentioned above. Accordingly, many organizations have bene-fitted from adopting this discipline within which new ideas, tools, techniques and strategies are constantly being conceived, developed and profitably employed. Having formulated the case for COMADEM – as with all human industrial activity – for sustainability, this potentially beneficial and problem solving multi-discipline must be economically justifiable, and it is to that area we now turn.

1.3 ECONOMIC JUSTIFICATION AND BENEFITS

The outcome of the DTI's report on managing maintenance into the 1990s has already been mentioned earlier in this chapter. The Tribology action campaign, launched in March 1992 by the Institution of Mechan-ical Engineers (I.Mech.E) and the DTI, suggested that British industry can save a collective £1.5 billion through encouraging the application of Tribological principles to production processes of all kinds, to mainten-ance and to design. Tribology is concerned with the reduction of wear and controlling friction for rubbing and rolling surfaces. The type and percentage of wear processes that are encountered in industry are shown in Table 1.4, with the potential energy savings in four major areas of Tribology being shown in Table 1.5. The high proportion of abrasive and rubbing associated wear shown in the table is in accord with a Rolls-Royce observation that 70% of their unscheduled engine shutdowns and services were associated with some form of debris contamination. Vickers Systems came to a similar conclusion after undertaking a wide survey of hydraulic machinery, wherein over 70% of the failures were due to poor fluid condition.

The economic benefits of controlling lubricant and hydraulic fluid cleanliness are therefore obvious, and will result in longer component life, improved fuel efficiency, plant reliability and reduced maintenance costs. In the UK, manufacturing turnover in 1990 was in the region of £150 000 million, while the total cost of maintaining quality for business

Table 1.4 Percentage of wear encountered in industry

Type of wear	Percentage
Abrasive wear	50
Adhesive wear	15
Erosive wear	8
Fretting wear	8
Chemical wear	5

Table 1.5 An evaluation of potential energy savings in four major areas of tribology and the estimated cost of achieving them. An R&D benefit ratio is evaluated for each area. (From ASME Book No. H00109, *Strategy for Energy Conservation through Tribology*, 2nd edn.)

R&D programme area • Technology	Type of energy used	Potential energy savings		Estimated R&D cost millions of 1980 dollars	Benefit ratio*
		% US consumption	Billions of 1980 dollars per year		
Road transportation • Adiabatic diesel • Transmissions • Piston rings • Lubricants	Oil	2.20	8.88	16.12	55
Power generation • Bearings • Seals • Materials and wear	All types	0.23	0.93	3.15	30
Turbomachinery • Bearings • Seals • Materials and wear	All types	1.10	4.44	7.80	57
Industrial machinery and processes • Materials and wear • Metal processing • Lubricants	All types	1.80	7.20	7.65	93
Total		5.33	21.45	34.72	62

*Benefit ratio = Savings/(10 × R&D cost).

was estimated as being between 4% and 15% of this figure. The cost of failures were estimated as being around 50% of the total quality costs – and much higher if insufficient funds were spent on failure prevention. Thus it is likely that roughly £6000 million was wasted in failures and defects, and a 10% improvement in failure costs would have released into the economy approximately £600 million. The cost of preventing these failures was estimated to be 1% of the total turnover and therefore around £1500 million. A quarter of the money wasted!

In the UK, energy consumption is on a prodigious scale. Savings of up to 20% on the consumption bill could be realized by employing efficient energy monitoring and management strategies. In the USA, for example, new legislation will save $28 billion by outlawing inefficient appliances by the year 2000, thus saving the cost of building 25 large power generating stations. Similar regulations in the UK could save up to 600–700 megaWatts (mW) of power. The additional examples which follow are all taken from UK companies, and show the financial benefits to be gained from the judicious application of COMADEM tools, techniques and strategies.

- A 25-year-old flour mill implemented a planned and condition-based maintenance strategy and achieved a 43% saving within 12 months.
- An estimated benefit of £2.175 million has been reported by Imperial Chemical Industries (ICI) Plc after implementing permanent vibration monitoring systems at a number of sites.
- Significant savings in repair costs have been reported by a shipping company, after the introduction of a vibration analysis programme.
- A survey carried out in several British collieries has revealed many benefits after implementing a condition monitoring programme.
- In a period of four years, the number of plant stoppages in a paper plant was reduced from 300 to 30 via the use of condition monitoring. A considerable saving, given that the cost per stoppage was approximately £6000.
- In the British Steel Corporation (BSC), condition monitoring is encouraged, and its application is based upon a rigorous rationale aimed at cost reduction rather than blanket coverage.
- British Coal believes that a total approach to machine cost/benefit is the pragmatic way forward for reducing mining machine downtime and hence cost. They have demonstrated that the cost/benefit of applying routine condition monitoring can be measured.
- On-line non-invasive condition monitoring and diagnostic techniques have been successfully applied by the North Sea Oil industry to detect airgap eccentricity, high-voltage winding insulation degradation and broken bars in high-voltage squirrel cage induction motors.
- The cost benefit achieved by condition monitoring, in identifying possible blade failure in a chemical plant turbine, amounted to £700 000.
- A systematic approach to condition-based maintenance in a company has provided a financial cost benefit to the plant operators. This approach has ensured that condition monitoring is driven by financial, operational and safety requirements, not by technology.
- The Yorkshire Water Authority has made a significant reduction in its energy costs in real terms by implementing an active energy management programme.
- The on-line monitoring of fluid system cleanliness in a company has resulted in a profitable and cost effective operation.
- The successful implementation of an overall condition monitoring plan by British Petroleum (BP) on one site alone has saved a considerable amount of money.
- Texaco's Pembroke refinery saved nearly £500 000 per year by implementing an effective energy monitoring and management programme.
- A recent UK DTI–British Computer Society joint award for the best expert system application in manufacturing brought to light a number of practical software packages in regular use, and these were contributing significant savings to industrial companies.

Table 1.6 The power of predictive maintenance. 'Judicious use of condition monitoring can yield 10 to 20 times the initial outlay *within the first year.*' (Works Management Boardroom Report, *Into the Late 90s*, July 91)

Industry	Application	Savings due to predictive maintenance (PM)
Defence	Navy	The Canadian Navy estimates average savings of $2m/annum through the use of PM across its fleet of destroyers
Metals	Aluminium mill	ALCOA saved $1.1m in 1992 in motor repairs alone
Metals	Steel works	Armco Steel saves some $12m/annum through PM
Metals	Steel mill	Unplanned repair of a failed 1000 hp motor bearing cost $79k before PM; planned repair of same motor after PM identified potential bearing failure: $1.6k
Petrochemical	Oil production	Introduction of PM reduced gas turbine compressor maintenance outages by 20% and eliminated the associated lost production cost of 1100 barrels of crude oil per hour
Petrochemical	Oil refinery	An oil refinery produced nearly $1m/year savings by reducing maintenance costs by 29% on 100 major and 3900 minor machines
Power	Co-generation	On average, maintenance of co-generation plant costs $7/hp; one western Texas facility has reduced theirs to $3.5/hp through PM
Power	Nuclear	Following installation of PM in 1985, a nuclear power plant estimated its first-year savings were $2m, second-year savings were $3.5m
Power	Utilities	An Electric Power Research Institute study compared the actual costs of maintenance in N. American utilities: run-to-failure, $18/hp; periodic, $13/hp; predictive, $9/hp
Pulp and paper	Paper and board converter	Company specializing in high-value coating of plastic film and paper saved $40k within three months of installing PM
Pulp and paper	Paper and board mill	Georgia Pacific Paper saved $72k on one machine outage when PM detected a pump problem

Sources: Schlumberger.

Thus, the growth of predictive maintenance over the past decade has led to major improvements in the productivity of a very wide range of machinery, from the petrochemical industry to food packaging equipment. It has progressively replaced breakdown maintenance as part of an integrated machinery reliability programme, allowing many maintenance groups to control costs much more closely by providing accurate fault diagnosis. There are many similar stories to those recorded above reported throughout the world (Rao, 1993), and in a newly released report by the United States-based Thomas Marketing Information Centre, on predictive maintenance, the following significant findings are among those detailed:

● The majority of US firms are currently employing some form of predictive maintenance.
● The larger the company the greater the usage level of predictive maintenance techniques.
● Less than 10% of the study respondents see their current predictive maintenance programme as 'strong', and over 70% will be increasing their commitment to these technologies and services within the next five years.
● While overall maintenance spending is up by 12% in the period 1993 and in 1995, predictive maintenance spending is up 27%, better than twice the industry rate for maintenance in general.

Further savings, and the latent potential in the use of predictive maintenance is shown in Table 1.6, which again illustrates the effectiveness and cost benefits to be gained by employing this technique.

1.4 MARKET RESEARCH

Given the wide-ranging examples quoted above, in respect of the cost savings and financial justification for using COMADEM techniques, it is perhaps worthwhile, to look at the results of some market research relating to this area. Figure 1.3 shows the total European Economic Community (EEC) condition monitoring equipment market in 1989, as predicted by Frost & Sullivan Ltd in their report E1217. This report indicated that condition monitoring equipment could soon be as widely used as automation and process control systems are used today. In addition, Figure 1.4 gives the total European sensor market revenue forecast, country by country, for the years 1992, '93, '95 and '97. This shows Germany to be the leading sensor user with a market of nearly $1000 million, and the UK trailing a poor second, with a market of only half the size at around $500 million. Table 1.7 reinforces the impression of emerging high growth in the sensor market as uncovered by Frost & Sullivan's market intelligence.

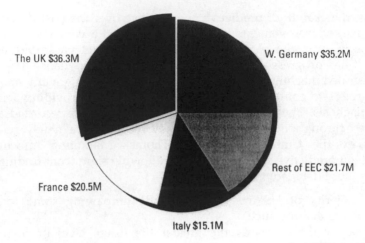

Source: Frost and Sulivan Ltd, Report #E1217   Total 1989 Market: $128.8 Million

Fig. 1.3   The total EEC condition monitoring equipment market in 1989.

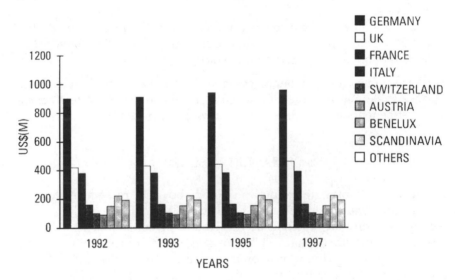

Fig. 1.4   Total European sensor market revenue forecast by country: 1992, 1993, 1995, 1997.

The increased demand for improved safety, performance and comfort has led to strong markets for all types of sensor systems. Accordingly, demands for better performance and advanced capabilities will soon make sensors not a luxury but a necessity on all industrial systems. According to the Frost & Sullivan market intelligence report 2636-30, US

Table 1.7 Total emerging sensors market: unit
shipment and revenue forecasts (world), 1989–99

Year	Units (thousand)	Revenue ($ million)	Revenue growth rate (%)
1989	47 560	1146.8	—
1990	55 412	1395.8	21.7
1991	62 354	1589.0	13.8
1992	67 374	1760.3	10.8
1993	73 773	1935.1	9.9
1994	84 937	2262.0	16.9
1995	102 434	2815.7	24.5
1996	121 164	3442.7	22.3
1997	141 792	4156.0	20.7
1998	165 745	4972.8	19.7
1999	193 533	5904.8	18.7

Compound annual growth rate (1992–9): 18.9%.
Note: All figures are rounded.
Source: Frost & Sullivan/Market Intelligence.

sales of Non-Destructive Test (NDT) equipment will grow from $407 million in 1992 to $512 million in 1997. In particular, Artificial Intelligence (AI) and expert systems will have an increasing impact on NDT methods, particularly in the ultrasonic and acoustic emission areas. As NDT computerization spreads, expert sysems will be developed to analyse computer resident information.

In a complementary report 927-10, Frost & Sullivan also predict that condition monitoring equipment sales will grow worldwide from $348 million in 1992 to $711 million in 1999. This growth being spurred on by increasing manufacturer concern with improving their competitive position and avoiding costly mishaps. Table 1.8 indicates the trend in world machine condition monitoring equipment markets, covering the following product areas:

- vibration monitoring equipment;
- oil and wear particle analysis equipment;
- thermography equipment;
- corrosion equipment.

Also included in the report is an analysis of the software used on condition monitoring applications. The petrochemical industry being identified as the largest purchaser of such software, and a moderate to strong growth rate is predicted in this area over the coming years. The report further indicates that the process industries represent a huge potential market for both condition monitoring equipment and software.

Table 1.8 Total condition monitoring equipment market: unit shipment and revenue forecasts (world), 1989–99

Year	Units (thousand)	Revenues ($ million)	Revenue growth rate (%)
1989	28.7	309.4	—
1990	31.3	336.6	8.8
1991	34.0	364.8	8.4
1992	35.8	383.6	5.2
1993	37.5	402.9	5.0
1994	40.2	433.0	7.5
1995	43.7	472.6	9.1
1996	48.0	521.0	10.2
1997	52.9	577.0	10.7
1998	58.4	640.5	11.0
1999	64.4	711.3	11.1

Compound annual growth rate (1992–9): 9.2%.
Note: All figures are rounded.
Source: Frost & Sullivan/Market Intelligence.

It is important, therefore, to note exactly what is currently available in the marketplace, with respect to the tools and techniques employed in the field of condition monitoring and diagnostic technology, and accordingly this will now be reviewed.

The selection of the right sensors for any particular application is one of the keys to effective condition monitoring, for without the ability to acquire accurate information, the quantitative monitoring and controlling of process machinery would be impossible. In the process industry, increasing use of unmanned production resources dictate a requirement for on-line, real-time actual condition knowledge. Such information is required in several areas of this type of plant, and because the operating environment can be harsh or difficult these sensing systems need to be (Davies, 1990):

- very fast, accurate and self correcting;
- simple in design and rugged in construction;
- non-intrusive and preferably non-contacting;
- highly reliable, and minimizing system complexity;
- economically competitive.

A variety of sensors exist to monitor and control the various on-line process parameters encountered in manufacturing industry effectively. These include the following:

- mechanical transducers such as, displacement, location or position, strain, motion, pressure and flow sensors;

Table 1.9 Some techniques of measurement in the process industries

Temperature	thermocouple, resistance thermometer, liquid and gas expansion, bimetal, radiation pyrometer
Pressure	bourdon tube, diaphragm, bellows, piezoresistance, strain gauge
Flow	orifice plate, venturi tube, nozzle, pilot tube, flume, weir, electromagnetic, ultrasonic, vortex, turbine, variable area, positive displacement, Coriolis mass flow, Ball prover
Level	float, capacity probe, conductance probe, ultrasonic echo, pressure difference, optical refraction, nucleonic
Other physical	• viscosity: drag plate, ultrasonic • load: strain gauge • force: strain gauge • displacement: LVDT (see also level) • consistency: pressure difference • thickness: nucleonic • turbidity: photoelectric • colour: photoelectric • density: vibrating cylinder • humidity: animal skin, photoelectric (dew point), chemical absorption/conductance, quartz oscillator
Analytical	pH, selective ion electrodes, electrolytic conductance, thermal conductance, chromatography, IR/visible/UV absorption, magnetic susceptibility, flame ionization, spectrophotometric, colorimetry, oxidation/reduction potential, X-ray fluorescence, mass spectrometry, solid electrolytes

- optical transducers, such as photodetectors, pyrometers, lasers, optical fibre systems, and solid-state sensors;
- thermal transducers, such as thermocouples and thermistors;
- environmental transducers such as, spectrometers, ph indicators, air, water, and soil pollution monitors.

Extensive research into the electrical properties of silicon has led to the discovery of several potentially beneficial effects. These have allowed a reduction of sensor size and permitted rugged construction, and have the potential to reduce costs. Silicon sensors are now employed in the monitoring of gas and chemical environments and within the foreseeable future silicon biosensors will be commercially available. Table 1.9 shows some of the measurement techniques and sensors currently in use in the process industry (Rao, 1988), and as indicated in the market research above, the industrial application of all types of sensor systems is rapidly expanding. As automation takes hold in manufacturing industry, it is anticipated that this trend will increase, as will that of device micro-miniaturization and distributed 'intelligent' instrumentation.

Obtaining information via sensor systems is not the whole story, however, and a judiciously selected condition monitoring and diagnostic scheme is also required to complement any data-acquisition system. If properly applied, this should make it possible to detect, diagnose, predict and control any impending machinery failures effectively and efficiently. Condition monitoring ensures that all decisions are made on substantive and corroborated diagnostic information, thereby providing a basis for cost-effective and logical decision-making. Thus, condition monitoring and diagnostic technology is nowadays extensively employed in numerous industrial applications, including petrochemical refining, steelmaking, aerospace and nuclear engineering, shipping and power generation.

One point, which is important to note now, and which becomes obvious when reading later chapters in this book, is that condition monitoring and diagnostic technology is a multi-disciplinary subject area. It requires a team approach to implementation and is both proactive and predictive in nature. The technology involved must be economically justifiable and effectively integrated into a wider maintenance system. It is also important to note that on no account should it be regarded as an isolated technical discipline, only performed by specialist staff, for this attitude could well lead to difficulties in implementation.

Thus, it may be seen that the condition monitoring of process control systems require two preconditions to ensure a successful application. First, the host system should be stable in normal operation, and this stability reflected in the parameters under surveillance. Secondly, accurate continuous/periodic, automatic or manual measurement of the process parameters must be possible for the system to be monitored. Provided these conditions are met, then any change from the normal behaviour of the system can easily be monitored, and via trend or various other analyses, the presence of any potential failure can be revealed.

As will be seen in later chapters, such performance monitoring methods cover the following areas:

- Measuring the variations in, and or the absolute values of system output in terms of quality and quantity.
- Measuring the system's input/output relationship.
- Measuring and simultaneously comparing two output parameters within a standard set of operating conditions.

Table 1.10 illustrates the wide range of currently available condition monitoring and diagnostic technology techniques. It amply outlines the wide diversity of methods used in condition monitoring by many companies worldwide and also indicates the application potential of many of these techniques. Table 1.11 presents some of the currently available maintenance software which is being used by many process-

Table 1.10 Some available condition monitoring and diagnostic techniques

A. Vibration monitoring:
(a) Overall monitors
(b) Spectral analysis
(c) Discrete frequency monitoring
(d) Shock pulse monitoring
(e) Kurtosis method
(f) Cepstrum analysis
(g) Signal averaging

B. Wear debris analysis:
(a) Inductive sensors
(b) Capacitive sensors
(c) Electrically conducting filters
(d) Existing and special filter systems
(e) Optical oil turbidity monitor and level sensors
(f) Magnetic plugs
(g) Centrifuges
(h) Particle counters
(i) Ferrography
(j) Rotary Particle Depositor (RPD) and Particle Quantifier (PQ)
(k) Spectrography

C. Visual inspection:
(a) Borescopes and fibrescopes
(b) Stroboscopes
(c) Dye penetrants
(d) Thermographic paints and crayons
(e) Infrared thermography
(f) Radiography
(g) Laser systems
(h) Magnetic flux
(i) Electrical resistance
(j) Eddy current
(k) Ultrasonics
(l) Stress wave sensors
(m) Corrosometer

D. Noise monitoring:
(a) Sound pressure monitoring (microphones)
(b) A-weighting
(c) Damage risk criteria
(d) Equivalent continuous energy level monitoring (Leq)
(e) Impulsive noise monitoring
(f) Spectral analysis
(g) FFT/Zoom FFT
(h) Infrasonic noise monitoring
(i) Sound intensity monitoring

E. Environmental pollution monitoring:
(a) Air pollution monitoring
(b) Water pollution monitoring
(c) Earth pollution monitoring

Table 1.11 Some currently available maintenance softwares

Name of software	Suppliers
MP2 for Windows maintenance management	Datastream Systems Inc, USA
COMPASS–Predictive maintenance	Bruel & Kjaer (UK) Ltd
PMIS Computerized maintenance management system	BC Computing Ltd, UK
ADRE (Automated Diagnostics for Rotating Machinery) for Windows+DAIU (Data Acquisition Interface Unit)	Bentley Nevada, USA
TELEVIEW–Maintenance and asset management systems	Dynamic Logic Ltd, UK
Planet XL Maintenance system	FDS Advanced Systems, UK
JOBWISE Maintenance resource planning and scheduling system	Insight Logistics Ltd, UK
LIPS Image processing system	Land Infrared, UK
ROTALIGN Laser shaft alignment system	Pruftechnik (UK) Ltd
FLEXIMAT Corrosion monitoring system	AEA Technology, UK
PHOCUS Machinery health advisor	AES Ltd, UK
IRWIN PRO Thermal analysis and report software	AGEMA Infrared Systems Ltd, UK
BERL Bond motor management system	British Electrical Repairs Ltd, UK
MAINTRACKER Integrated maintenance management system	Cotec Computing Services, UK
CASP Maintenance planning system	Delta Catalytic (UK) Ltd
SCOFTROL Vibration analysis	
CUI Corrosion Under Insulation	
Q4CAMM Condition monitoring and asset management system	Engica Technology Systems International Ltd, UK
Q4RCM Reliability centred maintenance software	
EMONITOR & EMONITOR for Windows predictive maintenance softwares	Entek Scientific Corp, USA
CONTROLLER Computerized maintenance management system	ESBI Computing Ltd, UK
MCS-II, RAPIER & PM3 Oracle based integrated maintenance management systems	Kvaener AM Ltd, UK
COMO Condition monitoring system	Southampton Institute, UK
AMETHYST, VIOLET Expert systems	IRD Mechanalysis (UK) Ltd
IQ 2000 Oracle-based maintenance management systems	

based companies worldwide and Table 1.12 provides a comparison of some of the most popular condition monitoring systems and software (Unger, 1994). Some of the diagnostic methods presently used in manufacturing industry are outlined in Table 1.13 and a condition monitoring method selection chart is presented in Table 1.14.

Table 1.12 Comparison of condition monitoring systems and condition monitoring software

Sensors:
A — Accelerometer
CL — Current loop 4-20mA
D — Displacement
E — Eddy current
F — Frequency
L — LVDT
P — Pressure
S — Strain gauge
T — Thermocouple
V — Voltage

Company or product name	Sensors	ADC resolution	ADC speed in kHz	No. of channels	Input range	Memory on board	Communication	Signal processing	Digital filters	Numerical analysis	Statistical analysis	Regression analysis	MS-Windows interface	
									Data analysis					
Amplicon Liveline	T, CL	12 / 16	312.5 / 4	8D 18SE 1	±20mV ±200mV 2V 20V	1MB	RS232 RS485 IEEE	DSP 68001	—	—	—	✓	—	Software DIGIS DAP
Bentley Nevada	E, D	n/a	n/a	2040	n/a	—	RS422	—	✓	✓	n/a	n/a	n/a	System Trendmaster 2000
Biodata	T, CL, F / S	12 / 16	200 / 40	16	100mV ±250mV 2.5V ±5V 10V	up to 1MB	IEEE RS485	✓	✓	✓	✓	✓	✓	Software FAMOS 1.5
Calex Instr./Analog Devices	T, S, F / L	12 / 16	50 / 250	8D 18SE	±5V ±10V ±10V	25k x 16 25k x 12	RS232 RS422 RS485	✓	✓	✓	✓	—	—	Compatible to most Analysis software packages
COMO	A, D, L / S, T	14		258 SE	±5V	—	IEEE	—	✓	✓	✓	—	—	Developed at Southampton Institute
CP Instruments	T, CL, S	16	250	6	0.3mV to 10V	—	RS232 IEEE LAN	✓	✓	✓	✓	—	—	Software Instatrend
CSI	n/a	n/a	n/a	n/a	n/a	—	RS232	—	✓	✓	✓	—	—	System Mastertrend
Data Translation	T, CL	12 / 16	750 / 100	4D	±5V ±10V	—	RS232	✓	✓	✓	—	—	—	Boards DT2827, DT2841L Software GlobalLab 3.0
Digitron Instrumentation	T, CL, V / P	n/a	1Hz	4	0-2.5V 0-10V	—	RS232	—	—	—	—	—	—	Hand-held data logger SF12
Endevco / SKF	T, F, V / A	n/a	20	1	250mV to 25V	12k	RS232 LAN	✓	✓	✓	✓	—	—	Hand-held system SKF CVMA 4 Software PRISM[2]
GfS	n/a	n/a	n/a	n/a	n/a	—	RS232 IEEE	✓	✓	✓	✓	✓	—	Software DIA / DIGO
IRD Mechanalysis	n/a	n/a	n/a	n/a	n/a	—	RS232	✓	✓	✓	✓	✓	✓	Software IQ 2000
Micromeasurement Group	T, CL, S / V	12 / 16	10	128	±10mV to ±10V	58k	RS232 IEEE	—	—	—	—	—	—	Basic data acquisition systems mainly for strain gauges and thermocouples
National Instruments	T, CL, S / V	12 / 16	1000 51.2	4 / 2	±5V ±2.8V	—	RS232 RS485 IEEE	—	—	—	—	—	—	Boards EISE A 2000 AT DSP 2200
SciTech / HuDe	T, CL, F / V	16	50	230 47SE	±1.25V ±1V	—	RS232	—	✓	—	—	—	—	Hardware UPC 607 Software EasySense DAQ
Validyne	T, CL, F / S, L, V	14	50	8D 18SE	±10mV to 10V	—	n/a	—	—	—	—	—	—	Interfacer 4

Table 1.13   Some diagnostic methods currently used

Diagnostic methods (columns 1–20):
1. Visual inspection
2. Thermography
3. Optical metrology and holography
4. Liquid penetrant inspection
5. Magnetic particle inspection
6. Eddy current testing
7. Magnetic flux leakage methods
8. Potential drop crack sizing
9. Radiography
10. Television fluoroscopy- real time radiography
11. Neutron radiography
12. Ultrasonic flaw detection
13. Ultrasonic thickness gauging
14. Acoustic methods
15. Acoustic emission methods
16. Leak testing
17. Plant condition monitoring
18. Stress measurement
19. Coating thickness measurement
20. Other methods

● Indicates good prospects   O Indicates some prospects

Materials	Inspection task	1	2	3	4	5	6	7	8	9	10	11	12	13	14	15	16	17	18	19	20
Metals	Surface opening cracks	●			●	●	●	●	●	●	O		●		O	●	O	●			
	Surface corrosion pits etc	●			●	●		●	●	●	O		O								
	Severe corrosion thinning	●	O	O		●		●	●	●	O		●								
	Internal cracks						O	●	O	O	●		O	O		●					
	Porosity								●	●			●			O					
	Lack of fusion defects								O	O			●								
	Internal voids inclusions						O		●	●	O		●		O						
	Defect sizing	O		O		O			●	O	O		●								
	Thickness measurement		O	●			O	O		●			●	●							
	Microstructure variation	O					●						O		O				O		O
	Stress/strain measurement			●						O			O						●		
Coated metals	Coating thickness measurement			O			●	●			O	O	●							●	●
	Coating delamination	O		O			O						●	O							
	Coating 'pin holes'	O				●												O			
Composite materials	Detaminations and disbonds	O	O	O			O						●	●	●	O					
	Fibre/matrix ration evaluation						O							O	O						
	Incomplete cure of resin											O		O	O						
	Internal porosity								●	O											
Concrete	Concrete thickness measurement								●					O	O						
	Reinforcing-bar corrosion								O	O						O					
Ceramics	Surface cracks	O				●			O	O			O		O	O					O
	Internal cracks porosity								●	O			O		O	O	O				O
Any	Assembly verification	●	●						●	O	●				O						
	Sorting						●	●							●						●

Source: British Institute of NDT

Fairly obviously, good process control is fundamental to the effective operation of all production plant, and therefore automation ultimately provides a means of solving most safety, environmental and economic issues. However, the benefits of advances in process control and automation are not realized until they have been carried through into

Table 1.14 Condition monitoring method selector

	Vibration analysis	Noise analysis	Acoustic emission	On-line debris monitoring	Debris analysis	On-line oil condition monitoring	Oil condition analysis	Water in oil detection	Electrical motor insulation and winding monitoring	Optical detection systems	Optical alignment systems	On-line pressure monitoring	On-line temperature monitoring	Thermal imaging	Stress/strain analysis	Erosion/corrosion monitoring	Performance monitoring	Orifice restriction monitoring
Bearings	•	•		•	•	•	•			•			•	•	•	•		
Belts																	•	
Blowers/fans	•	•											•	•			•	
Boilers/heat exchangers			•						•			•	•	•		•	•	
Brazing/welding equipments				•					•	•		•	•				•	
Casting/forging machines									•			•					•	
Compressors/pneumatic machines	•	•			•	•	•	•	•			•	•	•	•		•	
Couplings	•	•													•		•	
Guilotines/cutting machines	•	•		•	•	•	•				•	•	•	•			•	
Earthmoving/excavating plant	•	•		•	•	•	•	•			•	•	•	•			•	
Electric motors generators	•	•				•	•	•	•			•	•	•	•	•	•	
Elevators/hoppers/conveyors	•	•							•			•	•	•	•	•	•	
Escalators	•								•									
Filters/separators/valves	•	•		•	•	•	•	•		•		•	•	•	•	•	•	•
Gearboxes	•	•		•	•	•	•	•				•	•	•		•	•	
Vacuum equipment	•			○	○	○	○	○		•		•	•	•		•	•	
Incinerators/furnaces/autoclaves				○	○	○	○	○		•		•	•	•		•	•	○
Internal combustion engine				•	•	•	•	•	•			•	•	•		•	•	
Loaders/stackers				•	•	•	•	•									•	
Machine tools mechanical	•	•		•	•	•	•	•				•	•	○	•	•	•	
Machine tools hydraulic	•	•		•	•	•	•	•				•	•	○		•	•	
Pressure vessels/accumulators			•									•	•	•	•	•	•	
Pumps	•	•		○	○	○	○	○				•				•	•	○
Structures/rigging	•	•		•											•			
Transformers						•	•	•					•	•			•	
Turbines/aero engines	•	•		•	•	•	•					•	•	•	•	•	•	
Wire/cable making																○	•	
Winding/lifting machinery			•												•	•	•	

Source: British Institute of NDT

Table 1.15 Some well-known control systems, software suppliers and some useful information on wear debris and particle monitoring equipments used in industry

Control systems, software and equipment	Suppliers
Self-tune controllers supervisory control systems	ABB Kent-Taylor Ltd, UK
Process control and instrumentation	Able Instruments & Controls Ltd, UK
Process and safety related systems	Allison Engineering Ltd, UK
Network 3000 family of digital controllers	Bristol Babcock Ltd, UK
Micro and mini computers, PLCs and softwares	Communication & Technology (1987) Ltd, UK
Precision Industrial PID controllers	Eurotherm Ltd, UK
Control systems for processing Industry	Hartmann & Brqun (UK) Ltd
SCADA, Emergency shutdown systems	ICS Scotland Ltd, UK
MODBUS protocol for process control	Precia Industries Ltd, UK
Networking and Communications	The SCADA centre, UK
Expert systems	Honeywell Control Systems, UK
Fuzzy logic systems	Omron Electronics, Japan
Neural net and expert systems	Gensym Ltd, UK
ACTIPROBE–uses thin layer activation technique	Cormon Ltd, UK
AEROMETRICS PHASE DOPPLER PARTICLE ANALYSER–uses optical-phase/ Doppler scatter/light scattering interferometry	Aerometrics Inc, USA
ON-LINE CONTAMINANT MONITOR– uses silting technique	BHR Group Ltd, UK
PARTICLE SIZER BI-90 & BI-DCP–uses optical-photon correlation spectroscopy and disc centrifuge mass separation techniques respectively	Brookhaven Instruments Corp., USA
TECALERT–CONTINUOUS Debris Monitor– uses magnetic flux path change technique	Ranco Controls Ltd, UK
DANTEC PARTICLE DYNAMICS ANALYSER– uses optical-phase/Doppler scatter technique	Dantec Elektronik, Denmark
DEBRIS TESTER–uses inductance technique	Staveley NDT, UK
FLUID CONDITION MONITOR–uses filter blockage technique	Lindley Flowtech Ltd, UK
FRITSCH ANALYSETTE 22 LASER PARTICLE SIZER–uses optical-Fraunhofer diffraction technique	Fritsch GmbH, Germany
LIQUID CONTAMINATION monitor (LCMII)– uses filter blockage technique	Coulter Electronics Ltd, UK
MAGNETIC CHIP COLLECTORS–uses magnetic attraction technique	Muirhead Vactric Component Ltd, UK
OILCHECK–uses Dielectric constant technique	UDC International Ltd, UK
QUANTITATIVE DEBRIS MONITOR QDM– uses magnetic attraction technique	Vickers Inc, USA
SPI-WEAR–uses surface activation technique	Spire Corp., USA

industrial practice. Advanced control techniques must be implemented if the total system is to cope with the complex interactions and varying dynamics of the process operations involved. Thus, since around 1980 onwards, process-control technology has undergone a major revolution, in conjunction with the rapidly emerging area of information technology. Significant advances have been made in the following areas:

- Microprocessor Based Control (MBC) systems;
- Supervisory Control and Data Acquisition (SCADA) systems;
- Computer Aided Control System Design (CACSD);
- Plant area networks;
- Artificial intelligence systems;
- Computer Aided Software Engineering (CASE).

Table 1.15 outlines some of the best-known and utilized commercial control systems and their associated software. In a survey conducted in 1989, the development of expert systems for diagnosis and maintenance was found to be widespread throughout industry. The findings being

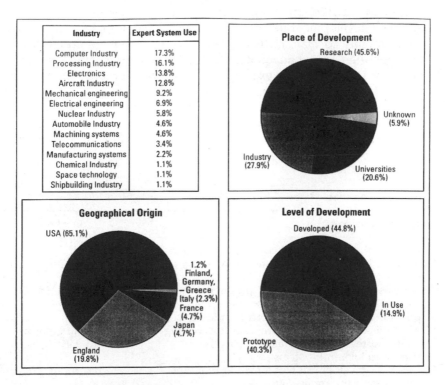

Industry	Expert System Use
Computer Industry	17.3%
Processing Industry	16.1%
Electronics	13.8%
Aircraft Industry	12.8%
Mechanical engineering	9.2%
Electrical engineering	6.9%
Nuclear Industry	5.8%
Automobile Industry	4.6%
Machining systems	4.6%
Telecommunications	3.4%
Manufacturing systems	2.2%
Chemical Industry	1.1%
Space technology	1.1%
Shipbuilding Industry	1.1%

Place of Development
Research (45.6%)
Unknown (5.9%)
Industry (27.9%)
Universities (20.6%)

Geographical Origin
USA (65.1%)
1.2% Finland, Germany, Greece Italy (2.3%)
France (4.7%)
Japan (4.7%)
England (19.8%)

Level of Development
Developed (44.8%)
In Use (14.9%)
Prototype (40.3%)

Fig. 1.5 Expert systems survey results, Majstorovic (1990).

summarized in Figure 1.5 and Table 1.16, which shows some of the currently available knowledge based tools.

Developments are taking place extremely rapidly in this area, and one of the newest technologies emerging within the AI field is that of neural networks. Artificial Neural Networks (ANNs) are now finding new and exciting applications in many areas including the fields of condition monitoring and control engineering. The inspiration for neural networks lies with the human central nervous system, which is literally composed of billions of neurons. Although the basic concept of a neuron is simple, the way they work in a large network can be very complex, a neuron being effectively a switch, with a number of weighted inputs.

In principle, when the inputs of the neuron have values exceeding a preset threshold, then the output of the neuron is switched on. Each neuron can have a large number of inputs and each are connected to other neurons to form a large network. The input for each neuron will have a strength and weighting factor associated with it, and the neuron will itself turn on when there is enough input. That is when the sum of the weighted inputs of the individual inputs exceed a certain value. Table 1.17 presents a comparison of expert systems and artificial neural networks, while Table 1.18 indicates some application areas of ANNs in the fields of condition monitoring and control engineering.

Another development in this area is that of fuzzy logic. This derives its applicability to manufacturing from the idea that traditional true/false logic often cannot adequately deal with situations which present a number of ambiguities or exceptions. It allows a process specialist to describe, in everyday language, how to execute decisions or control actions without having to specify the process behaviour in complex equations. It is often used for complex control applications with multiple inputs and outputs, either as an alternative to traditional Proportional,

Table 1.16   Available knowledge-based tools

Knowledge-based tools	Suppliers
Ilog Rules C^{++}–Object-oriented inference engine	Ilog Ltd, France
Goldworks III–A KBS which runs on PC using Microsoft Windows	Goldhill Corp, USA
CLIPS 6.0–C language integrated production system runs on a number of platforms	COSMIC, USA
Kappa–PC runs on a PC Microsoft Windows	Intellicorp Ltd, UK
RTWorks is a family of software tools for creating intelligent real-time monitoring, analysis, display and control systems	Scientific Computers Ltd, UK
G2 is a KBS development tool aimed at real-time process industry application	Gensym Ltd, UK

Table 1.17 Comparison of expert systems and ANNs

Rule-based systems	*Neural-network systems*
Excellent explanation capability	Little or no explanation capability
Requires an articulate expert to develop	Requires many examples, but no expert is needed
Many turnkey shells are available	Few turnkey shells available; most must be customized for your application
Average development time is 12 to 18 months	Development time is as little as a few weeks or months
Preferred system when examples are few and an expert is available	Preferred system when examples are available or an expert is not available
Many successful, fielded systems are available for public reference	Few successful, fielded systems are available for public reference
Large systems can be unwieldy and difficult to maintain if not carefully developed and designed	Large networks can not be built today; smaller networks can be hierarchically linked for more complex problems, making them more maintainable
Systems built through knowledge extraction and rule-based development	Systems built through training using data examples
Accepted validation procedures for completed systems	Validation of completed system is dependent on statistical analysis of performance
Work fine on ordinary digital computers	For all but the smallest networks, best performance comes from use with accelerator-assisted or specialized parallel chip boards

Source: Caudill (1991).

Table 1.18 Application areas of artificial neural nets

Products	*Suppliers*
NeuDesk 2, NeuRun, NeuModel, NeuralDesk, NeuSpring, Optional Algorithms	Neural Computer Sciences, UK
DataSculptor is available on PCs running Windows	Scientific Computers Ltd, UK
NT5000 Neural network controller	Neural Technologies, UK
NeurOn-Line, is a graphical, object-oriented software that enables users to easily build ANNs and integrate them into G2 Real-time Experts Systems	Gensym Ltd, UK

Application areas include: data validation, advanced control, quality management, process optimization, fault diagnosis, pattern recognition, maintenance management etc.

Table 1.19   Application areas of fuzzy logic

Products	Suppliers
Fuzzy controllers	Allen-Bradley, UK
Fuzzy chips	Texas Instruments Inc., USA
Fuzzy chips	Motorola, USA
Fuzzy controllers	Yokogawa, Japan
Fuzzy temperature controllers	Omron Electronics, Japan
Fuzzy systems	Sony, Mitsubishi, Japan

Application areas include, manufacturing planning, oil recovery techniques, personnel detection, configuring a digital filter, simulation of traffic flow and control, pattern recognition, controlling a robot arm, power generation control, resin curing, flight control, tuning fuzzy logic controllers, etc.

Integrative or Derivative (PID) control, or as a way to automate the control actions of a skilled process operator. Fuzzy logic is also used in the field of condition monitoring and diagnostic engineering management, as shown in Table 1.19, and as such is another burgeoning technique with potentially a large market in this area.

1.5   COMADEM EDUCATION

Thus, it may be seen from the discussion above that by judiciously selecting the correct COMADEM technology and techniques, it is possible to carry out the right tasks at the right time, safely, reliably and profitably throughout the life-cycle of industrial plant and equipment. As previously mentioned, in today's highly competitive environment, a company's profitable survival is the only factor of importance which underlies all other business activities. Only through profitable company projects and activities can survival be achieved. Hence, a key factor in any survival strategy must be an effective yet cost-efficient means of controlling maintenance activities, by use of the most appropriate management techniques that are available.

As stated above, only a few people recognize and really understand our total dependence on the wealth-creation aspects of industry. Thus, we need of necessity to re-emphasize the value and 'goodness' of those things which our industry manufactures, and which also enhance the quality and standard of our lives. It is only from the value-added or the profits made by manufacturing industry that we can pay for all the services that our society requires, including government, defence, health and education. Although a revolution has taken place in the field of education within the last 20 years, it still does not relate in the main to the way in which the nation earns its living. This is a failing and it should be recognized as such. Unfortunately, education is not seen as a life-long process, despite the recent attempts to introduce CPD. While industry, if it is to remain competitive in an age of automation, requires, and is totally dependent upon, highly educated and trained personnel.

Although this fact is stressed by all political parties, by the Trades Union Congress (TUC) and by the Confederation of British Industry (CBI), continuing professional development is not yet accepted by our society. While it is a truism that 'good engineers and managers are made and not born', the process industry, in particular, still suffers from an inadequacy in human resource planning as a consequence of this attitude. Thus, to have the right people in the right place at the right time with the right skills, knowledge and attitudes to achieve world class company performance is still a rarity in UK. As both the process and manufacturing industries still suffer from an inadequate supply of properly educated and trained personnel, most especially in the following areas:

- condition monitoring and diagnostic technology;
- maintenance management;
- health and safety management;
- environmental technology management;
- accident and risk management;
- clean technology and waste disposal management;
- energy technology and conservation management.

The lack of adequately trained human resources in these vitally important inter-disciplinary fields is now costing industry dearly, for many firms now compete in a marketplace where excellence in manufacturing is regarded as normal. Although the benefits of condition monitoring are well-known and reported widely, much of its success can be attributed to the skills of a few dedicated individuals. This knowledge and expertise has been, to a large extent, either self-taught or supported by equipment manufacturers' training programmes. More recently, a number of consultancy firms have also started to offer education and training programmes in this area. In addition, higher educational establishments have begun to offer courses in condition monitoring and maintenance management.

At Southampton Institute for example, a successful MSc programme in condition monitoring was pioneered by the author and has been running for quite some time. This course has been specifically designed to meet the demands of the process and manufacturing industries worldwide. The programme is flexible and is offered on both a full-time and part-time basis. The course is modular in structure and consists of

Table 1.20   Available training courses in condition monitoring

Courses and programmes	*Offered by*
A. *Product vendor courses:*	
Vibration and noise monitoring, computer-aided-maintenance management, machinery monitoring and diagnostics, acoustic emission monitoring	Bruel & Kjaer, Diagnostic Instruments, Entek Corporation, IRD Mechanalysis, SKF, CSI, ENDEVCO
B. *Commercial training:*	
Corrosion monitoring, reliability centred maintenance, condition monitoring, maintenance management, electrical machinery monitoring, total productive maintenance, rotating machinery monitoring and diagnosis, vibration monitoring	Monitron Ltd, RCM Ltd, Wolfson Maintenance, Electrical Power Research Institute, WCS International, Update International, COMADEM International, Boyce International, Vibration Institute, IRD Mechanalysis Intelligent Information Technology Ltd.
C. *Short- and long-term courses:*	
Instrumentation and control, condition monitoring, vibration and noise monitoring, maintenance management, oil debris monitoring, environmental monitoring, industrial tribology, signal processing	Institute of Measurement & Control, Institution of Mech. Engrs, Institution of Plant Engineers, British Institute of NDT, Institution of Electrical Engrs
D. *Academic programmes:*	
M.Sc Condition Monitoring/ M.Phil/Ph.D M.Sc Industrial Maintenance Asset Management System Operational Effectiveness	Southampton Institute, Glasgow-Caledonian University, University of Manchester, Robert Gordon University, Sunderland University, Glamorgan University, Exeter University, Monash University, University of Toronto
Various post-graduate programmes	ISVR, UMIST, UWE, Brunel, Swansea, Cranfield, MIT, IIT

the following five modules:

- Principles and Applications of Condition Monitoring Technology;
- Principles and Applications of Diagnostic Technology;
- Principles and Applications of Integrated Maintenance Management;
- Project and Technology Management;
- Student Project.

The feedback from both students and employers in respect of the course is very complimentary, and demand for the course is growing year by year. Students from many parts of the world are participating in this unique post-graduate programme and hopefully will one day occupy positions of influence in industry. As a further example of the interest and education currently available in this area, Table 1.20 outlines some of the worldwide training programmes and courses in the field of condition monitoring and diagnostic engineering management.

1.6 CONCLUSIONS

This introductory chapter started by asking the question:

Why is condition monitoring so vital to the integrity of industrial systems?

Hopefully, the chapter has answered the question by addressing and making a case for the following issues.

Firstly, that there is a need for a new inter-disciplinary science, covering the areas of condition monitoring and diagnostic engineering management and, further, that there is sufficient economic justification and proven benefits to industry for such a science to be adopted by companies wishing to become world-class manufacturers.

Secondly, that there is a strong international market which supports, and fully exploits, advances in this cost-effective and potentially beneficial technology and, in addition, that there are tried and tested techniques available to effectively monitor, diagnose, maintain and control manufacturing process plant.

Finally, that research, education and development in the fields of sensor engineering, condition monitoring and diagnostic technology are vital to the national interest, and should be encouraged to develop at an accelerating pace, thereby providing the users of this science with a much needed, profitable and competitive edge in the marketplace via highly trained personnel.

1.7 REFERENCES

Anon (1993) *British Institute of Non-Destructive Testing Year Book.*

Carson, P.A. and Mumford, C.J. (1979) An analysis of incidents involving major hazards in the chemical industry, *Journal of Hazardous Materials*, **3**, 149–65.

Caudill, M. (1991) Expert networks, *BYTE*, **16**, October, 108–16.

Davies, A. (1990) *Management Guide to Condition Monitoring in Manufacture*, IEE publication, London, UK.

Department of Trade and Industry (1989) *Managing Maintenance into the 1990's*, DTI Report, London, UK.

Frost & Sullivan Market Research Report No. E1217.

Frost & Sullivan Market Research Report No. 972-10.

Frost & Sullivan Market Research Report No. 2636-30.

Garrison, W.G. (1988) *100 Large Losses: A Thirty Year Review of Property Damage Losses in the Hydrocarbon Chemical Industry*, Report by M&M Consultants, Chicago, USA.

Goverts-Lepicard, M. (1990) *Major Chemical Disasters – Medical Aspects of Management*, ed. V. Murray, Royal Society of Medicine Services, London, UK.

Hunt, T.M. (1993) *Handbook of Wear Debris Analysis & Particle Detection in Liquids*, Elsevier Applied Science, London, UK.

Majstorovic, V.D. (1990) Expert systems for diagnosis and maintenance: the state of the art, *Computers in Industry*, **15**, 43–68.

Rao, B.K.N. (1988) *COMADEM 88*, Proceedings of the 1st UK Seminar on Condition Monitoring and Diagnostic Engineering Management, Chapman & Hall, London, UK.

Rao, B.K.N. (1993) *Profitable Condition Monitoring*, Kluwer Academic Publishers, London, UK.

Unger, T. (1994) *Intelligent Multi-Sensor Nodes for Condition Monitoring of an FMC Environment*, MPhil/PhD Transfer Document (unpublished), Southampton Institute, UK.

CHAPTER 2

Condition based maintenance

R. Keith Mobley

President and Chief Operating Officer,

Integrated Systems Inc., 215 South Rutgers Avenue,

Oak Ridge, Tennessee, 37830, USA

2.1 INTRODUCTION

Condition monitoring is perhaps the most misunderstood and misused of all the plant improvement programmes. Most users define it as:

A means to prevent catastrophic failure of critical rotating machinery.

Others define condition monitoring as:

A maintenance scheduling tool that uses vibration, infrared or lubricating oil analysis data to determine the need for corrective maintenance actions.

A few share the belief, precipitated by vendors of monitoring systems, that it is:

The panacea for critically ill plant and machinery.

One common theme of all these definitions is that condition monitoring is solely a maintenance management tool. Because of these misconceptions, the majority of established programmes have not been able to achieve a marked decrease in maintenance costs, or a measurable improvement in overall plant performance. In fact, the reverse is too often true. In many cases, the annual costs of repairs, repair parts, product quality and production have all dramatically increased as a direct result of the programme.

Handbook of Condition Monitoring
Edited by A. Davies
Published in 1998 by Chapman & Hall, London. ISBN 0 412 61320 4.

Condition monitoring is much more than a maintenance scheduling tool and accordingly it should not be restricted to maintenance management. As part of an integrated, total plant performance management programme, it can provide the means to improve the production capacity, product quality and overall effectiveness of our manufacturing and production plants. Condition monitoring is not a panacea for all the factors limiting total plant performance. In fact, it cannot directly affect plant performance. Condition monitoring is thus a management technique that uses the regular evaluation of the actual operating condition of plant equipment, production systems and plant management functions, to optimize total plant operation.

The output of a condition monitoring programme is data. Until action is taken to resolve the deviations or problems revealed by the programme, plant performance cannot be improved. Therefore, a management philosophy committed to plant improvement must exist before any meaningful benefit can be derived. Without the absolute commitment and support of senior management and the full co-operation of all plant functions, a condition monitoring programme cannot provide the means to resolve poor plant performance. Condition monitoring technology can be used for much more than just measuring the operating condition of critical plant and machinery. The technology permits accurate evaluation of all functional groups, such as maintenance, within the company. Properly used, condition monitoring can identify most, if not all, of the factors limiting the effectiveness and efficiency of the total plant.

One factor limiting the effective management of industrial companies is the lack of timely, factual data that defines the operating condition of critical production systems and the effectiveness of critical plant functions such as purchasing, engineering and production. Properly used, condition monitoring can provide the means to eliminate all the factors limiting plant performance. Many of these problems are outside the purview of maintenance and accordingly must be corrected by the appropriate plant function. High maintenance costs are the direct result of inherent problems throughout the plant, not just ineffective maintenance management. Poor design standards, purchasing practices, improper operation and outdated management methods, contribute more to high production and maintenance costs than do delays caused by the catastrophic failure of critical plant machinery.

Because of the breakdown mentality, and myopic view of the root cause of ineffective plant performance, too many companies restrict condition monitoring to the maintenance function. Expansion of the programme to include regular evaluation of all factors that limit plant performance will greatly enhance the benefits that can be derived. In a total plant performance mode, predictive technology can be used to measure the effectiveness and efficiency accurately of all company functions, not just machinery. The data generated by regular evaluation can isolate specific limitations in skill levels, inadequate procedures and poor management methods as well as incipient machine or process system problems.

As a maintenance management tool, condition monitoring can provide the data required to schedule both preventive and corrective maintenance tasks on an as-needed basis instead of relying on industrial average life statistics, such as Mean Time To Failure (MTTF), to schedule maintenance activities. Condition monitoring uses direct monitoring of the operating condition, system efficiency and other indicators to determine the actual MTTF or loss of efficiency for each machine train and system within the plant. At best, traditional time-driven methods provide a guideline to normal machine train life-spans. The final decision in preventive or run-to-failure programmes, on when to repair or rebuild a machine, must be made on the basis of intuition and the personal experience of the maintenance manager. The addition of a comprehensive condition monitoring programme, can and will, provide factual data defining the actual mechanical condition of each machine train and operating efficiency of each process system. This data provides the maintenance manager with factual information that can then be used to schedule maintenance activities.

A condition monitoring programme can minimize unscheduled breakdowns of all mechanical equipment in the plant, and ensure that repaired equipment is in an acceptable mechanical condition. The programme can also identify machine train problems before they become serious. Most problems can be minimized if they are detected and repaired early. Normal mechanical failure modes degrade at a speed directly proportional to their severity. Thus, if the problem is detected early, major repairs can be prevented in most instances. To achieve its goals, the programme must correctly identify the root cause of incipient problems. Many of the established programmes do not meet this fundamental requirement. These, precipitated by the claims of condition monitoring system vendors, are programmes established on simplistic monitoring methods, that identify the symptoms rather than the real cause of problems. In such instances derived benefits are greatly diminished. In fact, many of these programmes fail because maintenance managers lose confidence in the programme's ability to detect incipient problems accurately.

A condition monitoring programme cannot function in a void. To be an effective maintenance management tool, it must be combined with a viable maintenance planning function that will use the data generated to plan the appropriate repairs. In addition, it is dependent on the skill and knowledge of maintenance craftsmen. Unless proper repairs or corrective actions are made, the data provided by the programme cannot be effective. Both ineffective and improper repairs will severely restrict programme benefits.

Condition monitoring utilizing vibration signature analysis for example is predicated on two basic facts:

- 'All common failure modes have distinct vibration frequency components that can be isolated and identified.'

- 'The amplitude of each distinct vibration component will remain constant unless there is a change in the operating dynamics of the machine train.'

Condition monitoring utilizing process efficiency, heat loss or other non-destructive techniques can quantify the operating efficiency of non-mechanical plant, equipment or systems. These techniques, used in conjunction with vibration analysis, can therefore provide maintenance managers or plant engineers with all the factual information that will enable them to achieve optimum reliability and availability from their plant.

Condition monitoring can also be an invaluable production management tool. The data derived from a comprehensive programme can provide the information needed to increase production capacity, product quality and the overall effectiveness of the production function. As production efficiency is directly dependent on a number of machine-related factors. Condition monitoring can provide the data needed to achieve optimum and consistent reliability, capacity and efficiency from critical production systems. While these parameters are viewed as maintenance responsibilities, many of the factors directly affecting them are outside the maintenance function. For example, standard operating procedures or operator errors can directly influence these variables. Unless production management uses regular evaluation methods to determine the effects of these production influences, optimum production performance cannot be achieved.

Product quality and total production costs are other areas where condition monitoring can benefit production management. Regular evaluation of critical production systems can anticipate potential problems, that would otherwise result in reduced product quality and an increase in overall production costs. While the only output of a condition monitoring programme is data, this information can be used to correct a myriad of production problems directly affecting the effectiveness and efficiency of the production department. As most product quality problems are the direct result of production systems with inherent problems, poor operating procedures, improper maintenance or defective raw materials. Condition monitoring can isolate this type of problem and provide the data required to correct many of the difficulties which result in reduced product quality.

2.2 VIBRATION MONITORING

There are a variety of technologies that can and should be used as part of a comprehensive, total company programme. Since mechanical systems or machines account for the majority of plant equipment, vibration monitoring is generally the key component of most condition monitoring programmes. However, vibration monitoring cannot provide all the information that will be required for a successful programme. This technique is limited to monitoring the mechanical condition of equipment, and not other critical parameters required to maintain the reliability and efficiency of machinery. It is therefore, a very limited tool for monitoring critical process and machinery efficiencies, and other parameters that can severely limit productivity and product quality. Thus, a comprehensive programme must include other monitoring diagnostic techniques aside from vibration monitoring. These techniques include thermography, tribology, process parameter monitoring, visual inspection and many other non-destructive testing techniques.

Vibration analysis is the dominant technique used for condition monitoring programmes. Consequently, since the greatest population of typical plant equipment is mechanical, this technique has the widest application and benefits in a total plant programme. The technique uses noise or vibration created by mechanical equipment and, in some cases, by plant systems to determine their actual condition. Using vibration analysis to detect machine problems is not new. During the 1960s and '70s the US Navy, petrochemical and nuclear electric power generating industries invested heavily in the development of analysis techniques based on noise or vibration that could be used to detect and identify incipient mechanical problems in critical machinery. By the early 1980s, the instrumentation and analytical skills required for noise-based condition monitoring were fully developed.

These techniques and instrumentation had proven to be extremely reliable and accurate in detecting abnormal machine behaviour. However, the capital cost of instrumentation and the expertise required to acquire and analyse noise data precluded general application of this type of programme. As a result, only the most critical equipment in a few select industries could justify the expense required to implement a noise-based condition monitoring programme. Recent advances in microprocessor technology, coupled with the expertise of companies specializing in machinery diagnostics and analysis technology, have evolved the means to provide vibration-based condition monitoring that can be cost-effectively used in most manufacturing and process applications. These microprocessor-based systems have simplified data acquisition, automated data management and minimized the need for vibration experts to interpret data.

Commercially available systems are capable of routine monitoring, trending and evaluation of the condition of all mechanical equipment in a typical plant. This type of programme can be used to schedule maintenance on rotating, reciprocating and most continuous process mechanical equipment. Monitoring the vibration from plant machinery can provide a direct correlation between the mechanical condition and the recorded vibrational data of each machine in the plant. Any degradation of the mechanical condition within plant machinery can be detected by using vibration monitoring techniques. Used properly, vibration analysis can identify degrading machine components or the failure mode of plant machinery before serious damage occurs. Most vibration-based programmes rely on one or more trending and analysis techniques. These techniques include broadband trending, narrowband trending and signature analysis.

The broadband trending technique acquires overall or broadband vibration readings from selected points on a machine train. This data is compared to either a baseline reading taken from a new machine or to vibration severity charts to determine the relative condition of the machine. Normally an unfiltered broadband measurement that provides the total vibration energy between 10 and 10000 Hertz (Hz) is used for this type of analysis. Broadband or overall Root Mean Square (RMS) data is strictly a gross value or number that represents the vibration of the machine at the specific measurement point where the data was acquired.

It does not provide any information pertaining to the actual machine problem or failure mode. At best, broadband trending can be used as a simple indication that there has been a change in either the mechanical condition or operating dynamics of the machine or system. Consequently, this technique can be used as a gross scan of the operating condition of critical process machinery. However, broadband values must be adjusted to the actual production parameters, such as load and speed, to be effective even in this reduced role. Changes in both the speed and load of machinery will have a direct effect on the overall vibration levels of the machine.

Narrowband trending, like broadband, monitors the total energy for a specific bandwidth of vibration frequencies. Unlike broadband, narrowband analysis used vibration frequencies representing specific machine components or failure modes. This method provides the means to quickly monitor the mechanical condition of critical machine components, not just the overall machine condition. The technique provides the ability to monitor the condition of gear sets, bearings and other machine components without manual analysis of vibration signatures. As in the case of broadband trending, changes in speed, load and other process parameters will have a direct, often dramatic, impact on the vibration energy produced by each machine component or narrowband. To be meaningful, narrowband values must be adjusted to the actual production parameters.

Unlike the two trending techniques above, signature analysis provides a visual representation of each frequency component generated by a machine train. With training, plant staff can use vibration signatures to determine the specific maintenance required by plant machinery. Most vibration-based condition monitoring programmes use some form of signature analysis in their programme. However, the majority of these programmes rely on comparative analysis rather than full root cause techniques.

This failure limits the benefits that can be derived from this type of programme. A more detailed explanation of vibration analysis is given in later chapters of this book.

2.3 THERMOGRAPHY

Thermography is a technique that can be used to monitor the condition of plant machinery, structures and systems. It uses instrumentation designed to monitor the emission of infrared energy, that is temperature, to determine their operating condition. By detecting thermal anomalies or areas which are hotter or colder than they should be, an experienced surveyor can locate and define incipient problems within the plant. Infrared technology is predicated on the fact that all objects having a temperature above absolute zero emit energy or radiation. Infrared radiation is one form of this emitted energy. Infrared emissions are the longest wavelengths of all radiated energy and are invisible without special instrumentation.

The intensity of infrared radiation from an object is a function of its surface temperature. However, temperature measurement using infrared methods is complicated because there are three sources of thermal energy that can be detected from any object. These are the energy emitted from the object itself, energy reflected from the object, and energy transmitted by the object. Only the emitted energy is important in a condition monitoring programme. Reflected and transmitted energies will distort raw infrared data. Therefore, the reflected and transmitted energies must be filtered out of acquired data before a meaningful analysis can be made.

The surface of an object influences the amount of emitted or reflected energy. A perfect emitting surface is called a 'blackbody' and has an emissivity equal to unity. These surfaces do not reflect. Instead, they absorb all external energy and re-emit it as infrared energy. Surfaces that reflect infrared energy are called 'greybodies' and have an emissivity less than unity. Most plant and equipment fall into this classification. Careful consideration of the actual emissivity of an object improves the accuracy of temperature measurements used for condition monitoring. To help users determine emissivity, tables have been developed to serve as guidelines for most common materials. However, these guidelines are not absolute emissivity values for all machines, plant or equipment.

Variations in surface condition, paint or other protective coatings, and many other variables, can affect the actual emissivity factor for plant or equipment. In addition to reflected and transmitted energy, the user of thermographic techniques must also consider the atmosphere between the object and the measuring instrument. Water vapour and other gases absorb infrared radiation. Airborne dust, some lighting and other variables in the surrounding atmosphere can distort measured infrared radiation. As the atmospheric environment is constantly changing, using thermographic techniques requires extreme care each time infrared data is acquired. Most infrared monitoring systems or instruments provide special filters that can be used to avoid the negative effects of atmospheric attenuation of infrared data.

However, the plant user must recognize the specific factors that will affect the accuracy of the infrared data, and apply the correct filters or other signal conditioning required to negate that specific attenuating factor or factors. Collecting optics, radiation detectors and some form of indicator are the basic elements of an industrial infrared instrument. The optical system collects radiant energy and focuses it upon a detector, which converts it into an electrical signal. The instrument's electronics amplifies the output signal and processes it into a form which can be displayed. There are three general types of instruments that can be used for condition monitoring; these are infrared thermometers or spot radiometers, line scanners and imaging systems.

Infrared thermometers or spot radiometers are designed to provide the actual surface temperature at a single, relatively small point on a machine or surface. Within a condition monitoring programme, the point of use infrared thermometer can be used in conjunction with many of the microprocessor-based vibration instruments to monitor the temperature at critical points on plant, machinery or equipment. This technique is typically used to monitor bearing cap temperatures, motor winding temperatures, spot checks of process piping temperatures and similar applications. It is limited in that the temperature represents a single point on the machine or structure. However, when used in conjunction with vibrational data, point of use infrared data can be a valuable tool. Line scanning provides a single dimensional scan of comparative radiation. While this type of instrument provides a somewhat larger field of view, that is an area of machine surface, it is limited in condition monitoring applications.

Unlike other techniques, thermal imaging provides the means to scan the infrared emissions of complete machines, processes or equipment in a very short time. Most of the imaging systems function much like a video camera. The user can view the thermal emission profile of a wide area by simply looking through the instrument's optics. There are a variety of thermal imaging instruments on the market ranging from relatively inexpensive, black and white scanners to full colour, micro-processor-based systems. Many of the less expensive units are designed strictly as scanners and do not provide the capability to store and recall thermal images. The inability to store and recall previous thermal data will limit a long term condition monitoring programme.

Inclusion of thermography into a condition monitoring programme will enable the user to monitor the thermal efficiency of critical process systems relying on heat transfer or retention, electrical equipment, and other parameters that will improve both the reliability and efficiency of plant systems. Infrared techniques can be used to detect problems in a variety of plant systems and equipment, including electrical switchgear, gearboxes, electrical substations, transmission lines, circuit-breaker panels, motors, building envelopes, bearings, steam lines and process systems relying on heat retention or transfer. A more detailed treatment of the equipment involved is presented in a later chapter of this book.

2.4 TRIBOLOGY

This is a general term referring to the design and operating dynamics of the bearing, lubrication and rotor support structure of machinery. Several tribology techniques can be used for condition monitoring and these include lubricating oil analysis, spectrographic analysis, ferrography and wear particle analysis. The first of these, lubricating oil analysis – as the name implies – is an evaluation technique that determines the condition of lubricating oils used in mechanical and electrical equipment. It is not a tool for determining the operating condition of machinery. Some forms of lubricating oil analysis will provide an accurate quantitative break-down of individual chemical elements, both oil additive and con-taminants, contained in the oil. A comparison of the amount of trace metals in successive oil samples can indicate wear patterns of oil-wetted parts in plant equipment and will provide an indication of impending machine failure.

Until recently, tribology analysis has been a relatively slow and expensive process. Analyses were conducted using traditional laboratory techniques and required extensive, skilled labour. Microprocessor-based systems are now available that can automate most of the lubricating oil and spectrographic analysis, thus reducing the manual effort and cost of analysis. The primary applications for spectrographic or lubricating oil analysis are in quality control, reduction of lubricating oil inventories, and determination of the most cost-effective interval for oil change. Lubricating, hydraulic and dielectric oils can be periodically analysed, using these techniques, to determine their condition. The results of this analysis can be used to determine if the oil meets the lubricating requirements of the machine or application. Based on the results of the analysis, lubricants can be changed or upgraded to meet the specific operating requirements.

In addition, detailed analysis of the chemical and physical properties of different oils used in the plant, in some cases, allows consolidation or reduction of the number and types of lubricants required to maintain plant equipment. Elimination of unnecessary duplication can reduce required inventory levels and, therefore, maintenance costs. As a predictive maintenance tool, lubricating oil and spectrographic analysis can be used to schedule oil-change intervals based on the actual condition of the oil. In mid-size to large plants, a reduction in the number of oil changes can amount to a considerable annual reduction in maintenance costs. Relatively inexpensive sampling and testing can show when the oil in a machine has reached a point that warrants change. The full benefit of oil analysis can only be achieved by taking frequent samples and trending the data for each machine in the plant. It can provide a wealth of information on which to base maintenance decisions. However, major payback is rarely possible without a consistent programme of sampling.

Oil analysis has become an important aid to preventive maintenance. Laboratories recommend that samples of machine lubricant be taken at scheduled intervals to determine the condition of the lubricating film that is critical to machine train operation. Typically, the following ten tests are conducted on lube oil samples.

- **Viscosity** This is one of the most important properties of lubricating oil. The actual viscosity of an oil sample is compared to an unused sample to determine the thinning or thickening of the sample during use. Excessively low viscosity will reduce the oil film strength, weakening its ability to prevent metal to metal contact. Excessively high viscosity may impede the flow of oil to vital locations in the bearing support structure, reducing its ability to lubricate.
- **Contamination of oil by water or coolant** This can cause major problems in a lubricating system. Many of the additives now used in formulating lubricants contain the same elements used in coolant additives. Therefore, the laboratory must have an accurate analysis of new oil for comparison.
- **Fuel dilution of oil in an engine** This weakens the oil film strength, sealing ability and detergency. It may be caused by improper operation, fuel system leaks, ignition problems, improper timing or other deficiencies. Fuel dilution is considered excessive when it reaches a level of 2.5–5%.
- **Solids content** This is a general test. All solid materials in the oil are measured as a percentage of the sample volume or weight. The presence of solids in a lubricating system can significantly increase the wear on lubricated parts. Any unexpected rise in reported solids is cause for concern.

- **Fuel soot** This is an important indicator for oil used in diesel engines and is always present to some extent. A test to measure fuel soot in diesel engine oil is important since it indicates the burning efficiency of the engine. Most tests for fuel soot are conducted by infrared analysis.

- **Oxidation of lubricating oil** This can result in lacquer deposits, metal corrosion, or thickening of the oil. Most lubricants contain oxidation inhibitors, however, when additives are used up, oxidation of the oil itself begins. The quantity of oxidation in an oil sample is measured by differential infrared analysis.

- **Nitration** This results from combustion in engines. The products formed are highly acidic and they may leave deposits in combustion areas. Nitration will accelerate oil oxidation. Infrared analysis is used to detect and measure nitration products.

- **Total Acid Number (TAN)** This is a measure of the amount of acid or acid-like material in the oil sample. Because new oils contain additives affecting the TAN number, it is important to compare used oil samples with new, in used oil of the same type. Regular analysis at specific intervals is important to this evaluation.

- **Total Base Number (TBN)** This indicates the ability of an oil to neutralize acidity. The higher the TBN, the greater is its ability to neutralize acidity. Typical causes of low TBN include using the improper oil for an application, waiting too long between oil changes, overheating and using high sulphur fuel.

- **Particle count tests** These are important in anticipating potential system or machine problems. This is especially true in hydraulic systems. The particle count analysis, made as part of a normal lube oil analysis, is quite different from wear particle analysis. In this test, high particle counts indicate that machinery may be wearing abnormally or that failures may occur as a result of temporarily or permanently blocked orifices. No attempt is made to determine the wear patterns, size and other factors that would identify the failure mode within the machine.

Spectrographic analysis allows accurate, rapid measurements of many of the elements present in lubricating oil. These elements are generally classified as wear metals, contaminants or additives. Some elements can be listed in more than one of these classifications. Standard lubricating oil analyses do not attempt to determine the specific failure modes of developing machine train problems. Therefore, additional techniques must be used as part of a comprehensive condition monitoring programme. Normal spectrographic analysis is limited to particle contamination with a size of ten microns or less. Larger contaminants are ignored, a fact which can limit the benefits that may be derived from the technique.

Table 2.1 Five types of wear

Type	Description
Rubbing wear	Result of normal wear in machine
Cutting wear	Caused by one surface penetrating another machine surface
Rolling fatigue	Primary result of rolling contact within bearings
Combined rolling and sliding wear	Results from moving of contact surfaces within a gear system
Severe sliding wear	Caused by excessive loads or heat in a gear system

Wear particle analysis is related to oil analysis only in that the particles to be studied are collected through drawing a sample of lubricating oil. Where lubricating oil analysis determines the actual condition of the oil sample, wear particle analysis provides direct information about the wearing condition of the machine train. Particles in the lubricant of a machine can provide significant information about the condition of the machine. This information is derived from the study of particle shape, composition, size and quantity. Wear particle analysis is normally conducted in two stages. The first stage used for wear particle analysis is routine monitoring and trending of the solids content of the machine lubricant. In simple terms, the quantity, composition and size of particulate in the lubricating oil is indicative of the mechanical condition of the machine. A normal machine will contain low levels of solids with a size less than 10 microns. As the machine's condition degrades, the number and size of the particulate matter will increase.

The second wear particle stage involves analysis of the particulate matter in each lubricating oil sample. Five basic types of wear can be identified according to the classification of particles, as shown in Table 2.1 these are, rubbing wear, cutting wear, rolling fatigue wear, combined rolling and sliding wear and severe sliding wear. Only rubbing wear and early rolling fatigue mechanisms generate particles predominantly less than 15 microns in size.

- Rubbing wear is the result of normal sliding wear in a machine. During the normal run-in of a wear surface, a unique layer is formed at the surface. As long as this layer is stable, the surface wears normally. If the layer is removed faster than it is generated, the wear rate increases and the maximum particle size increases. Excessive quantities of contaminants in a lubrication system can increase rubbing wear more than an order of magnitude without completely removing the shear mixed layer. Although catastrophic failure is unlikely, these machines can wear out rapidly. Impending trouble is indicated by a dramatic increase in wear particles.

- Cutting wear particles are generated when one surface penetrates another. These particles are produced when a misaligned or fractured hard surface produces an edge that cuts into a softer surface, or when abrasive contaminants become embedded in a soft surface and cut an opposing surface. Cutting wear particles are abnormal and are always worthy of attention. If they are only a few microns long and a fraction of a micron wide, the cause is probably a contaminant. Increasing quantities of longer particles signal a potentially imminent component failure.

- Rolling fatigue is associated primarily with rolling contact bearings and may produce three distinct particle types. These are fatigue spall particles, spherical particles and laminar particles. Fatigue spall particles are the actual material removed when a pit or spall opens up on a bearing surface. An increase in the quantity or size of these particles is the first indication of an abnormality. Rolling fatigue does not always generate spherical particles and they may be generated by other sources. Their presence is important in that they are detectable before any actual spalling occurs. Laminar particles are very thin and are thought to be formed by the passage of a wear particle through a rolling contact. They frequently have holes in them. Laminar particles may be generated throughout the life of a bearing, but at the onset of fatigue the spalling quantity increases.

- Combined rolling and sliding wear results from the moving contact of surfaces in gear systems. These larger particles result from tensile stresses on the gear surface, causing the fatigue cracks to spread deeper into the gear tooth before pitting. Gear fatigue cracks do not generate spheres. Scuffing of gears is caused by too high a load or speed. The excessive heat generated by this condition breaks down the lubricating film and causes adhesion of the mating gear teeth. As the wear surfaces become rougher, the wear rate increases. Once started, scuffing usually affects each gear tooth.

- Severe sliding wear is caused by excessive loads or heat in a gear system. Under these conditions, large particles break away from the wear surfaces, causing an increase in the wear rate. If the stresses applied to the surface are increased further, a second transition point is reached. The surface breaks down and catastrophic wear ensues.

Ferrography is a technique similar to spectrography but there are two major exceptions. First, ferrography separates particulate contamination by using a magnetic field rather than burning a sample as in spectrographic analysis. Because a magnetic field is used to separate contaminants, this technique is primarily limited to ferrous or magnetic particles. The second difference is that particulate contamination larger than 10 microns can be separated and analysed. Normal ferrographic analysis will capture particles up to 100 microns and provides a better representation of the total oil contamination than spectrographic techniques.

There are three major limitations with the use of tribology analysis in a condition monitoring programme, these are the equipment costs, acquiring accurate oil samples and the interpretation of the data. The main factor which severely limits the benefits of tribology is the acquisition of accurate samples representing the true lubricating oil inventory in a machine. Sampling is not a matter of opening a port somewhere in the oil line and catching a pint sample. Extreme care must be taken to acquire samples that truly represent the lubricant that will pass through the machine's bearings.

One recent example is an attempt to acquire oil samples from a bullgear compressor. The lubricating oil filter had a sample port on the clean, downstream side. However, comparison samples taken at this point and one taken directly from the compressor's oil reservoir indicated that more contaminants existed downstream from the filter than in the reservoir. Which location actually represented the oil's condition? Neither sample was truly representative of the oil condition. The oil filter removed most of the suspended solids, that is metals and other insolubles, and was, therefore, not representative of the actual condition. The reservoir sample was also not representative, since most of the suspended solids had settled out in the sump.

Proper methods and frequency of sampling lubricating oil are critical to all condition monitoring techniques that use lubricant samples. Sample points consistent with the objective of detecting large particles should be chosen. In a recirculating system, samples should be drawn as the lubricant returns to the reservoir and before any filtration. Do not draw oil from the bottom of a sump where large quantities of material build up over time. Return lines are preferable to reservoir as the sample source, but good reservoir samples can be obtained if careful, consistent practices are used. Even equipment with high levels of filtration can be effectively monitored as long as samples are drawn before oil enters the filters. Sampling techniques involve taking samples under uniform operating conditions. Samples should not be taken more than 30 minutes after the equipment has shut down.

Sample frequency is a function of the Mean Time To Fail (MTTF) from the onset of an abnormal wear mode to catastrophic failure. For machines in critical service, sampling every 25 hours of operation is appropriate. However, for most industrial equipment in continuous service, monthly sampling is adequate. The exception to monthly sampling is machines with extreme loads. In this instance, weekly sampling is recommended. Understanding the meaning of analysis results is perhaps the most serious limiting factor. Most often results are expressed in terms totally alien to plant engineers or technicians. Therefore, it is difficult for them to understand the true meaning in terms of oil or machine condition. A good background in quantitative and qualitative chemistry is beneficial. As a minimum, plant staff will require training in basic chemistry and specific instruction on interpreting tribology results. Further specific information on techniques in this area is given in later chapters of this book.

2.5 OTHER MONITORING TECHNIQUES

Many plants do not consider machine or systems efficiency as part of the maintenance responsibility. However, machinery that is not operating within acceptable efficiency parameters severely limits the productivity of many plants. Therefore, a comprehensive condition monitoring programme should include routine monitoring of process parameters. As an example of the importance of process parameter monitoring, consider a process pump that may be critical to plant operation. Vibration-based condition monitoring will provide the mechanical condition of the pump and infrared imaging will provide the condition of the electric motor and bearings. Neither provide any indication of the operating efficiency of the pump. Therefore, the pump can be operating at less than 50% efficiency and the condition monitoring programme would not detect the problem.

Process inefficiency, like the example, is often the most serious limiting factor in a plant. The negative impact on plant productivity and profitability is often greater than the total cost of the maintenance operation. However, without regular monitoring of process parameters, many plants do not recognize this unfortunate fact. If your programme included monitoring of the suction and discharge pressures and the 'amp' load of the pump, then you could determine the operating efficiency. The brake horsepower formula shown below for example:

$$BHP = \frac{GPM \times TDH \times Sp.Gr}{3960 \times Efficiency} \tag{2.1}$$

where   TDH = Total Dynamic Head,
        GPM = the actual flow,

could be used to calculate operating efficiency of any pump in the programme. By measuring the suction and discharge pressure, the total dynamic head can be determined. A flow curve, used in conjunction with the actual total dynamic head, would define the actual flow and an ammeter reading would define the horsepower. With this measured data, efficiency can be calculated.

Process parameter monitoring should include all machinery and systems in the plant process that can affect its production capacity. Typical systems include the following, heat exchangers, pumps, filtration equipment, boilers and other critical items. Inclusion of process parameter surveillance in a condition monitoring programme can be done in two ways, by manual or microprocessor-based systems. However both methods will normally require installing instrumentation to measure the parameters indicating the actual operating condition of plant systems. Even though most plants have installed pressure gauges, thermometers and other instruments that should provide the information required for this type of programme, many of them may no longer be functioning. Therefore, including process parameter monitoring in your programme will require an initial capital cost to install calibrated instrumentation.

Data from the instrumentation can be periodically recorded using either manual logging or a microprocessor-based data logger. If the latter is selected, many of the vibration-based microprocessor systems can also provide the means of acquiring process data. This should be considered when selecting the vibration monitoring system that will be used in your programme. In addition, some of the microprocessor-based condition monitoring systems provide the ability to calculate unknown process variables. For instance, they can calculate the pump efficiency used in the example above. The ability to calculate unknowns based on measured variables will enhance a total plant condition monitoring programme without increasing the effort required. In addition, some of these systems include non-intrusive transducers that can measure temperatures, flows and other process data without the necessity of installing permanent instrumentation. This further reduces the initial cost of including process parameter monitoring in your programme.

Evaluation of electric motors and other electrical equipment is critical to a total plant condition monitoring programme. To some extent, vibration data can isolate some of the mechanical and electrical problems that can develop in critical drive motors. However, vibration cannot provide the comprehensive coverage required to achieve optimum plant performance. Therefore, a total plant condition monitoring programme must include data acquisition and evaluation methods specifically designed to identify problems within motors and other electrical equipment. Insulation tests are important, although they may not be conclusive, since they can reveal flaws in insulation, poor insulating material, the presence of moisture and a number of other problems. Such tests can be applied to the insulation of electrical machinery from the windings to the frame, and also to underground cables, insulators, capacitors plus a number of other auxiliary electrical components.

Normally these tests are conducted using a megger, wheatstone bridge, Kelvin double bridge or a number of other instruments. A megger provides the means to measure the condition of motor insulation directly. This method uses a device which generates a known output, usually 500 volts, and directly measures the resistance of the insulation within the motor. When the insulation resistance falls below the prescribed value, it can be brought to required standards by cleaning and drying the stator and rotor. The accuracy of megging and most insulation resistance tests varies widely with temperature, and cleanliness of the parts. Therefore, they may not be absolutely conclusive. A complete condition monitoring programme should include all testing and evaluation methods required to regularly evaluate critical plant systems. As a minimum, a total plant programme should also include dielectric loss analysis, gas oil analysis, stray field monitoring, high-voltage switchgear discharge testing, resistance measurements, Rogowski coils and rotor bar current harmonics.

Regular visual inspection of the machinery and systems in a plant is a necessary part of any condition monitoring programme. In many cases, visual inspection will detect potential problems that will be missed using the other condition monitoring techniques. Even with the predictive techniques discussed, many potentially serious problems can remain undetected. Routine visual inspection of all critical plant systems will augment the other techniques and ensure that potential problems are detected before serious damage can occur. Most of the vibration-based condition monitoring systems include the capability of recording visual observations as part of the routine data acquisition process. Since the incremental costs of these visual observations are small, this technique should be incorporated in all condition monitoring programmes. All equipment and systems in the plant should be visually inspected on a regular basis. The additional information provided by visual inspection will augment the condition monitoring programme regardless of the primary techniques used.

The ultrasonic monitoring technique uses principles similar to vibration analysis. Both monitor the noise generated by plant machinery or systems to determine their actual operating condition. Unlike vibration monitoring, ultransonics monitors the higher frequencies, that is ultrasound, produced by unique dynamics in process systems or machines. The normal monitoring range for vibration analysis is less than 1 to 20 000 Hertz. Ultrasonic techniques monitor the frequency range between 20 000 and 100 000 Hertz. The principle application for ultrasonic monitoring is in leak detection. The turbulent flow of liquids and gases through a restricted orifice, that is a leak, will produce a high-frequency signature that can be easily identified using ultrasonic techniques. This technique is ideal for detecting leaks in valves, steam traps, piping and other process systems.

Two types of ultrasonic systems are available that can be used for condition monitoring. These are structural and airborne. Both provide fast, accurate diagnosis of abnormal operation and leaks. Airborne ultrasonic detectors can be used in either a scanning or contact mode. As scanners, they are most often used to detect gas pressure leaks. Because these instruments are sensitive to ultrasound, they are not limited to specific gases as are most other leak detectors. In addition, they are often used to locate various forms of vacuum leaks. In the contact mode, a metal rod acts as a waveguide. When it touches a surface, it is stimulated by the high frequencies, ultrasound, on the opposite side of the surface. This technique is used to locate turbulent flow and or flow restriction in process piping.

Some of the ultrasonic systems include ultrasonic transmitters that can be placed inside plant piping or vessels. In this mode, ultrasonic monitors can be used to detect areas of sonic penetration along the container's surface. This ultrasonic transmission method is useful in quick checks of tank seams, hatches, seals, caulking, gaskets or building wall joints. In a typical machine, many other machine dynamics will also generate frequencies within the bandwidth covered by an ultrasonic instrument. Gear-meshing frequencies, blade pass and other machine components will also create energy or noise that cannot be separated from the bearing frequencies monitored by this type of instrument. The only reliable method of determining the condition of specific machine components, including bearings, is vibration analysis. The use of ultra-sonics to monitor bearing condition is thus not recommended.

The operating dynamics analysis method is driven by machine design and is not limited to traditional analysis techniques. The diagnostic logic is derived from the specific design and operating characteristics of the machine train or production system. Based on the unique dynamics of each machine train or system, all parameters that define optimum operating condition are routinely measured and evaluated. Using the logic of normal operating condition, operating dynamics can detect, isolate and provide cost-effective corrective action for any deviation from the optimum. Operating dynamics analysis combines traditional condition monitoring techniques into a holistic evaluation technique that will isolate any deviation from optimum condition of critical plant systems. This concept uses raw data derived from vibration, infrared, ultrasonics, process parameter monitoring and visual inspection but applies a unique diagnostic logic to evaluate critical plant systems. More detailed information on many of the techniques mentioned above is included in other chapters of this book.

2.6 CONCLUSIONS AND FINANCIAL IMPLICATIONS

There are numerous other non-destructive techniques that can be used to identify incipient problems in industrial plant, equipment or systems.

However, these techniques either do not provide a broad enough application or are too expensive to support a condition monitoring programme. Therefore, these additional techniques are normally used as the means of confirming failure modes identified by the main condition monitoring techniques mentioned in this chapter. Other techniques supporting condition monitoring include acoustic emission, eddy current, magnetic particle, residual stress and most of the traditional non-destructive methods, some of which are outlined in later chapters of this book.

The initial and recurring costs required to establish and maintain a comprehensive condition monitoring programme will vary with the technology and type of system selected for plant use. While the initial or capital cost is the more visible, the real cost of a programme is the recurring labour, training and technical support required to maintain a total plant programme. The capital cost for implementing a vibration-based condition monitoring programme can range from about $8000 to more than $50 000, with costs depending on the specific techniques desired.

Training is critical for condition monitoring programmes based on vibration monitoring and analysis. Even programmes relying strictly on the simplified trending or comparison techniques require a practical knowledge of vibration theory so that meaningful interpretation of machine condition can be derived. More advanced techniques, that is signature and root-cause failure analysis, require a working knowledge of machine dynamics and failure modes.

2.7 BIBLIOGRAPHY

Mobley, R.K. (1990) *An Introduction to Predictive Maintenance*, Van Nostrand Reinhold, New York, USA.

PART TWO

Techniques for Visual Inspection

CHAPTER 3

Visual inspection systems

A. Davies

Systems Division, School of Engineering, University of Wales College of Cardiff (UWC), P.O. Box 688, Queens Buildings, Newport Road, Cardiff, CF2 3TE, UK

3.1 INTRODUCTION

This technique is probably the simplest and most cost-effective method of condition monitoring. Essentially it consists of the visual inspection of machinery or a product, using some form of optical/computer-based assistance, or the unaided eye. The computer system being used for data acquisition, information logging or analysis. One obvious advantage of the technique is that at its most unsophisticated, it is a cheap and easy method to employ, and one which provides the rudiments of a condition monitoring scheme. Another is that in most cases a direct and immediate indication of machinery condition is available, via the use of what is basically a simple technique.

The employment of personnel in this way, to provide flexible inspection/monitoring, is also attractive to maintenance managers. For the inspectors, who could also be production operators, may only require a fairly low level of skill and/or experience, with possibly some simple specialist training, to become very effective ancillary maintenance personnel. The disadvantages associated with this type of condition monitoring are, firstly that it may be limited to items of equipment which are stationary at the time of inspection, and secondly that it usually requires direct optical or sensor access. Normally, because it is in essence a condition checking technique, where interpretation is undertaken during the inspection, a record or measurement of the equipment state or component surveyed may not necessarily be noted.

However, if automatic electronic data acquisition, analysis and trend monitoring is also being done in conjunction with visual inspection, records may be kept via the use of a variety of techniques. These can

Handbook of Condition Monitoring
Edited by A. Davies
Published in 1998 by Chapman & Hall, London. ISBN 0 412 61320 4.

include portable data or text collectors, still photography, sound tape or video recording, surface printing or casting, and witness indenting. A wide range of techniques and equipment are now available to help implement effective visual inspection. Some of the most common are outlined in the sections that follow, together with a commentary on their usefulness in this form of condition monitoring. A look into the future is also provided in the sections on automated visual inspection, and the state of the art, where new and exciting technologies are shown to be under development. Many of these have implications for condition monitoring, together with a potential application in industrial inspection and maintenance.

3.2　UNSOPHISTICATED LOW-COST AIDES

Borescopes and fibrescopes are essentially optical systems for inspecting components in complex machinery which are normally inaccessible without major plant stripdown. They are available in a rigid or flexible form, with or without magnification and/or lighting attachments; and can be very useful, particularly where small access holes have been included in the equipment design. They are widely used throughout industry as a relatively low-cost aid to the internal visual surveillance of machinery, and a typical use of this equipment is the inspection of aircraft engine components during pre-flight or routine maintenance. Some systems have the capability for use with photographic or closed-circuit television equipment, thereby allowing trend monitoring to take place by providing a means by which component condition records may be kept (Davies, 1990).

Stroboscopes are another useful tool for visual inspection. They are essentially a lamp and associated apparatus which allow the direct visual inspection of moving parts by producing a rapidly flashing bright light. The rate at which the light flickers may be adjusted and in so doing the equipment under inspection can be made to appear stationary. By operating the strobe at a slightly different frequency to the rotating or reciprocating movement generated by the equipment, any pronounced vibrational characteristic, imbalance or gap stretching can be observed. This may then be corrected, and the equipment subsequently re-inspected after the maintenance is carried out, to ensure that the repair is satisfactory. The use of stroboscopes in industry is widespread, and involves their application to a variety of rotating and reciprocating machinery, for which they are a low cost and quite valuable inspection tool (Davies, 1990).

Crack detection is another area in which visual inspection is heavily involved, and several low-cost aides are available to improve the efficiency of this task over simple eyeball examination. Dye penetrants for example are cheap and useful for the detection of surface cracking in machine components, typical applications being on support castings and on fabricated items to check the integrity of welded joints etc. Other low-cost systems for surface crack detection include those based on a variation of magnetic flux or electrical resistance. However, some skill is required in the operation of both these systems for efficient crack detection, and the magnetic flux equipment is fairly obviously limited to finding cracks in magnetic materials. In addition, both techniques are sensitive to crack orientation and clean surfaces, thus care is required during use or the crack will be missed. The electrical resistance method can also give some indication of crack depth (Dressel *et al.*, 1992).

Sub-surface defects, where no visual access is possible, can be detected by manual inspection if two other relatively low-cost techniques are used. Skill is essential when using equipment based on the principle of a variation in eddy current, for the proximity of the probe to the surface of the component under inspection can affect the results. However, a wide range of material discontinuities such as near surface cracks and inclusions can be detected by a skilled operator. In a similar fashion, equipment based on the use of ultrasonics requires skill to operate successfully. This type of detection is directionally sensitive and searches may be lengthy. It does, however, detect cracks anywhere in the component providing there is access through a clean smooth surface (Dressel *et al.*, 1992).

Corrosion is another area which is often the subject of visual inspection. A simple method of detection in component materials is by use of a corrosometer, which is basically a low-cost electrical potentiometer. The meter does require some skill for effective use and detection levels down to 1 micron are claimed. Equipment temperature levels may also be checked visually via low-cost thermographic paints, crayons and tapes. Typical applications are on equipment such as electric motors, which may prior to failure exhibit an increase in temperature. The usual range of temperature measuring devices such as thermometers and thermocouples etc. can also be used effectively, depending upon the circumstances (Davies, 1990). In addition, low-cost hand-held electronic sensing and digital readout systems such as the infrared-red 'Pyroscan' are now readily available as effective aids to visual inspection (Anon, 1995a).

3.3   PORTABLE INFORMATION LOGS

Two useful and customized computer aids, which are available to support visual inspection, are portable data-loggers and pen-based computer systems. Both, in their modified form, are specifically intended for use with computer-based Maintenance Management Information

Table 3.1 Attributes, advantages and disadvantages of portable data logging systems

Attributes	Advantages	Disadvantages
• 286 processors	Specifically tailored to application	Cost versus non-specific application systems
• DOS-based	Rugged systems designed for field use often for specific environment	Inflexibility, usually linked to specific application and company product
• Analogue and digital data acquisition	Linked to specific application software running on desk-top PCs	
• Output channels to control external devices or sensors	DOS and PC programming/ application software compatible	
• 4 MByte internal RAM	Menu-driven and easy to use	
• Memory card drive	Wide range of sensors available to use with the system	
• Shock impact hardened	Can run 'Windows' type programmes on some machines	
• Waterproof	Cost-effective data collection	
• Temperature hardened		
• Plug in application modules		
• Lightweight		

Systems (MMIS). In addition, there are various 'palmtop' computers, 'organizers' and 'notebooks' available, which are personal computer (PC) compatible, and thus may be used with an integrated MMIS system. As the essence of a condition monitoring programme is an organized schedule of periodic inspections, one of the major problems of a manual system, that of the accurate collection of information relating to the plant under inspection, is now close to solution, via the use of these relatively low-cost computer aids.

Table 3.1 outlines the various main attributes, advantages and disadvantages of the wide range of portable data loggers which are now available for use in maintenance. Many of these systems are linked to a specific application and/or the implementation of a particular condition monitoring technique, the collection of machinery vibration data via this type of system, being a good example of their use in practice. Figure 3.1

Fig. 3.1 Typical portable vibration data collector/analyser.

illustrates a typical portable vibration data collector/analyser. As a result of the many advances in computer technology, self-contained sensing units for data acquisition, and the use of sophisticated software, several of these systems can now analyse data on site, thereby assisting in the rapid inspection of plant or equipment. More sophisticated data analysis, for deep equipment diagnostic checks and trend monitoring, can then be undertaken on a larger office-based system, when data is downloaded from the portable recording unit (Anon, 1994a).

Aside from the vibration monitoring role, there are portable units available for use in a number of other applications, such as measuring the water content in oil (Figure 3.2) (Anon, 1995b), particle counting in a variety of fluids (Figure 3.3) (Anon, 1994b), and temperature measurement. Other, non-specific portable data-loggers are also available. These feature a number of analogue/digital input/output channels which may be used to record or implant equipment information. Pen computers, for example (Figure 3.4), are an interesting development in this area. Basically, they are portable computer systems which have a pen interface rather than a keyboard, and this permits both text and graphical information capture. Table 3.2 outlines their main attributes, advantages and disadvantages. They have been field-tested in typical maintenance environments, and the results are claimed to support their use in this application (Billson, 1992, 1993).

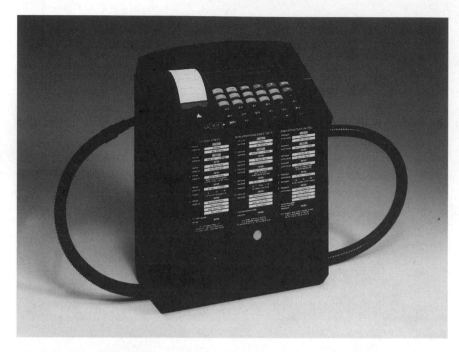

Fig. 3.2 Portable unit for measuring the water content in oil. (By kind permission of UCC International Ltd)

Fig. 3.3 Portable unit for particle counting in a variety of fluids. (By kind permission of UCC International Ltd)

Fig. 3.4 Examples of pen computers. (By kind permission of Kerry Technology Ltd)

Table 3.2   Attributes, advantages and disadvantages of pen computers

Attributes	Advantages	Disadvantages
• Simple pen-driven interface	Up to date on site repair information available with high mobility	Product still under development to achieve its full potential
• Simple data comms to MMIS database	Standardization and accuracy of data collection–low error rate	Cost when compared with paper systems
• Access to text and diagrams from maintenance manuals	Reduction in paperwork and transcription time or errors to MMIS database	Compatibility with existing MMIS hardware or software
• May incorporate digitized speech recognition and photography in the future	Reduction in time to enter field data hence more efficient	Time to design special data entry screens
• Lightweight, battery-powered with strong rugged construction	Security of data entry via signature capture and verification	Careful systems management required for security, data transfer, battery-charging etc.
• Modem and FAX modem available as options	Operator goodwill via use of hi-tech tool and the availability of data	
• Optional backlighting	Lower costs of data transfer, standardization interpretation and analysis	
• Fluid spill proof	Quick copy and editing	
• Graphical information capture	Easy to use big display	
• 3D digitizer available		

Currently, pen computers are still under development for niche-market applications such as maintenance, and one obvious addition to their capabilities for this area should be the automatic receipt/transfer of electronic data via inbuilt systems. The use of palmtop computers/ personal organizers in maintenance is also increasing (Figure 3.5), and in many ways they overlap the functions of dedicated portable data-loggers at a reduced cost. This is obviously attractive within a limited maintenance budget, plus the fact that there are a wide variety of such units available, with differing capabilities, to suit every purse (Anon, 1993). Table 3.3 outlines their main attributes, advantages and disadvantages for use in maintenance. One obvious disability is their cost when compared to notebook PCs and the requirement/time to 'tailor' them for a specific application.

Fig. 3.5 Palmtop computer (personal organizer).

Table 3.3 Attributes, advantages and disadvantages of palmtop computers and organizers

Attributes	Advantages	Disadvantages
• Fully DOS capable	Application driven product	High cost versus capability when compared to notebook PCs
• 640k RAM (max)	Software compatibility with desktop PCs	Limited memory capacity
• XT class processors	Reduced cost of data collection	Not 'Windows' capable
• Alkaline batteries supplied as standard	Improved data integrity	Display technology screen resolution low
• On-board data transfer	Easy to use	Low battery life
• Serial link to desk top PC (cable or infrared)		
• Rechargeable NiCAD batteries available on some systems		

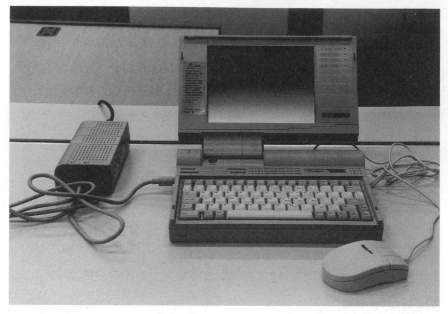

Fig. 3.6 Typical Notebook PC.

Notebook PCs on the other hand are now highly capable systems, with some units having facilities which rival the desktop PC at an equivalent or lower cost. Figure 3.6 illustrates a typical unit. Being general-purpose computer systems and fully Windows™-capable, their use in maintenance is only limited by the imagination of the user to design suitable application programmes. With judicious enhancement, these units can perform the functions of data-loggers, pen computers, palmtops and organizers. In all probability, and subject to the maintenance sector's providing sufficient market stimulus, a single computerized aid will evolve based on these systems, and enhanced by the best features of the other portable units mentioned above. Table 3.4 outlines the main attributes, advantages and disadvantages of the notebook-type PCs (Anon, 1995c).

3.4 MORE SOPHISTICATED SYSTEMS

Thermography involves the use of a colour thermographic camera to extend human vision into the infrared region of the electromagnetic spectrum. Infrared radiation is invisible to the human eye at temperatures below 525 degrees Celsius and the use of this technique permits the presentation of 'real-time' images to the operator. These can be supplemented by quantitative analysis to give accurate temperature

Table 3.4 Attributes, advantages and disadvantages of 'Notebook' PCs

Attributes	Advantages	Disadvantages
• 486 processors	General purpose PC	Cost/benefit depends on m/c and application
• 50–100 MHz clock speed	Full 'Windows' capability	Small screen tiring for extensive data entry
• 8 MB internal RAM	DOS-based	Not hardened for an industrial use
• £1000–£3000 price tag	Desktop compatible	Fairly short battery life
• NiMh batteries	Existing MMIS compatible	3–4 kg weight range
• Hard/floppy/CD disk drives fitted	Cost over other systems	Customized software may be required
• Good expansion potential	Reduced cost of data entry	Security – m/c theft possible
• Good network connectivity	Improved data integrity	Management of data transfer
• Colour screens	Easy to use	Management of battery charging
• 2–3 hour battery life	Standard software available	Power supply, mouse and portable printer may be required
• Built-in printer possible	In-house programming possible	
• Lightweight and portable	Pentium chip notebooks available	
• Modem and fax modem available		
• Power supply, mouse and portable printer available		

measurements and the facility to identify isotherms. The equipment is expensive, but portable, highly reliable, non-contacting, and single operator-based.

These are all characteristics which are useful when equipment is used in high temperature or dangerous environments. Typical applications include the detection of rogue hot/cold spots in equipment and heat losses from buildings, hot vessels and pipelines. They have also been used on electrical switchgear, printed-circuit boards, control equipment, motors and hydraulic systems. The technique does require some training in the use of the system, together with an appreciation of the effects of different target surface emissitivities and the cooling characteristics of draughts on the target surface.

Radiography has also been used as an aid to visual inspection. The techniques available involve either the X-ray examination of the condition of internal components, or the Gamma-ray examination of hollow machinery and components. Both are expensive requiring considerable skill and may involve the need to access both sides of the equipment under investigation. X-ray equipment tends to be large and not easily portable, while the Gamma-ray source is small and can often be placed inside equipment. Security for both techniques is essential due to the radiation hazard and both may be limited by section size. A typical application is the inspection of aircraft engine turbine blades for defects in cooling passages.

The use of laser systems to assist in the visual inspection of machinery and equipment is in certain circumstances cost-effective, and on the increase, as this type of equipment finds more and more application niches. They are basically precision surveying instruments which can measure distance and angle very accurately. However, they are expensive but also portable, reliable and non-contacting systems which require some skill in set up and operation. Machine-tool spindle/slideway accuracy testing after assembly is a typical application for this technique, although it has been used in a number of other novel situations such as to measure wear in refractive linings on steelmaking plant.

Sophisticated illumination chambers are another approach to assisting the human eye for the purposes of visual inspection. Some of the latest technology in this respect, has been used to permit the examination of high gloss paint finishes on motor vehicles and thereby identify surface blemishes. For this application, the walls and ceiling of an inspection tunnel were clad with specially designed illumination panels. These provide a banded light pattern which highlight areas of poor paintwork, and thus allow human inspectors to readily spot defects in the paint finish. The curvature of surfaces to be inspected does not limit the performance of this system and any low contrast surface distortions are ignored. A variation in the light patterns used allow different types of defect to be targeted and the system can be applied to any product with gloss surface finish (Anon, 1995d).

3.5   AUTOMATED VISUAL INSPECTION

Although we accept the gift of sight as something quite natural for human beings, together with the ability it confers on us to differentiate between 'good' and 'bad' in many situations, automating this process requires that we technologically mimic what is a substantial visual ability. This is not easy given our current level of technology, because most industrial inspection tasks require some level of flexibility. Accordingly, the hallmark of current automated visual inspection applications is their specialization. Normally, the item under surveillance is highly constrained and the camera systems used to perform the inspection are not generic. However, advances are being made via the use of multiple imaging, colour and range data. In addition, computer processing is less expensive in terms of time, memory and cost, allowing desktop PCs to be used for many industrial applications (Newman and Jain, 1995).

In a condition monitoring context, inspection can be defined as:

Identifying the deviation of a part, object, signal, item or product from a given set of specifications or values on many different types of equipment.

Accordingly, it usually involves the assessment of specific features, in both a qualitative and quantitative sense. Plus incorporating a 'testing' element which may be defined as:

The active examination of specific equipment operational functions for deviation from nominal conditions.

In addition, 'recognition' and 'diagnostic' factors are present which are defined as:

The ability to detect deviations and to link cause with effect.

It may therefore be seen that the flexibility requirement involved in automating this process presents some quite considerable technological challenges. Indeed, without human intervention, it is unlikely to be achieved, except in some highly constrained cases.

Figure 3.7 illustrates the linkage between condition monitoring and the various inspection stages in a typical manufacturing company environ-

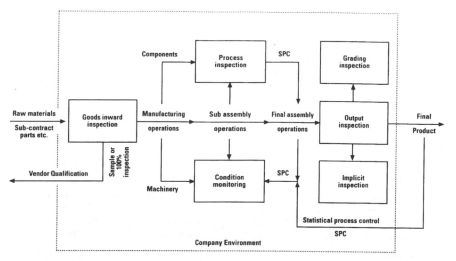

Fig. 3.7 Relationship between condition monitoring and the various stages of industrial inspection.

ment. It may be seen from this diagram that human and automated visual inspection have a part to play in the monitoring of the production process. The visual inspection of machinery being part of the condition monitoring role normally undertaken by maintenance, and the inspection of components being traditionally part of the quality-assurance process. A tight linkage is shown in the diagram between process inspection and condition monitoring. This linkage is an essential adjunct to condition monitoring, for Statistical Process Control (SPC) information is extremely valuable data, which can often help in the diagnosis of equipment failure. Unfortunately, this linkage is sometimes not present in an industrial situation to the detriment of the manufacturing system.

The use of SPC data, which can often be gathered by automated means, gives a new dimension to the application of visual inspection for the condition monitoring of machinery. Figure 3.8 illustrates the enhancement possible when process inspection is linked to traditional monitoring. As may be seen from the examples given in the diagram, the range of checks is broadened and includes both qualitative and quantitative inspection activities. The use of process and final inspection data as a

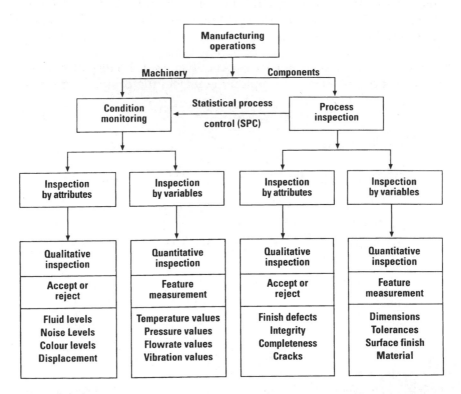

Fig. 3.8 Relationship between condition monitoring and process inspection.

facet of plant condition monitoring is increasing, especially in the context of automated machining, fabrication and assembly systems for the detection of tool wear and breakage. Systems with automatic adjustment to compensate for tool wear, in real time via adaptive control and based on such inspection data, are also increasing in popularity (Davies, 1995). In addition, for a good many components, the development and use of Just-In-Time manufacturing procedures have also increased the sensitivity of product quality, and thereby the requirement for 100% process inspection plus effective plant condition monitoring.

As human visual inspection is estimated to be only 80% effective when implemented via a rigidly structured checklist, and the fact that many inspection tasks are both time-consuming and boring (Smith, 1993). The attractiveness of automated visual inspection becomes clear, especially when linked to condition monitoring, and with the knowledge that inspection tasks can account for 10% or more of total labour cost in some applications (Mair, 1988). In addition, and as shown above, many part or machinery defects are too subtle for detection by unaided personnel. There being a limit to what can be achieved in this respect even when an active Total Quality/Productive Maintenance (TQM/TPM) programme is in effect. This is particularly so where the manufacturing environment may be inherently unsafe for regular personnel access. Automated visual inspection, being in general a non-contact technique, does not suffer from these disadvantages, and accordingly the risk to personnel, product or machinery of damage during surveillance is minimal.

The main defect associated with automated visual inspection is that such systems are expensive and time-consuming to develop, plus the fact that in general they are not adaptable to new situations, tending to be customized in respect of lighting requirements, part orientation and image analysis for effective operation. In addition they do not have the flexibility of human beings to handle unexpected component positions, orientations or objects. Even when linked to Intelligent Knowledge Based Systems (IKBS), their ability to handle difficult value judgements has to be suspect (Wright and Bourne, 1988). However, as previously indicated, the great advantage of these systems for condition monitoring lies in the fact that they can accurately record error rates for each defect. This makes machinery diagnosis easier and workpart redesign possible if this is required to reduce plant failure. It also has the effect of integrating maintenance more closely with production.

3.6 STATE-OF-THE-ART

There is a vast range of literature relating to automated visual inspection systems, much of it concentrating on the reporting of specific applica-

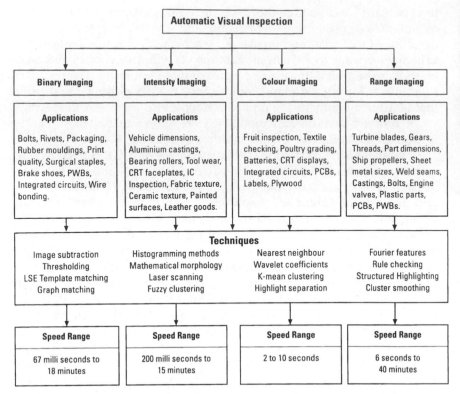

Fig. 3.9 Main approaches, applications, techniques and speed in AVI.

tions or techniques. Figure 3.9 illustrates the main subdivisions within this area of scientific research, along with the known applications and techniques used. It may be seen in the figure that there are four basic sensing arrangements, that is, binary, intensity, colour and range imaging, together with some systems which use alternative sensing modes. In general, the literature appears to reflect a two pronged approach to inspection problems. Either a concentration on developing specialized techniques, that allow the inspection of a single workpart, or an attempt to develop more generic inspection methodologies, which are applicable in a wider inspection environment.

It should also be noted that many of the applications quoted in the literature employ imaging constraints and special lighting arrangements. Thus the surveillance task is simplified, by reducing computational demands, increasing inspection speed and lowering system cost. As indicated in Figure 3.9, there are a large number of commercial surveillance systems in use and these cover a wide range of inspection tasks. Although most of these refer to product inspection, there are applications in the condition monitoring area. One example is a robot system which visually inspects for steam or water leakage in a high radiation area of a nuclear power plant (Yamamoto, 1992). Another, is the inspection of cutter tips for tool wear (Wright and Bourne, 1988).

Binary imaging is the most common technique used in commercial systems, although it produces images of only limited resolution and tends to be used to detect the presence or absence of a workpart. In some cases binary information is sufficient to inspect workparts via their silhouette images. Flat objects without surface features are typical examples, where it may be possible to use cheap sensors and a high-contrast background or backlighting to achieve the required results. Binary vision systems have the advantage of reducing the amount of data that needs to be processed, thus helping to reduce inspection time and total cost. Unfortunately, due to their lighting requirements, it is usually not possible to use orientation, texture or reflectance information in the inspection procedure (Chin, 1992). It is also important to note that binary images are inadequate for the inspection of surface characteristics or three-dimensional shapes.

Grey-level intensity imaging is another popular approach in the development of automated inspection systems, although at present only a few of the systems quoted in the literature are capable of operating in a normal industrial environment. Dirt, smoke and fumes tend to degrade their efficiency, along with a lack of specialist lighting conditions. In a condition monitoring sense, however, they have been used to inspect bearing rollers, detecting cracks, nicks, scratches and rust in a difficult environment, where surface oil and grease smears have been present (Xian *et al.*, 1988). Another application, recorded in the literature, is in the inspection of drill bits for tool wear (Liu, 1990). In this case worn areas of the tool were found to have a higher reflectivity than the unworn surfaces. It is possible to inspect for texture using intensity levels, but as with binary imaging, it may not be appropriate for looking at surface characteristics or three-dimensional shapes.

Colour imaging is a useful approach in automated visual inspection where defects are revealed by a change in this important factor. One advantage, which is pushing the introduction of these systems, is that human inspection of coloured objects is suspect, as apparently we have a poor memory for colour (Daley and Rao, 1990). The disadvantages holding back the use of colour systems for industrial inspection applications are the requirements for extensive computing power, intricate optics and controlled lighting. Other factors include light source intensity, colour temperature, angle of illumination, lens aperture and magnification. These are all due to the sensitivity of the system and the energy distribution from the light source (Daley and Rao, 1990).

In contrast to binary, intensity or colour; range imaging has the advantage that it explicitly represents surface information and this allows the shape data to be used for the efficient matching of three-dimensional features (Marshall, 1989). Unfortunately, few if any, three-dimensional industrial standard inspection systems for complex objects are available (Chin, 1992). Although this approach, plus the use of structured light, has been used for industrial component inspection (Pryor, 1982). This type of system tends to be highly optimized, and constrained for use in a single application. Efforts are being made to overcome some of these disadvantages, by investigating automatic task learning and rule generation for these systems (Freeman, 1989).

As outlined earlier, optical inspection is not necessarily confined to that part of the spectrum which is visible to the human eye. For example, considerable work has been done on the automated X-ray inspection of industrial parts (Hedengren, 1986) and other techniques such as scanning electron microscopy (Shu *et al.*, 1988), ultrasound imaging (Casasent and Carender, 1989) and infrared sensing (Durrani *et al.*, 1987). X-rays have the advantages of high contrast and a potential for real time imaging, although safety considerations need to be born in mind for any potential application. Gamma-rays are similar to X-rays and can be used on denser materials, with images obtainable in real time but of a lower contrast to that of X-rays. Safety considerations are of course paramount in the case of Gamma-ray systems which cannot be switched off (Chan *et al.*, 1990).

Magnetic Resonance Imaging (MRI) has been used for the inspection of food. However, it is a very expensive process and at present not really viable for an industrial application. Ultrasound imaging has also been used for food inspection, and this technique has the advantage of being a safe process. Unfortunately, it also has the disadvantage of requiring special transducers which are not well suited to an industrial environment. Thermal imaging using the infrared part of the spectrum has also seen application in this context. The sensors used being flexible and easily calibrated, with no safety considerations to worry about. They have the disadvantage, however, of limited sensitivity, and images they produce may require preprocessing. Ultraviolet light has also been used to produce a fluorescent analysis of food. There are safety considerations, however, and the colour processing required does add some complexity to the system (Newman and Jain, 1995).

In a more industrial context, automated visual inspection systems have used some of the defect enhancement techniques mentioned earlier, such as magnetic particle and liquid penetrant testing. These methods were originally developed to aid the human eye, but have been adapted for use in automatic inspection. In applications on cast or forged parts, for example, surface defects have been highlighted for automatic inspection systems by using coloured or fluorescent dyes. The magnetic particle method has been used to determine the position of near-surface non-metallic inclusions in castings. Both inspection systems use ultraviolet light to enhance the contrast of defective areas and thus improve the detection rate of the automatic system. The magnetic particle technique has the advantage of being less affected by surface geometry, roughness and material structure. Accordingly, it has been used to inspect a wide range of industrial components including casting rods, coil springs, axle housings, turbine blades and wheel hubs (Newman and Jain, 1995).

In general, automated inspection systems fall into two separate categories. The grouping depends upon the approach used for a particular application, and each may be summarized as follows:

- the automatic comparison of a sensed image of the object under inspection, with a synthetic image or template of a defect free model of the component;
- the automatic comparison of features extracted from the sensed image of an object under inspection, with a description or list of rules that describe an ideal model of the component.

Several different schemes can be used to implement these approaches including pixel by pixel matching, and use of Computer Aided Design (CAD) models. The pixel by pixel comparisons are quite time-consuming and the matching strategies used are heavily researched in order to improve efficiency. This type of matching becomes very slow if the orientation, position and scale of the image is unknown (Van Gool *et al.*, 1991). In like fashion, feature extraction can also be a slow process, although rule-based schemes have the advantage of not requiring an extensive database of templates. The disadvantages of rule-based schemes are that they are less adaptable to design changes than the pixel-based systems, and more complex in terms of formulating the rules from, say, a CAD-based model. In addition, they may require complex software to eliminate false defect detection. Hybrid arrangements of these two approaches have also been used in practical applications.

3.7 CONCLUSIONS

As outlined above, optical inspection covers a very wide area. It ranges from the simple, unaided 'policing' of plant and equipment for obvious defects, to sophisticated automatic visual inspection, using camera and computer technology. In between these extremes, there exists a wide range of equipment which is available to assist the human eye in the inspection task. Much of the equipment is well established in niche markets, and linked to condition monitoring either directly, or via process inspection and the use of SPC data.

In future, it is likely that, with the increasing use of computerization and the networking of information, the use of this type of equipment to aid inspection/condition monitoring tasks will increase. More PC-based application solutions are likely to become available via proprietary software, and costs will reduce as computer technology and usage improve. In the case of automated systems, further research should improve their flexibility, speed and capability, making them more useful in a generic sense for process inspection and condition monitoring.

3.8 REFERENCES

Anon (1993) Palm readings, *Computer Shopper*, February.

Anon (1994a) *Condition Monitor*, **95**, 12.

Anon (1994b) *Condition Monitor*, **86**, 12.

Anon (1995a) *Condition Monitor*, **98**, 9.

Anon (1995b) *Condition Monitor*, **100**, 12.

Anon (1995c) Rainbow road warriors, *PC Magazine*, March, 103–63.

Anon (1995d) Graduated illumination improves surface inspection, *Eureka on Campus*, **7**(1), 3.

Billson, C. (1992) Field information systems for maintenance management, *Maintenance*, **7**(4), 22–9.

Billson, C. (1993) Pen-&-paper versus pen-based computer–a comparison of methods used for asset data collection, *Maintenance*, **8**(2), 3–7.

Casasent, D. and Carender, N. (1989) An acousto-optic processor for inspecting cigarette labels, *Proceedings of the SPIE Conference on Automated Inspection and High Speed Vision Architectures III*, Philadelphia, USA, Vol. 1197, 181–90.

Chan, J.P., Batchelor, B.G., Harris, J.P. and Perry, S.J. (1990) Intelligent visual inspection of food products, *Proceedings of the SPIE Conference on Machine Vision Systems Integration in Industry*, Boston, USA, Vol. 1386, 171–9.

Chin, R.T. (1992) Automated visual inspection algorithms. In C. Torras (ed.), *Computer Vision Theory and Industrial Applications*, Springer-Verlag, New York, USA, 377–404.

Daley, W. and Rao, T. (1990) Colour vision for industrial inspection, *Proceedings of SME Vision Conference 90*, Detroit, USA, 12.11–24.

Davies, A. (1990) *Management Guide to Condition Monitoring in Manufacture*, IEE, London, UK.

Davies, A. (1995) *Machine Tool Breakdown Diagnostics and Maintenance*, invited paper for the Institution of Electrical Engineers (IEE) East Wales Area Lecture, University of Wales College of Cardiff, UK, 11 April 1995. UWCC Technote No EP 188.

Dressel, M., Heinke, G. and Steinoff, V. (1992) Inspect for maintenance with NDT methods, *Condition Monitoring and Diagnostic Technology*, **2**(3), 30–5.

Durrani, T.S. *et al. (1987)* Real time inspection of thermal image sequences for real time inspection of composites. In C. Cappellini (ed.), *Time Varying Image Processing and Moving Object Recognition*, Elsevier, New York, USA, 230–7.

Freeman, H. (1989) Development of a trainable machine vision inspection system. In Cantoni *et al.* (eds), *Proceedings of the 5th International Conference on Image Analysis and Processing*, Positano, Italy.

Hedengren, K. (1986) Automatic image analysis for inspection of complex objects: a systems approach, *Proceedings of the 8th International Conference on Pattern Recognition*, Paris, France, 919–21.

Liu, T.I. (1990) Automated inspection of drill wear, *Proceedings of Manufacturing International 90*, Atlanta, USA, Vol. 5, 115–19.

Mair, G.M. (1988) *Industrial robotics*, Prentice-Hall, New York, USA.

Marshall, A.D. (1989) The Automatic Inspection of Machined Parts using Three Dimensional Range Data and Model Based Matching Techniques, Ph.D dissertation, Dept of Computing Mathematics, University of Wales, Cardiff, UK.

Newman, T.S. and Jain, A.K. (1995) A survey of automated visual inspection, *Computer Vision and Image Understanding*, **61**(2), 231–61.

Pryor, T. (1982) Optical inspection and machine vision, *Proceedings of the IEEE Workshop on Industrial Applications of Machine Vision*, Research Triangle Park, NC, 3–20.

Shu, D.B. *et al.* (1988) A line extraction method for automated SEM inspection of VLSI resist, *IEEE Transactions on Pattern Analysis and Machine Intelligence*, **10**(1), 117–20.

Smith, B. (1993) Making war on defects: six sigma design, *IEEE Spectrum*, **30**(9), 43–7.

Van Gool, L. *et al.* (1991) Intelligent Robotic Vision Systems, in S. G. Tzafestas (ed.), *Intelligent Robotic Systems*, Marcel Dekker, New York, USA, 457–507.

Wright, P.K. and Bourne, D.A. (1988) *Manufacturing Intelligence*, Addison-Wesley, New York, USA.

Xian, W. *et al.* (1988) Unrolled problem of rotary objects and automatic inspection of appearance defects of bearing rollers, *Proceedings of the SPIE Conference on Automated Inspection and High Speed Vision Architectures II*, Cambridge, MA, USA, Vol. 1004, 69–76.

Yamamoto, S. (1992) Development of inspection robot for nuclear power plant, *Proceedings of the IEEE Conference on Robotics and Automation*, Nice, France, 1559–66.

CHAPTER 4

Thermal monitoring using infrared thermography

D.A.G. Dibley

AGEMA Infrared Systems Ltd, Arden House,
West Street, Leighton Buzzard, Bedfordshire, LU7 7DD

4.1   INTRODUCTION

This chapter will begin with a look at the basic principles of temperature measurement, defining the meaning of temperature and considering the need for measurement. After a brief overview of the history of temperature measurement, the article will provide a summary of the different measurement techniques available. More detailed information on the development of thermal imaging systems and their construction will also be provided.

The second half of the chapter will be devoted to application information, explaining briefly how thermal imaging techniques can be applied in different industries for condition monitoring. This section will concentrate mainly on case studies, giving practical information on how thermal imaging systems have been used by various companies for plant condition monitoring. The chapter will then conclude with a few words on the future of infrared thermography within the condition monitoring market, and will describe a new generation of systems which is emerging in this area.

Initially it is worth asking ourselves: what is temperature? Temperature can be defined in a number of ways. One of the most common definitions says that:

> Temperature is proportional to the random kinetic energy of the molecules or atoms making up a body.

Alternatively, it can be said that:

> If two bodies are in thermal equilibrium such that no thermal energy is exchanged, then these bodies are at the same temperature.

Handbook of Condition Monitoring
Edited by A. Davies

Both definitions are equally applicable and help us to understand 'thermometry' which is the measurement of temperature.

Well, why do we need to measure temperature? In almost all aspects of life, temperature provides an indication of whether a body is behaving normally. In the human body, for example, many forms of illness are accompanied by a rise in body temperature. In industry, there are numerous examples of faults being accompanied by an unexpected increase or decrease in temperature. In fact such faults are often preceded by a change in temperature before any physical damage occurs.

One example is the overheating of motor bearings and electrical connections and missing refractory lining from furnaces. Clearly it is an advantage therefore to be able to detect such changes in temperature before a catastrophic failure occurs. Regular monitoring of temperature consequently allows companies to devise well-informed and scheduled maintenance programmes and thus avoid expensive and unscheduled stoppages.

In terms of history, the science of temperature measurement preceded the discovery of the infrared spectrum by some 200 years. Around 1595 the invention by Galileo of the 'thermoscope', a primitive and somewhat unreliable gas thermometer which utilized atmospheric air expansion, provided valuable measurement experience, particularly in the medical field. Within 50 years the 'thermoscope' was replaced by thermometers using alcohol and consisting of a hermetically sealed bulb with a capillary tube attached.

The next turning point came when Gabriel Fahrenheit invented a new method of cleaning mercury by distillation, thus overcoming the problem of mercury's sticking to the walls of capillary tubes. The substitution of alcohol by mercury with a low freezing point, $-39\,^{\circ}$C, a high boiling point of $356\,^{\circ}$C and equal expansion at different temperatures, enabled higher temperatures to be measured. The alcohol thermometer, however, is still used for low-temperature measurements because at atmospheric pressure alcohol does not solidify, even at very low temperatures.

With the new mercury thermometry came the Fahrenheit scale which defines the freezing point of water to be $32\,^{\circ}$F and the boiling point to be $212\,^{\circ}$F. Although useful in medical work this scale was not ideal for scientific use and in 1742 Ander Celsius, the Swedish astronomer, proposed a more practical scale with 100° between the freezing and boiling points of water. This became known as the Centrigrade or Celsius scale.

In the early part of the nineteenth century greater precision in temperature measurement was again sought, which led Henry Regnault in France to develop gas thermometers accurate to $0.1\,^{\circ}$C by using a constant-volume hydrogen thermometer. With the aid of this apparatus, Regnault established that the rate of expansion of hydrogen at approximately $0\,^{\circ}$C is one part in 273. This work complemented that of Lord Kelvin in England and others, who were thus able to define absolute zero on the gas thermometer as $-273.15°$ on the Centigrade scale. This is now known as the Kelvin Scale where $273.15\,\text{K}$ equals $0\,°\text{C}$.

4.2 SIMPLE TEMPERATURE MEASUREMENT TECHNIQUES

The most commonly used instruments for temperature measurement today are as follows.

- **Liquid in glass thermometers** Nearly everyone has used a liquid in glass thermometer at some point in their life. Usually filled with mercury or alcohol, these thermometers depend on manufacturing quality, calibration accuracy and scale resolution for their measurement precision. This at best can reach in total immersion $0.01\,°\text{C}$ for lower temperatures, say $0–150\,°\text{C}$, and $1\,°\text{C}$ for higher temperature thermometers, say $300–500\,°\text{C}$. When used as a contact thermometer against a solid surface, accuracies are significantly reduced.
  Pressure can have an adverse effect on the reading of these thermometers when they are horizontal and vertical. The plasticity of the glass and the sensitivity of the reading to bulb volume, which is large, can cause a typical $0.1\,°\text{C}$ change in reading for a one atmosphere change in pressure.
  It is obvious from the nature of these thermometers, that finite time is required for their readings to stabilize, and hence these devices have relatively slow response times. Their application in industry is therefore somewhat limited, and we must look to other types of temperature measurement instruments to provide the required accuracy and speed of response.
- **Thermocouple thermometers** These work on the principle that when two different metals are joined at one point, an electrical potential difference, dependent on the temperature of the junction, is obtained at the open end. This is known as the 'Seebeck' effect. As simple, inexpensive and flexible temperature sensors, they have a wide number of applications throughout industry. Compared to liquid in glass thermometers, they have relatively fast response times, in some cases as high as 1 metre/sec. They are also relatively accurate, typically $0.1\%–1\%$ of absolute temperature, and durable.
  The inherent problem of contact devices like thermocouples is, first, that the contact resistance must be zero, which in practice, is difficult to ensure. Even with zero resistance, heat will be conducted along the device and hence will lower the temperature of the device at the point of measurement. One advantage of the thermocouple over the traditional mercury thermometer is its easy integration into automatic monitoring and recording systems which use electrical input signals.
- **Bimetal thermometers** These take advantage of the differences in thermal expansion properties of dissimilar metals to enable tempera-

ture measurement. The angular position of the bent strip, with pointer attached, can be calibrated to give a temperature readout. While they are quite rugged and inexpensive, they only provide a visual temperature reading and not an electrical signal suitable for automatic monitoring and control.

- **Resistance thermometers (RTDs)** These use the principle that any material whose electrical resistance changes with temperature in a constant and repeatable manner can be used as a thermometer. The relationship between resistance and temperature, not necessarily linear, can be calibrated to enable temperature measurement. Typical materials include platinum, nickel and copper.

    The accuracy of these thermometers can vary from 0.1 °C to ±3 °C at higher temperatures. For best performance they are designed to have a high sensing element to sheath resistance. This is important, because it is in parallel with the sensing element resistance and thus the shunting effect is negligible. In practice, the shunt resistance can be reduced significantly due first to moisture, secondly to contamination and finally high temperature operation, where the resistance of the insulating material decreases with temperature. Other effects such as thermal shock and mechanical vibration can cause long-term instability.

- **Thermistors** Similar in operation to resistance thermometers, thermistors are usually semiconductors, metal oxides, with large temperature coefficients of resistance. The main differences are much higher resistances, together with changes in resistance per degree change in temperature being larger, typically ten times higher. Also thermistors used for temperature measurement typically decrease in resistance as temperature increases, the reverse of RTDs, plus having a very non-linear relationship between temperature and resistance. In addition, temperature errors due to self-heating are more significant than RTDs.

    The accuracy of thermistors is influenced by much the same factors as RTDs. Interchangeability is a problem due to the composition of the thermistor material which is difficult to control during manufacture and can lead to inconsistencies in the resistance/temperature characteristics. Some degree of instability can be caused by changes in the crystal structure of the thermistor material. Experience has shown that these thermistors lose their calibration with time and may need regular checking.

4.3 PYROMETRY AND INFRARED LINE SCANNERS

The temperature of a body can be determined by measuring the thermal radiation it emits. This has the great advantage of enabling non-contact and remote temperature measurement of objects. For most applications the range of temperatures can be covered utilizing wavelengths from the far infrared down to the visible spectrum – although work on plasma temperatures involves measurements in the ultraviolet range. There are three main types of pyrometers as follows.

- **Total radiation pyrometers** The most common of the pyrometer family, these measure the total radiation over a range of wavelengths. Ideally they should measure all wavelengths; however, components of the system, that is lenses, filters and the atmosphere, attenuate some wavelengths. The detectors used in total radiation pyrometers are somewhat similar to thermocouples except that they sense radiation which heats up the contact. A number of these thermocouples added together make up a thermopile detector.

  Since a 'blackbody', an object that wholly absorbs/emits radiation, is used to calibrate the pyrometer, it is necessary to know the average emissivity of the observed body over the wavelength range of the chosen pyrometer. Emissivity can be defined as the ratio of radiant power from an object compared to a 'blackbody' at the same temperature and wavelength. The electrical output of these pyrometers is usually arranged to give simple digital readings of temperature and/or voltage/current equivalents for use in control loops. Speed of response can vary from 1 second up to 80 milliseconds, at the expense of increased noise. Since the instrument is calibrated by covering the whole detector with incoming radiation, it is important to know the ratio of 'spot size' to distance for any particular application.

  There are many different total radiation pyrometers on the market today, Figure 4.1 showing a typical example. They range from long-wave infrared, 8–14 microns, through to short wavelength, 2–6 microns, and for high-temperature applications some are working in the near infrared, around 1 micron. Some applications, e.g. plastics or glass, require narrow band pyrometers to match the absorption characteristics of the material. This is generally done by filtering the broader band systems previously mentioned. The type chosen will depend entirely on the application and its temperature range.

- **Optical or brightness pyrometers** These pyrometers measure the radiation power in a narrow band of wavelengths, generally in the optical wavelength. The object is viewed as a background behind the filament of a lamp housed inside the instrument. The intensity of the brightness of the light is adjusted until the filament disappears against the background. The current to the lamp is measured, and the temperature determined from a calibration of lamp current versus temperature of a 'blackbody'. Clearly, response times for these pyrometers are a function of the operator's speed in adjusting the filament current. Since a human operator is required, the basic model does not lend itself to integration in control loops.

- **Two-colour, ratio pyrometers** These compare the spectral radiances at two wavelengths to identify the temperature. This is done to

Fig. 4.1 Typical example of a total radiation pyrometer.

attempt to reduce the effects of emissivity of the object. However, the two wavelengths selected must be quite close together to avoid normal emissivity changes with wavelength. Since this creates a narrow bandwidth of operation, the sensitivity is greatly reduced, rendering the instrument suitable for only high-temperature measurements. These devices are complex and therefore not in common use in process control applications.

- **Line scanners** These could be described as rotating radiometers that present the temperature output as an amplitude signal much like a graph. In practice, they use quite different detectors which are similar to thermal imaging systems. These are photon detectors, usually constructed of semiconductor material, which produce a voltage or current output proportional to the radiation impinging on them. These detectors are generally very small. Accordingly, they give tremendous advantages over radiometers in terms of their small measured spot size and their rapid time constants – typically less than 1 millisecond, as shown in Figure 4.2.

A disadvantage of these detectors has been the necessity for cooling to cryogenic, −196 °C, temperatures. If they are not cooled, they would sense their own ambient temperature creating a noisy signal where only temperatures above that of ambient could be detected.

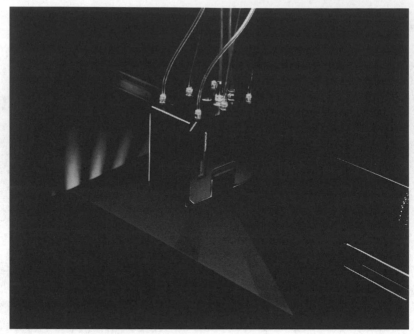

Fig. 4.2 Photon detector in action.

Cooling was initially achieved using liquid nitrogen or argon gas etc. However, nowadays detector technology enables thermoelectric cooling techniques to bring the temperature down to the required level.

In a typical line scanner, a rotating mirror receives infrared energy and reflects it onto a lens, which focuses it onto a thermoelectrically cooled detector. The detector receives energy from the target every quarter rotation or 90° of the mirror. During the remaining 270°, two internal temperature references are imaged onto the detector for self-calibration purposes. Because the scanning is in one plane only, these systems are ideally suited to looking at objects which are moving in a direction perpendicular to the scanning plane.

As non-contact measurement systems with fast response times and a range of electrical outputs, line scanners therefore lend themselves to process monitoring and control applications, such as strip steel passing through a mill, rotating cement kilns, plastic and paper webs. Modern processing techniques can also be used to combine a number of scanned lines to produce a thermal image of an object, although of course not in real time. While typical temperature measurement accuracies as high as ±1% can be achieved, it is essential to know the emissivity of objects at the wavelength of operation of the scanner, usually 2–6 microns and 8–14 microns.

4.4 THERMAL IMAGING OR THERMOGRAPHIC SYSTEMS

In the mid 1960s, the rapid development of semiconductor technology led to the introduction of passive thermal imaging systems for military use, as opposed to active imaging which requires an external infrared source. Since all objects emit infrared radiation, sensitive infrared detectors used in these systems can detect and convert incident energy into a picture which is independent of daylight. This led the military to develop a range of thermal imaging systems for night vision use. As a direct result of these developments, and with the addition of temperature measurement from image, thermographic systems were created.

Traditionally, a single element detector has been used to capture the incoming radiation from an object. To produce a two-dimensional picture, this requires the detector to be projected into the field of view as a single spot moving rapidly both horizontally and vertically. This is achieved using optical components such as rotating prisms and/or tilting mirrors driven by electrical motors. By using electro-mechanical scanning, as shown in Figure 4.3, these systems are very different to conventional TV systems which use electronic scanning techniques.

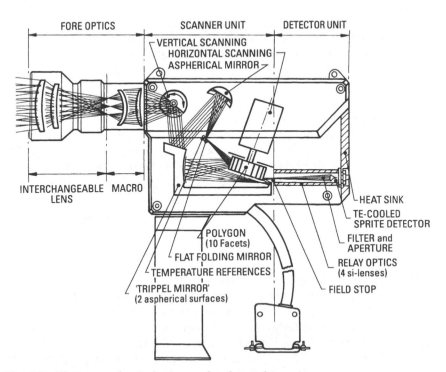

Fig. 4.3 Electro-mechanical scanner for thermal imaging.

The most commonly used detector in these thermal imaging systems is Cadmium Mercury Telluride (HgCdTe), commonly known as CMT or MCT. This can be used in either short, 2–6 micron, or long, 8–14 micron wavelengths. Since the sensitivity of CMT is lower in the shortwave, Indium Antimonide (InSb) is also used in the short wavelengths. These detectors require cooling, normally to cryogenic temperatures. Cooling of these detectors was traditionally achieved using liquid nitrogen or argon gas. Today, other methods are employed such as thermoelectric cooling, 'Peltier', for shortwave systems and a miniaturized 'Stirling' cooler for long wavelength systems.

While thermoelectric cooling has the advantage of no moving parts, no maintenance and short start-up times, say 15 seconds, it does not reach the same low temperatures as 'Stirling' coolers, and hence achieves lower sensitivity. Coolers of the 'Stirling' type, however, require regular refurbishment to replace worn seals and will take in excess of five minutes to reach a working temperature. The choice of operating wavelength is determined by a number of factors. Long wavelength systems are most sensitive to lower temperatures, ambient and below, and are also less affected by attenuation over long distances, because the atmosphere attenuates less at long wavelengths.

For these reasons, longwave systems are often preferred by the military. A short wavelength system is more sensitive to the higher temperatures found in industry, that is above ambient. For accurate temperature measurement, shortwave systems are generally restricted to measuring around 200 metres. At distances greater than this, at least 1 kilometre, imaging is still possible. Clearly this would be an unusual requirement for industrial temperature measurement.

The thermal sensitivity of these systems, whether short or long wave, is generally better than 0.1 °C at 30 °C. This is important, but should not be confused with the accuracy of the system. An equally important factor when choosing a thermal imaging system is the geometric or spatial resolution. This defines the smallest spot distinguishable in the image. Some confusion exists here because the traditional way of stating the spatial resolution of a military system was to define a slit response function at 50% modulation. This is calculated by measuring the slit size which gives a 50% reduction in energy when looking at a 'blackbody' radiation source.

Although this is good enough for military systems, that is recognizing differences in the image, it will obviously give temperature measurement errors. The correct way to define the spatial resolution for measurement, should be to specify the slit response function at 99% modulation where no signal deterioration is evident. Temperature measurement accuracies as high as $\pm 1\%$ can be achieved with these systems. Especially when internal 'blackbody' reference sources and accurate control of internal component temperatures are employed. It is important to remember when choosing a system that adequate compensation for external factors, such as emissivity, reflected ambient radiation and distance, is also provided.

Thermographic systems have different levels of 'built-in' temperature analysis. Typically, they will have functions such as spot temperature measurement, profiles, isotherms etc.

Isotherms are very useful as they highlight points of equal temperature within an image. By moving the isotherm over the temperature range of the image, it is possible to pinpoint the hottest or coldest points within an image. Additional statistical analysis, including functions such as histograms, is generally provided by dedicated software packages offering features such as report generation and dynamic linking within the Windows™ environment. A useful feature available in some packages is the facility to compare thermal images directly with visual images recorded on digital cameras and imported into the software package.

While the focus of this article is on condition monitoring, it is useful to mention some of the other application areas in which thermography is often employed. Historically, the first applications were in the medical sector, not as a clinical tool but for screening and research and development. In the same way as the temperature of industrial components can be used to detect abnormalities, the temperature of the human body fluctuates according to its physical condition. Originally used for breast cancer screening, today thermography is often applied in the control of inflammatory diseases such as rheumatism, arthritis, and back pain. In dentistry, thermography has been used recently to monitor the temperature in the mouth when lasers, rather than conventional electro-mechanical drilling tools, are used.

In industry, thermal imaging systems are often used for general research and development. When designing new materials/products a detailed knowledge of the thermal patterns which would be created under working conditions is often essential. A prime example is in the aerospace industry where the material temperatures of aircraft/spacecraft flying at very high altitudes is constantly affected by the sun and the extremely cold temperatures at those altitudes. The increased use of lightweight composite materials for these craft has produced a new application for thermography looking for voids, delaminations and inclusions below the surface. Within the aerospace industry, another important application is the thermal monitoring of both rocket and turbine engine performance.

The automative industry benefits from infrared thermography when designing and developing new parts such as disc brakes, tyres, and engines. The electronics industry, however, uses thermography not only for research and development, but also for quality assurance of Printed Circuit Boards (PCBs), electronic components and semiconductor devices. Finally, the military have a need for thermography in their research and development departments, quantifying infrared for such areas as target signature analysis, infrared camouflage, military clothing and the development of infrared missiles.

4.5   CONDITION MONITORING CASE STUDIES

Electrical systems

The largest users of thermographic instruments are the power generation, transmission and distribution authorities. Electrical transmission faults lend themselves to this kind of inspection because faults invariably produce heat, poor contacts, short circuits, overheating components and so on, or lack of heat, open circuits etc. This is particularly evident with high voltage/high current systems, the I^2R effect. Typical uses are in a substation or switch yard, where contacts or clamps overheat until the material melts and causes either the welding of the contacts together, or complete disconnection of the power line. Faults can be detected at a very early stage thermographically, thus allowing rectification before major breakdowns occur.

Many transmission authorities today use helicopters to patrol hundreds of kilometres of power line, with the many thousands of joints involved. The generating authorities also have many areas of use. An example is the inspection of stator cores during maintenance. With the rotor removed, the stator can be energized and thermographically will reveal such faults as shorted turns, and high resistance lamination contacts etc. Thermography is also invaluable for studying boiler insulation wear and the erosion/blocking of boiler tubes. Although distribution lines operate at lower voltages than transmission lines, they still require monitoring for the same reasons as transmission lines. Such monitoring is particularly important as distribution lines are frequently located in urban areas, where a line falling down poses a major hazard for the general public.

Two examples, which demonstrate the importance that power generation, transmission and distribution authorities place on infrared thermography for condition monitoring are the recent large orders for thermal imaging cameras from electricity authorities in India and Korea. Twenty-nine thermal imaging systems have recently been delivered to a state electricity board in India, for inspecting substations and overhead power lines. A further 31 systems have been ordered by the transmission division of a Korean power corporation, for condition monitoring of its transmission lines and substation equipment, including switchgear and transformers.

The inspection of overhead power lines along railway tracks is another example where thermal imaging systems can be usefully employed. A rail operator uses a thermal imaging system to monitor the condition of overhead lines on one of its main lines. The system is used to detect problems such as overheating clamped connections, before they have a chance to fail, and has averted a number of problems which, undetected, would have caused lengthy and costly delays. The system was deployed for overhead electrification inspection, as a means of introducing a method of preventive maintenance in their current carrying high voltage overhead line system. Overheating failures can develop in some clamped and friction held fittings over a period of years, until they manifest themselves as a failure, causing train delays costing many thousands of pounds and involving delays to strategic passenger services.

Infrared thermography enables maintenance engineers at the rail company to survey an entire overhead power system and pinpoint just those areas which require further investigation. As an electric locomotive passes along a line, electric current is drawn from the overhead power lines and various joints will heat up. A poor connection in one of these joints would significantly raise its temperature to the point at which it would appear as a hot spot in a thermal image of the joint. By pointing the camera at various sections along the overhead power line just after a train has passed by, it is therefore possible to detect these connections where some fault might be present.

Thermal images taken on the camera can be stored on the system's 'built-in' floppy disk drive. Following a thermographic survey, the company's electrification maintenance engineers will send copies of relevant images on disk back to the service department. Using a propri- etary thermal analysis software, it is then possible to take a closer look at the thermal images of those connections where some overheating is apparent. According to the company, a 2° or 3°C rise above normal is quite acceptable, but anything up to a 10° or 15°C rise would probably require further investigation. Depending on the seriousness of the problem, the engineer is then able to plan a schedule of maintenance which will cause fewer track isolations and minimum disruption to the operation of the line.

In one survey of overhead power lines a potentially serious fault was identified. Figure 4.4 shows the thermal image of an overhead line switch, in which it was discovered that the connection between the jaws and the blade, was about 15°C higher than the surrounding equipment. Figure 4.5 shows the switch in more detail. Clearly there has been some major erosion in the jaws, probably caused by electric arcing, which is causing a poor connection between the jaws and blade. If this had not been detected, the chances are that the switch would have failed open, in which case it would not have been possible to bring this section of line back into operation after a period of isolation; or it would have failed closed, in which case it would not have been possible to isolate the current from this section of track.

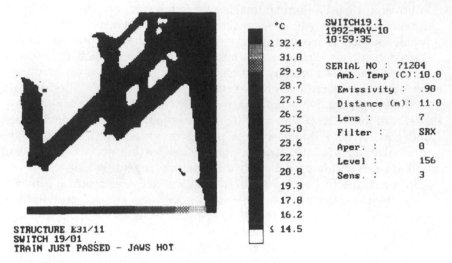

```
                                      °C      SWITCH19.1
                                              1992-MAY-10
                                   ≥ 32.4     10:59:35
                                     31.0
                                     29.9     SERIAL NO : 71204
                                                Amb. Temp (C):10.0
                                     28.7
                                              Emissivity :   .90
                                     27.5
                                              Distance (m): 11.0
                                     26.2
                                              Lens :          7
                                     25.0
                                              Filter :       SRX
                                     23.6
                                              Aper. :          0
                                     22.2
                                              Level :        156
                                     20.8
                                              Sens. :          3
                                     19.3
                                     17.8
                                     16.2
    STRUCTURE E31/11                ≤ 14.5
    SWITCH 19/01
    TRAIN JUST PASSED - JAWS HOT
```

Fig. 4.4   Thermal image of an overhead line switch.

Fig. 4.5   More detailed view of overhead line switch.

In either case, the cost in delays to the operation of the line could have been anything between £35 000 and £40 000 and, since the service department is charged for every disruption on the line, no matter how small, it was clearly in its advantage to locate this fault as soon as possible. In this event, the department was able to arrange an immediate electrical isolation, and to replace the faulty connection at a time which fitted in with the timetable of trains running through the affected area. The detection of this fault alone paid for the thermal imaging camera!

In another survey, internal burning was detected on a clamp fitting which was not crimped hard enough on the stranded conductor feeding it. If this had not been noticed, it was more than likely that the conductor would become so badly damaged that it would eventually fall out and break the connection altogether. The hot spot on the thermal image shown in Figure 4.6 shows the location of the internal burning, and Figure 4.7 shows the damage to the various parts which make up the clamp. In yet another survey, the thermal imager was able to detect a nut which had not been tightened sufficiently and was causing a connection to overheat by 10°C. Even problems as small as this are capable of causing major disruption if left undetected long enough.

The benefits of using infrared thermography within a planned schedule of maintenance for electrified railway systems are thus clear. On one line alone, the investment in a thermal imaging system has been more than justified with a number of potentially very serious faults being revealed. Undetected, these faults might have cost the company many

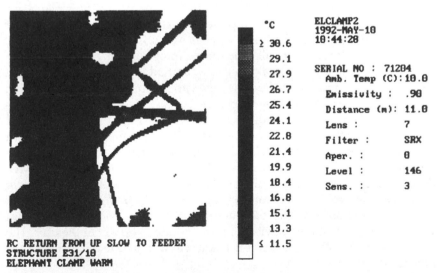

Fig. 4.6 The hot spot on thermal image showing location of the internal burning.

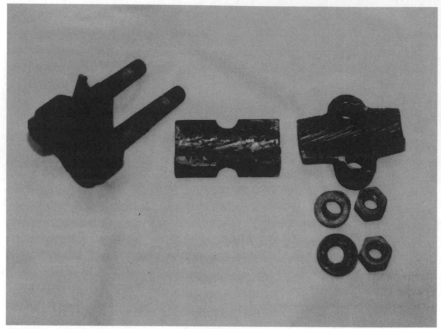

Fig. 4.7   Thermal image showing damage to the various parts making up the clamp.

thousands of pounds in lost revenue, man hours and replacement parts, not to mention the quality of the service to its customers. Thermal imaging techniques are also often applied at manufacturing plants for general electrical inspection. Components such as transformers, thyristor banks, circuit operating devices, switches, fuses, circuit-breakers, control equipment and motors are all susceptible to overheating.

One paper manufacturer averted a potentially very costly disaster at its plant. In a thermographic survey carried out by local condition monitoring consultants, a serious overheating problem in one of its circuit-breakers was revealed. Undetected, this might have failed and cost the company thousands of pounds in lost revenue and replacement equipment. Two years previously, the consultants had reported an unusually high temperature above ambient on the intake 11 kV Area Board's main supply circuit-breaker. The electricity board was called in to look at the problem but nothing unusual could be found. In the next thermographic survey, the temperature difference has risen to about 19°C and the consultants advised an immediate overhaul of the unit. The electricity board were called in once again and this time the extent of the problem was realized.

The removable pattern oil circuit-breaker contained six connecting rods. These engaged into a rose cluster which was supported by a brass cup. One of the roses was badly burnt and the connection could have failed causing the complete loss of site supply. Since these parts were no longer available, the paper company commissioned a switchgear contact manufacturer to manufacture a new set with spares. In the next planned shutdown at the plant, the replacement rose cluster assembly was installed. If this condition had remained undetected, it could have led to a major breakdown of contact materials and insulation within the breaker assembly. Since the time frame to rebuild, process and manufacture spare parts would have been many days, the result could have been several days' loss of all site production. For this company, it is clearly very important to carry out regular non-invasive maintenance checks on plant equipment.

Computer systems

Facilities management is a growing area for the application of thermal imaging techniques. Indeed one UK building society states that their central computer system has never 'gone down' since it purchased a thermal imaging system. The computer system, which services the accounts of hundreds of local branches, controls all the automated telling machines, all the individual customer accounts, in fact every aspect of the society's business. It runs 24 hours per day, 365 days a year, and if it were ever to fail, the cost to the society would be substantial. A continuous, smooth power supply is maintained using two Uninterruptible Power Supplies (UPSs), one for each of the computer rooms at the site. Once a year, the power supplies to each room are changed over. Immediately after the new but previously dormant supply has been switched in, and every three months thereafter, the live system is tested to ensure that no loose connections have been introduced by the change over.

The 'live system' is represented by 64 'boxes' linked together by numerous cables and all situated underneath the floor of the computer room. To check all the contactors on the 'live' boxes individually would not only be time-consuming but extremely dangerous. When first evaluating the best way of testing the system, the society estimated that it would take six men six weeks to complete the inspection. Over a four-hour period, this would equal $36 \times 4 = 144$ working hours paid at double time. Apply this cost to four inspections a year and the result was an expensive maintenance routine! Apart from the obvious cost impediment, the society could not guarantee that under these conditions a full and reliable inspection would be achieved. With so many connections to check, human error was inevitable. So the society looked for an alternative solution – infrared thermography.

According to the society, the results using their new thermal imaging camera were excellent. In its first inspection, the thermal imager detected over 20 potential problem areas, ranging from loose connections to poorly machined busbars, where heat was forming on the bends. Also significant was the time taken to complete an inspection after power changeover. It would take minutes rather than hours for the three men to conduct a complete survey of the computer room – and during normal operating hours at the normal hourly working rate. After a year using the thermal imaging camera, the society have been left in no doubt about the success of thermographic inspection. Their computer system has never 'gone down' and the number of faults detected during each investigation is rapidly diminishing.

Mechanical systems

Mechanical applications for infrared thermography generally involved rotating equipment. Faulty bearings, inadequate lubrication, misalignment, misuse and normal wear are just some of the problems which can cause overheating. Gears, shafts, couplings, V-belts, pulleys, chain drive systems, conveyors' air compressors, vacuum pumps and clutches are among the most common components to fail due to overheating. In most applications, infrared thermography is used to pinpoint a problem area while other inspection techniques such as vibrational analysis are used to find the cause of the problem.

Thermographic inspections of refractory and insulation materials rely on the principle that the exterior surface temperature of a vessel is proportional to the heat conduction through the insulation and external wall. If there is any moisture in the insulation or uneven wear in the refractory, non-uniform heat conductance will cause a hot spot to be detected by a thermal imaging camera. Typical components which can be inspected using infrared thermography are batch and continuous ovens, furnaces, heat treatment furnaces, ovens, dryers, kilns, boilers, ladles, hot storage tanks and insulated pipes. One Norwegian company using a thermal imaging system has now developed a technique for identifying the build-up of scale on process and production equipment offshore. If applied correctly, this technique could save operators millions of pounds in lost production.

Figure 4.8 shows a thermal image (thermogram) of a pipeline where scale has been allowed to build up. Figure 4.9 shows a profile of the build-up of scale. As the thickness of the scale decreases, the temperature of the outside surface of the pipe increases. The scale is therefore acting as an insulator, reducing heat flow from the inside of the pipe to its surface. The company maintains that they can now detect scale thickness down to as little as 5 mm. In this case they would expect a temperature difference of approximately 2–5°C across the length of the scale from its thickest to its thinnest point. Larger-scale thickness can bring about temperature differences in the region of 45°C.

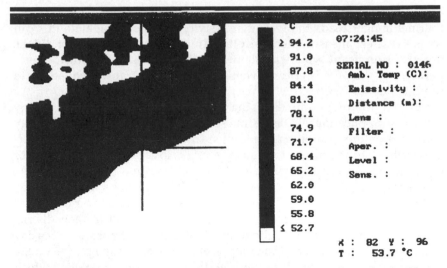

Fig. 4.8 Thermal image of a pipeline where scale has been allowed to build up.

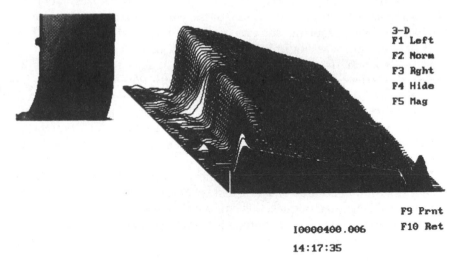

Fig. 4.9 Profile of the build-up of scale.

There is more to the technique than just pointing a thermal imaging camera at the pipeline and reading off temperature differences. The operator must first have a good understanding of petraphysics. By knowing what chemical reactions are taking place within the pipeline, for example, it is possible to get an idea of what to expect and how measurements should be taken to produce the best results. The physical characteristics of the pipe – its material content, diameter and thickness –

are also essential parameters in determining scale thickness. If the thermal imager reads the wrong emissivity value of a pipeline this can severely affect measurement results. Ambient temperature conditions are also significant when calculating scale thickness. The outside temperature of the pipe is critical when applying heat transfer equations to the calculation.

Infrared thermography is being used more and more for energy efficiency applications in large manufacturing plants, and for energy conservation applications in buildings. In the latter case, infrared thermography can be used to detect energy loss as a result of poor construction or moisture build-up. A new application in this field is the inspection of district heating networks, looking for hot spots caused by leaking pipes. A leading UK thermographic consultant company was called in by a London council, to try and locate a fault in their district heating network, which was allowing 12 000 gallons of hot water per day to escape from the system. Using a portable thermal imaging camera, the company took just a couple of hours to locate the leaking pipeline which was situated several feet under the ground in a parking bay.

The district heating system was serving over 3000 houses and flats over an area of 4 square miles. Some of the cast-iron pipework was over 20 years old and problems such as corrosion were not uncommon. The cost of corrosion was not only in the amount of water which was lost from the system. To ensure that each block of flats received water at about 30°C, the water which was pumped into the system had to be heated to at least 80°C. The cost of heating water which never reached the end customer was clearly expensive. In addition, the water which was escaping, could also cause further damage to the pipework, attacking the insulation which surrounded the pipes, resulting in further heat loss.

To maintain such a vast network, the council needed to be able to locate such faults accurately and quickly. Before the advent of thermographic inspection techniques, this could only be achieved using a rather crude technique whereby maintenance engineers would look for evidence of dry spots on the ground after a heavy rainfall – an indication of localized heat leakage warming up the ground. Alternatively, engineers would compare the inlet temperatures around the system for different blocks of flats – if any temperature was significantly lower than another, there was a chance that hot water might be escaping from the network just prior to entering the building. Either method was clearly not very reliable, and it was not long before the council turned to thermographic techniques to survey their district heating system.

The advantage of infrared thermography for the council is that it allows operators to scan wide areas of terrain very quickly, pinpointing localized hotspots caused by hot-water leakages several feet under the ground. By employing the services of a trained thermographic consultant, particularly one experienced in carrying out surveys of district heating systems, the council could be assured of receiving a full and accurate report on the quality of their system. Accompanied by a council employee with a plan of the district heating system, the consultant located two hotspots underneath two cars in a parking bay near a block

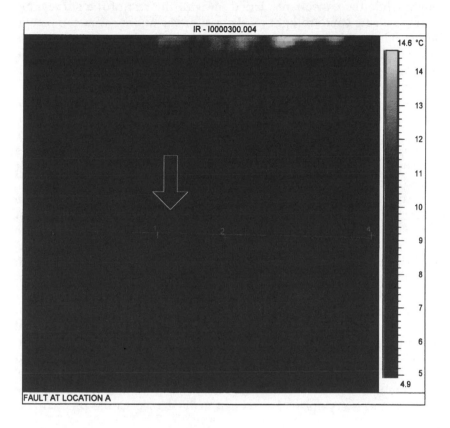

SPOT 1 = MAXIMUM TEMP OF SURFACE
SPOT 2 = AREA AROUND FAULT
SPOT 3 = AMBIENT TEMP OF SURFACE

Results						
°C	I0000300.004 10/02/94 11:31:41					
	Temp					
SP01	9.4					
SP02	8.1					
SP03	5.3					
SP04	9.2					
	Min	Max	Diff	Avg	Med	Sd
LI01	7.3	9.7	2.4	8.4	8.1	0.7
	Diff	Formula				
DI01	4.1	SP01 - SP03				

Fig. 4.10 Thermal image showing centre of hot spot (SPO1), another spot (SPO2) on the outer edge of the hot spot and a third spot (SPO3) further away.

of flats. After checking the temperature of the exhaust and the underside of the car engine, it was apparent that the hot spots were probably not being caused by heat reflected from a recently used car engine – a possibility that the inexperienced operator is not always aware of.

Nevertheless it was essential to make sure that the hotspots remained even after the cars had been removed. So the owners of the cars were sought while the consultant carried on with the rest of the survey. No other problems came to light as the rest of the survey was completed. On returning to the parking lot one of the cars had been removed by its owner and the consultant was able to carry out a fuller investigation. Using a thermal imaging camera, he produced a thermal image which showed the temperature at the centre of the hot spot to be 9.4°C. Figure 4.10 shows this spot (SPO1), along with two other points (SPO2) on the outer edge of the hot spot with a temperature of 8.1°C, and (SPO3) placed further away from the hot spot, and indicating an average temperature for the surrounding area of 5.3°C. The difference, 4.1°C, was more than likely to have been caused by hot water leakage under the ground at this point.

To confirm his prognosis, the consultant carried out an inspection underneath the other car which it had not been possible to remove. A temperature difference of 3.3°C was found in this case. In preparation for his report, he stored thermal images of the problem areas on disk, and took visual images of the corresponding areas using a digital still camera. Using a thermal imaging analysis and report software package, the consultant was able to provide a full report on his results for presentation to the council the next day. The report included visual and thermal images of the problem area, plus a brief analysis of the results. Following the report, the council requested a more in-depth inspection of the problem area, digging up the ground to find out if there was in fact a leak at this point. On finding this to be the case, they were then able to re-route the water away from this point, at a time convenient to everyone, and carry out maintenance on the pipework as necessary.

Fire detection in waste dumps

More and more waste companies are now recognizing the benefits of infrared thermography for preventing fires at incineration plants and waste dumps. As a non-contact temperature measurement system, the thermal imaging camera is able to scan the dump from a safe distance and detect the build-up of heat just before ignition. It is able to detect smouldering material caused for example, by burning cigarettes or ashes, the warming-up process caused by local fermentation in the dump, and the reaction of different chemicals within the waste. It can also detect the sparks caused by pieces of metal colliding too close to flammable materials – especially useful in smoky or dusty conditions.

A thermal imaging system has been successfully installed at a plant in Germany to look for hot spots in dumps where waste is stored prior to incineration. In the installation, the camera forms part of a total surveillance system designed to provide plant operators and fire crew with enough time to avert a major fire. To increase the chances of detection, the camera is mounted alongside an additional black and white video camera which has a spectral sensitivity between 600 nanometres and 2 micrometres. Both cameras are housed together in a dust-protected environmental enclosure which is mounted on a pan and tilt head, and controlled from a central 'cockpit' located about 40 m above the trash.

The thermal imager is specially adapted to automatically measure the highest temperature in a particular sector. If the temperature exceeds a particular alarm level, warning procedures come into play. At the highest level, for temperatures above 80°C, the fire department is automatically notified and internal warnings are issued within the plant. At each gate a map is printed out for the firemen indicating the best way to approach the fire. Various reports are also produced including thermal images of the problem sector. As soon as the operator receives notification of a hot spot, he is instructed to remove the burning waste from the dump and place it directly into the incinerator.

The alarm function is just one of five routines which form part of the system control software. The other routines include a 'teach-in' programme in which the system learns which sectors to measure and in which order, a self-test programme which alarms on detecting defective components or loss of power, a measurement programme which forms the major part of the software, and a small routine which allows for manual intervention by the operator, but which automatically 'times out' and reverts to the normal measurement mode if no input is received after a certain period of time.

The benefits of using infrared thermography in the detection of fires at waste dumps have already been proved at the German plant – since the installation, there have been no fires recorded at the incineration plant. It does not only help, however, to reduce the number of undetected fires and toxic emissions. It also gives rise to an increase in plant safety and a decrease in pollution. Plant reliability and efficiency can also be improved with the subsequent reduction in unexpected plant stoppages.

4.6 THE FUTURE OF THERMOGRAPHY, APPLICATIONS AND CONCLUSIONS

New technology in the form of Focal Plane Array (FPA) scanners is now being introduced which will significantly reduce the size and power of the thermal imaging systems, and at the same time increase the spatial and thermal resolution, thus improving the image quality. These 'staring' arrays are a matrix of detectors, typically 320 × 256 detectors and obviously do not require the opto-mechanical scanning technique necessary for the previous single element systems. The detectors used are commonly Platinum Silicide (PtSi) working in the short wavelength, 3–5 micrometres. Indium Antimonide (InSb) is also used, in the 3–5 micrometre band, and Cadmium Mercury Telluride in the long wavelength, 8–15 micrometre band.

All of these detectors require cooling to cryogenic temperatures and with a fairly recent development of miniature 'Stirling' coolers, plus their increasing reliability, this is now a viable proposition. The reduction in size and weight means that existing systems used in condition monitoring, typically the size of a professional outside broadcast TV-camera weighing some 7–10 kg including battery, as shown in Figure 4.11, will be reduced to the size of a modern camcorder weighing 2–3 kg including the battery. Image and data storage will be on considerably smaller PC-cards instead of the standard 3.5 in computer mini-disks and these will hold seven or eight times more images then the mini-disks. Figure 4.12 shows one of the new generation of Focal Plane Array cameras currently on the market.

Fig. 4.11 Thermal imager, typically the size of a professional outside broadcast TV camera.

Fig. 4.12   Focal Plane Array camera.

The continuing downward trend in the cost, in real terms, of thermal imaging systems will enable smaller plants to acquire these systems for the traditional condition monitoring functions of electrical and mechanical inspection. The reduction in the size and weight will also encourage wider use of these systems in plants where the large areas, which must be covered and which often require many stairs and restrictive catwalks to be negotiated, can dull the enthusiasm for the technology, especially if the instrument is heavy. The trend for more automation in industrial processing will also necessitate a major increase in the monitoring of temperature of such installations, not only manually but automatically, with remote controlled systems that can be alarmed and programmed to make the necessary adjustments to the process.

CHAPTER 5

Surface and internal defect detection

G. Hands[1] *and T. Armitt*[2]

[1]*Godfrey Hands Ltd, Unit 14, Hammond Business Centre, Hammond Close, Attleborough, Nuneaton, CV11 6RY, UK*

[2]*Lavender International NDT Ltd, Unit 7, Penistone Station, Sheffield, S30 6HJ, UK*

5.1 INTRODUCTION

This chapter deals with the detection of surface and internal defects on components and structures. To be able to use the component or structure after examination, obviously, the inspection should not affect the item involved, and must therefore be non-destructive. This is normally called Non-Destructive Testing (NDT), sometimes Non-Destructive Evaluation (NDE) or Non-Destructive Inspection (NDI).

NDT includes many different technologies, each suitable for one or more specific inspection tasks, with many different disciplines overlapping or complementing others. Thus the best technique(s), for any one application, should be decided by an expert. The following chapter, therefore, serves as a guide to the different technologies available, their advantages and disadvantages, and also outlines some of the limitations involved in their application. Accordingly, the chapter is intended to provide some guidelines for the non-expert, and should not be regarded as a technical manual.

Many NDT techniques can detect defects on the surface of components or structures. One major limiting factor of most techniques, however, is the surface finish of the area being inspected. Few normal inspection techniques can reliably detect defects that are smaller than twice the size of the largest features present on the surrounding surface. In other words, if one needs to detect defects that are 1 mm deep, the features

Handbook of Condition Monitoring
Edited by A. Davies
Published in 1998 by Chapman & Hall, London. ISBN 0 412 61320 4.

permissible on the surface of the inspected area should be not more than 0.5 mm deep. Some (computer assisted) techniques can improve on this, but the application of these methods is very specialized and possibly only suited to exotic applications that can justify the high inspection costs involved.

Some of the different NDT techniques that can detect surface defects are:

- Eddy Current Testing (ET);
- Electrical Resistance Testing (AC or DCPD);
- Flux Leakage Testing (see under MT);
- Magnetic Testing (MT) also known as Magnetic Particle Testing (MPT);
- Penetrant Testing (PT);
- Radiographic Testing (RT);
- Resonant Testing;
- Thermographic Testing (see Chapter 4);
- Ultrasonic Testing (UT);
- Visual Testing (VT) (see Chapters 3 and 6).

It should be noted that significantly fewer NDT techniques can detect internal defects. Here again, the condition of the surface where the inspection is performed, has an influence on the detectability of internal defects, with good surfaces normally giving better detectabilities. We should also note that the detection of surface features on the inside of

Table 5.1 Detection and measurement possibilities

Technique	Defect surf.*	Defect remote*	Corrosion	Wall thick measurement
Eddy current testing	50 μm	Yes	Yes	Yes (restricted)
Electrical resistance testing	300 μm	No	No	No
Flux leakage testing	50 μm	Yes	Yes	No
Magnetic testing	10 μm	No	No	No
Penetrant testing	25 μm	No	No	No
Radiography	150 μm	150 μm	Yes	No
Resonant testing	50 μm	50 μm	Yes	Yes*
Thermography	250 μm	No	Yes	No
Ultrasonic testing	200 μm	200 μm	Yes	Yes
Visual testing	100 μm	No	No	No

*Please note, the above detection limitations are severely dependent on the surface condition of the components under test, wall thickness, environment etc. Accordingly, they should be used as a guideline only.

inaccessible structures and components, ought to be placed in the internal defect detection classification.

Some of the different techniques for detecting internal defects are:

- Eddy Current Testing (ET) but only near surface defects;
- Magnetic Testing (MT) but only near surface defects;
- Radiographic Testing (RT);
- Resonant Testing;
- Thermographic Testing (only near surface defects);
- Ultrasonic Testing (UT).

Table 5.1 indicates the detection and measurement possibilities of the various techniques.

The remaining sections of this chapter go into more detail about the different techniques listed above with the exception of thermographic testing. Although considered a Non-Destructive Testing (NDT) technique, this subject is covered in detail in Chapter 4.

5.2 EDDY CURRENT TESTING

Eddy current testing is based around the physics of electromagnetic induction. Early physicists between 1775 and 1900, whose names include Coulomb, Ampere, Faraday, Oersted, Arago, Kelvin and Maxwell, investigated and discovered the fundamentals of present electromagnetic techniques. Unlike magnetic particle testing, which relies on the ability to magnetize a material, eddy currents are best suited to non-magnetizable materials. The one fundamental requirement, however, is that the test item has to be an electrical conductor. Eddy currents are generated by utilizing a coil excited by alternating current to produce an alternating magnetic field. This field, when placed in close proximity to a conductive material, induces a secondary current within the material, as shown in Figure 5.1.

The secondary current opposes the primary magnetic field by producing an opposing magnetic field, which is less than but proportional to the primary field strength. Such opposition is registered electronically by the eddy current flaw detector, and is used for initial calibration or balancing. Changes to the eddy current flow can be registered as changes in the primary field thus indicating a shift on an instrument scale. Therefore, factors within a material that constitute possible eddy current variations become important variables. Like all scientific tests, limiting the quantity of unknown variables is of paramount importance; therefore great care must be taken to ensure a systematic approach for reliable results. Some of the principle variables are subsequently outlined.

- **Permeability** This is the ease with which a material can be magnetized. Changes in metallurgical structure including alloy and heat

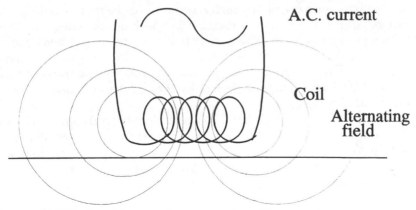

Fig. 5.1 Field induced into test item generates eddy currents.

treatment cause permeability changes. Even a single item may vary in magnetic permeability due to variations of metallurgical state. Magnetizable materials, therefore, require magnetic saturation to minimize the effect of permeability variations. This is usually accomplished locally using a secondary Direct Current (DC) coil within the test probe arrangement.

- **Conductivity** Electrical conductivity also varies with alloy and heat treatment state. This, however, can be utilized to advantage for material sorting, in the instance of two known alloys being mixed together, the sorting and verifying of heat treatment condition, or age hardness, using conductivity values which in certain materials are characteristic of the metallurgical state. Dedicated eddy current conductivity meters are generally used for this purpose.
- **Material thickness** Eddy currents generally do not usefully penetrate great depths into a material. However, thin sections will register thickness variations on an eddy current instrument. The frequency at which the test coil or probe is excited will govern the depth of eddy current penetration, which usefully may range from 1 mm to 10 mm.
- **Edge effect and end effect** The eddy currents produced within a test material travel in circular paths. These paths can be interrupted by the test component geometry when the test coil is moved up to or near a sharp change in section such as the plate edge, a machined slot or drilled hole. Instrument readout variations at these points may mask flaws, and, therefore, individual scans parallel to geometric changes with the instrument 'balanced' or calibrated specifically are necessary to eliminate edge effect and end effect from the test variables.
- **Lift off** This is usually an undesirable effect when the probe coil varies in distance from the test item. Eddy current density is rapidly changed with slight variations in probe contact with the test item, thus leading to irregular or false indications. For flaw detection and

conductivity measurements, such an effect must be minimized. Lift-off can be utilized for one particular test technique, namely coating thickness measurement. Non-conductive coatings can be measured by the lift-off effect achieved away from a conductive substrate material. As with all eddy current tests, suitable reference standards are mandatory to achieve quantifiable results through representative instrument calibration.

- **Fill factor** When encircling coils and bobbin coils are used, the degree to which an item fills the coil, or the coil fills the test item bore will dictate eddy current induction levels within the test material. Variations due to dimensional inaccuracies may cause fluctuations in test data, and as such require minimizing to enable consistent results.

Eddy current test coils fall into two main categories, absolute and differential. The absolute test arrangement is used as a coil solely on its own, without any simultaneous reference to a standard or different part of the material. The differential arrangement uses two coils wound in opposing direction for single probe applications, or two individual coil probes when using the reference standard technique, see Figure 5.2.

The single differential arrangement is often used for bar or tube inspection as an on-site examination method, and compares two sections of the one sample enclosed by the two parts of the coil. Whereas the separate reference test coils, are ideally suited to fixed installations or a

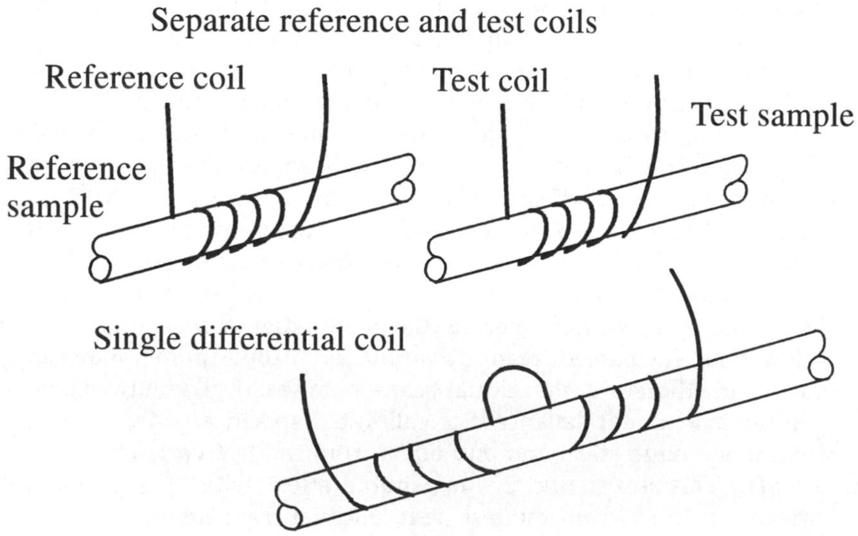

Fig. 5.2 Separate reference and test coils.

mass production test environment. Absolute coils usually are necessary for weld scans or variable geometry tests, such as machined components, aircraft service inspections and general condition monitoring.

Both absolute and differential coil arrangements can be in the form of a probe coil, encircling or bobbin types. The probe coil is manufactured in various styles and frequently custom-built coils are manufactured for specific applications. Tube or bar inspections utilize encircling coils through which the material is fed. However, tube bores can be inspected by inserting coil arrangements known as bobbin coils. Steam tubes in boiler tube plates are frequently monitored using this technique. The system sensitivity is dependent upon the control of variables and the size of the test coil. Discontinuities that exist sub-surface are more difficult to detect with increasing distance from the test surface.

Discontinuities that are surface-breaking can be detected with a wide range of frequencies, while lower frequency probes are necessary to penetrate material when searching for sub-surface flaws. Non-magnetizable material such as austenitic stainless steel, would be restricted to approximately 10–12 mm maximum flaw depth for detectability (ASM Metals Handbook, 1989). At such depths the induced flux lines would be very few, and thus reduced sensitivity prevails. Recently, new developments in corrosion detection and evaluation have been made. This technology is able to measure corrosion in tubing, especially in heat exchanger tubing, and the measured results can be evaluated to predict the remaining life of the components (Alferink, 1995). It is even possible to detect corrosion in pipes with insulation by means of Pulsed Eddy Currents (PEC) operating at low frequencies (Enters, 1995).

Eddy current techniques require consistency during a test, as many variables will lead to false indications. The test temperature must also remain consistent during a test. This is because temperature changes conductivity and therefore variations in temperature will dictate eddy current strength. Surface-breaking flaws are easiest to detect, but it is possible to detect flaws through a coated/painted surface. Thick non-conductive coatings reduce eddy current sensitivity due to lift-off effects. Therefore a limitation to coating thickness must be made. Simulated reference pieces, representative of a material, joint detail and surface condition are imperative to calibrate with. Natural or artificial flaws are generally used to establish sensitivity levels and functional ability. Non-magnetizable conductive materials are easiest to test. Ferromagnetic, magnetizable materials can cause difficulties due to magnetic permeability variations.

5.3   ELECTRICAL RESISTANCE TESTING (AC OR DCPD)

These are methods of measuring the depth of surface-breaking cracks on electrically conductive components and structures. They are not crack-

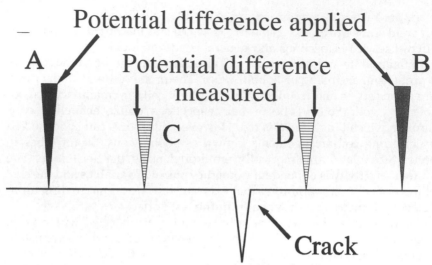

Fig. 5.3 Potential difference crack depth measurement.

detecting methods, so any cracks have to be detected with other techniques. The operating principle, as the name suggests, is electrical resistance. A potential difference is applied to the surface of the component over the crack from two electrodes, A and B and the Potential Difference (PD) is measured between two other electrodes C and D also aligned across the crack, as shown in Figure 5.3. When one has a deep crack, there will be more resistance between the electrodes than across a similar length of uncracked material.

If we suppose that the electrodes applying the PD, A and B, are 10 cm apart, and that 10 volts is applied to a perfect, defect free, homogeneous conductive plate, then it is reasonable to suppose that the PD will change by 1 volt per cm along the line directly between these two electrodes. If we now apply two further sensing electrodes, C and D at distances from A and B of 2 cm, there will be a PD between C and D of 6 volts.

If the effective electrical path between the two sets of electrodes, A,C and B,D is now increased, by a crack as shown in Figure 5.3 above, then the potential difference change per unit length across the conductive plate will also change. The current is forced to flow along a longer path between A and B, and therefore the PD will now change to less than 1 volt per cm, depending upon the depth of the crack. Electrodes C and D sense this.

The test system is calibrated against artificial defects of known depths, and then it is possible to adjust the system to give a direct reading in crack depth. Normally used calibration defects are one 'saw-cut' extending from perhaps 10 mm deep at the edge of a plate to 0 mm deep nearer to the other side of the plate. If the depth change is assumed to be linear along the saw-cut, then the depth will vary linearly from 10 mm to 0 mm. DC potential difference measurements were the first used for this technology. Alternating currents, (AC) however, do not penetrate so deeply as direct currents, and the current flow differences become more pronounced for smaller depths of crack. DC systems are simpler, and cheaper, for portable use.

Typical electrode spacings employed in commercially available instruments are about 1.5–2 cm for the two outside electrodes and about 8–10 mm for the inner electrodes. The technology is somewhat restricted in its application because the surfaces on either side of the crack must be similar, and without any grease, corrosion products, paint or other coatings, because these will cause extra resistance in the electrical circuits. Typical accuracy of this type of technology is about 10–20%, and crack depths from cracks shallower than a few tenths of a millimetre cannot accurately be measured.

5.4   FLUX LEAKAGE TESTING AND MAGNETIC TESTING

This technique is only applicable to ferromagnetic materials. The lines of magnetic flux run around a bar magnet in circular paths as shown in Figure 5.4. When a magnetic pole is brought in proximity to another, opposite, magnetic pole, there is a mutual attraction between the poles. There is also a concentration of magnetic flux between the poles. If these two poles are now brought closer together, until they are almost touching, there will still be a concentration of magnetic flux around the

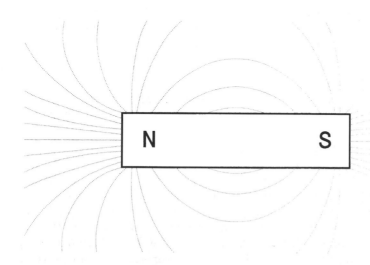

Fig. 5.4 Magnetic flux around a bar magnet.

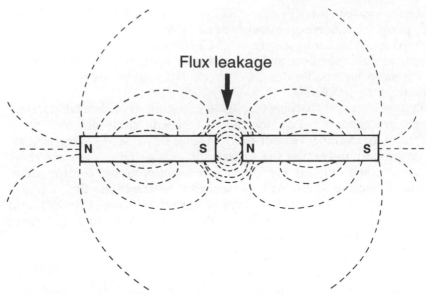

Fig. 5.5 Flux leakage around two adjacent magnets.

'split' between the two magnets as shown in Figure 5.5. This 'split' can be viewed as a crack. Detection of the concentration of flux 'leaking' out of what is now almost one, cracked, magnet will reveal this 'crack', as shown in Figure 5.6. Detection is normally effected via one of two general techniques:

- Magnetic Particle Testing (MPT)–that is the application of small particles that are then attracted to the flux leakage (such as iron powder). This is similar to the experiments you may have done at school with iron filings and a bar magnet.
- Flux Leakage Testing (FLT)–using electronic sensors such as Hall Effect probes.

There are several detection methods applicable to MPT. These all rely on very small particles 'flowing' over the surface, and being attracted to the flux leakage from the cracks. The particles can be suspended in water, water-based inks, a light hydrocarbon such as paraffin, (hydrocarbon-based inks), or dry powder applied in a gentle stream of air, (dry technique). These particles can also be coloured to contrast with the background, or black with a layer of white paint applied over the surface to be inspected for contrast enhancement. Alternatively, the particles can be coated with a fluorescent layer, and viewed under ultraviolet light with reduced ambient lighting. Each technique has advantages and disadvantages, dependent upon detectability, surface condition, surface temperature and economical viability.

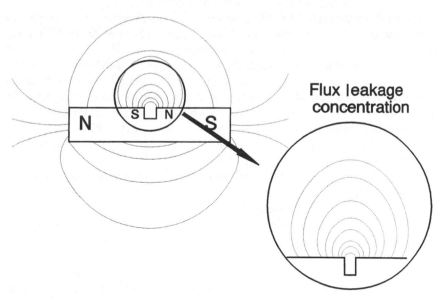

Flux leakage
concentration

Fig. 5.6   Flux leakage around a 'crack'.

The detection by all these indication technologies is 'manual', and depends upon the vigilance of the operator, environment, ambient lighting etc. Generally, approximately 80% to 90% of visible indications will be observed by the human operator (Juran, 1979), with the higher probability of detection for larger defects. Smallest detection limit, depending upon the surface condition, is about 0.5 mm long × 10 μm deep for a polished surface with the most sensitive techniques. There is a recent development of the fluorescent inspection technique becoming available for specific applications, using the same indication technology with the fluorescent inks, but using artificial viewing and intelligence for evaluation of the surfaces. This relies on very sensitive cameras connected to powerful computers to analyse the images.

It is more reliable, detecting 100% of all indications visible to the system, although the smallest detectable defect is about 3 mm long and 50 μm deep on a polished surface, i.e. about 5 times larger in each dimension than 'manual' detection. It is not very suitable for on-site applications, because the large amount of equipment involved brings handling and portability difficulties with it.

Indications from cracks will generally appear as sharp lines if the cracks are on the surface of the part being tested. Slightly subsurface cracks will tend to have a 'blurred' or 'fuzzy' appearance. When too much magnetic flux is used in the test, there will be a tendency for a high 'background' from the ink used, and this may mask some indications. Also, some sharp profile changes in a component will give rise to flux leakage concentrations around these profile changes. Beginning tests with a reduced flux, and then later increasing the flux, will help the operator to be able to distinguish between these conditions. A certain amount of experience is necessary before an operator can accurately interpret indications. Defect indications on weldments are difficult to interpret, especially around the sides of the weld, or undercuts. Here, 'dressing' of this area by grinding may be necessary before the test, to reduce the risk of any false indications, or of missing defects. Generally, the test specifications will define this.

Flux leakage testing is a technique suited to automation, which relies on a Hall Effect sensor to detect the flux leakage. A Hall Effect sensor produces a small voltage signal proportional to the magnetic flux that it is subjected to, and this signal can be the basis of a detection system. Here, the detection is by instruments, and much less subjective than the conventional operator-based MPT. A Hall Effect sensor consists of a small, thin plate that conducts an electric current from end to end. Electrodes are attached to the sides of the plate, such that the potential seen by the electrodes is the same on both sides. When the plate is subjected to a magnetic field, this will change the flow pattern of the electric current, thereby generating a small potential difference, or signal, between the electrodes. This signal is what then reveals the flux leakage.

This technology is more suited to mass production inspection of components than to condition monitoring, as the Hall Effect sensor is normally moved over the surface of the component under test mechanically. The mechanics for such an application tend to be very specific and expensive. Reliability is high, but detectability is less than that found in MPT, with a realistic lower detection limit being a few millimetres long and about 0.1 mm deep on an ideal surface. Recent developments in magnetic flux leakage corrosion detection, have been made with this technology. Piping, above 75 mm diameter, and plate with wall thicknesses of up to 15 mm can be inspected with reliable results. The technique is described in Figure 5.7. Here, permanent magnets are used, these frequently being fitted with rollers to assist positioning and movement (Enters, 1995).

There are two main techniques used for magnetization of components and local areas of structures. These are current flow and magnetic flow. There are also several ways of magnetizing the components or structure. The best methods to apply will be dictated by the application, required sensitivity, type location, orientation of expected defects and by economic viability. For structures, magnetization techniques are almost entirely restricted to current flow through prods or clamps, and to magnetic flow from a permanent or electromagnet.

Direct current magnetization is a technique that produces a steady magnetic field in one direction only. This penetrates deeper into the

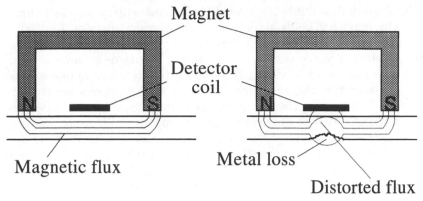

Fig. 5.7 Magnetic flux leakage detection technique.

material than AC magnetization, and allows the detection of defects up to a few millimetres under the surface. It is, however, difficult to remove the residual magnetism from components and structures that are sensitive or retentive to this residual magnetism. DC magnetization can also be produced in the test area with permanent magnets and current flow. Alternating current magnetization is a technique that does not penetrate so deeply into the material, so that the flux becomes more concentrated on the surface. This produces a little higher sensitivity, and the magnetic alternation will also help particle mobility over the surface. AC magnetization can be produced by electromagnets, yokes, current flow and induction, but not by permanent magnets.

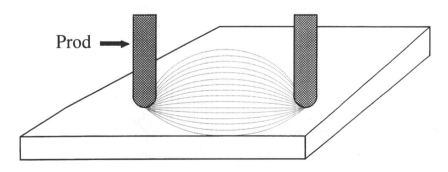

Fig. 5.8 Current flow with prod magnetization.

The Current flow method is probably the most widely used technique, using the magnetic fields associated with a current passing through a conductor for the inspection. A transformer, with or without rectification, provides a heavy current, from several hundreds to several thousands of amperes. This current is passed through the component to be tested, as shown in Figure 5.8, or through a small part of a structure to be tested. The flowing current generates a magnetic flux that is perpendicular to the direction of current flow, as shown in Figure 5.9. Similar effects are seen in cylindrical components, as shown in Figures 5.10 and 5.11, and in all cases, defects with orientation greater than about 45° from the magnetic flux flow have a possibility of being detected. If defects are suspected in other orientations, then for current flow with prods you have to reposition the prods to detect these other defects. For cylindrical components with circumferential defects, one should change the technique from current flow to magnetic flow.

For surface inspection of structures and larger components, typical prod spacings of 100 mm with about 800 amperes should be used. When 1000 to 1250 amperes are available, the spacing may be increased to about 125 mm. A general safe guide to typical values is 200 amperes for every 25 millimetres of prod spacing (Betz, 1967). Some inspection procedures will specify different values, and these should be used. The values given above are an indication of typical magnitudes to use. Hollow cylindrical components may be inspected by the 'threaded bar' technique, where a conductor carrying a large current is passed through the component, as shown in Figure 5.12. This produces a circular magnetization, suitable for the detection of axial defects. For the circular magnetization inspection of components using current flow techniques,

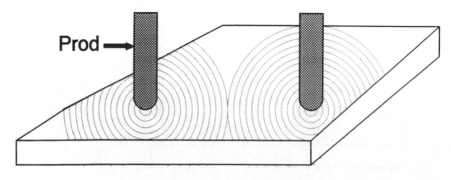

Fig. 5.9 Magnetic field with prod magnetization.

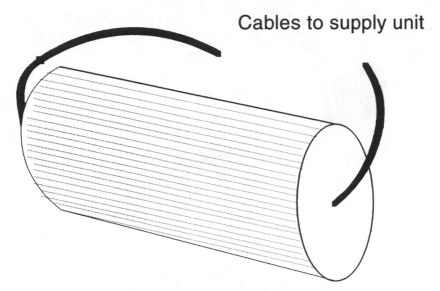

Fig. 5.10 Current flow with circular magnetization.

a current of approximately 1000 amperes per 25 mm diameter of the component is a good guideline. This may, however, give rise to some spurious indications due to the high flux density (Betz, 1967). For non-round components, one should use about 2 amperes for every square millimetre of section as a guide.

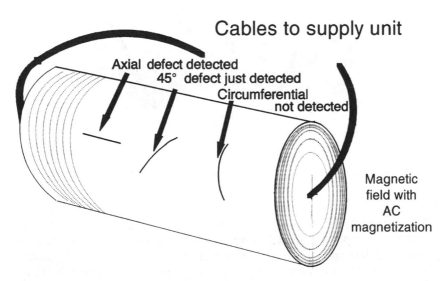

Fig. 5.11 Magnetic flux with circular magnetization.

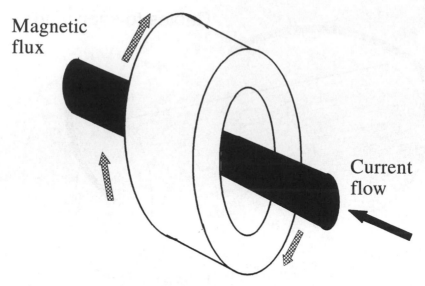

Fig. 5.12 Threaded bar technique of current flow.

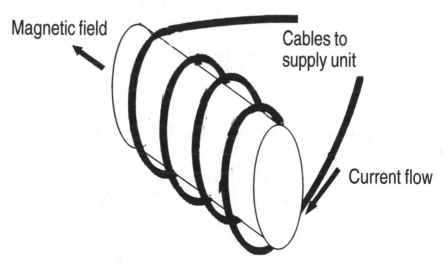

Fig. 5.13 Encircling coil method of magnetic flow.

There is generally one way to inspect with magnetic flow; that is to subject the component to a strong magnetic field. This field may be produced with permanent magnets or coils carrying AC or DC, as shown in Figure 5.13. As in current flow methods, defects are only detectable within 45° of perpendicular to the magnetic flow through a component or structure. Permanent magnets may be used, sometimes equipped with wheels to help movement over the surface of components. Coils may also be used on MPT test benches, often a test bench has a combination of current flow and magnetic flow capabilities. Here, a separate coil is installed on the test bench, and the component is inserted into the coil. Alternatively a coil from a flexible cable is wound around the component, or structure, being tested. The magnetic field from a coil is parallel with the axis of the coil.

5.5   PENETRANT TESTING

This is one of the simplest and oldest surface crack detection techniques. It may be applied to all non-porous materials, but can only detect cracks open to the surface. The surfaces also have to be free of any coating, grease, paint, scale or rust. Penetrant testing has a very simple principle. The surface to be inspected is coated with a film of a special liquid or penetrant. This is drawn into any surface breaking cracks by capillary action, and the surface is then cleaned with a suitable cleaner or remover. After this, a layer of developer – a chalk powder or similar absorbent material, suspended in a suitable solvent – is applied. This layer when dry, draws the liquid out of the crack like blotting paper, and spreads it over a larger area of the surface, thus making it more visible.

The liquid used as penetrant is frequently coloured to contrast with the chalk developer, or else a fluorescent penetrant is used that can be inspected under ultra-violet light. The earliest form of penetrant testing was early in the twentieth century when components were dipped into dirty penetrating oil for some time, washed in a solvent, then painted with a suspension of chalk in a solvent. Large cracks were visible with this technology many years before magnetic testing became a recognized technique in the 1930s.

There are different types of penetrant available today, each type with its own advantages and disadvantages. Some types are solvent-based, applied directly from aerosol cans and used with a solvent remover, (also applied from aerosols). Some types are water-washable penetrants, requiring an emulsifying liquid to help the water-based removal process. Similarly, there are different types of developer obtainable. Some developers may be applied by spraying, or directly from aerosol cans, some by dipping the component into a bath of developer, and some may be applied in a cloud of dry powder over the component being tested. With each type, there are different procedures to be applied in order to make a meaningful and repeatable test. The most commonly used, and most portable, type in condition monitoring of structures is the aerosol-based, coloured dye system. This is frequently used with a red coloured penetrant which contrasts well against the white chalk background.

The surfaces to be tested should be correctly prepared. They must be clean, dry and free from any loose scale, rust or dirt. Any paint, varnish or other layer on the surface **must** be removed before testing. Scale, paint and rust removal can be done with many different techniques. Wire-brushing, shot blasting with shot, sand, water or other suitable materials, grinding, electro-chemicals, or abrasive papers may be used. One must consider that the removal technique chosen must not close up any possible surface breaking cracks, especially in softer materials such as aluminium, and this preparation itself must be non-destructive. In some cases, the surfaces are blasted with the ground-up kernel of walnuts. This has a suitable abrasive effect without the risk of sealing cracks by peening over the surfaces.

The surfaces should also be dry and free from grease or similar materials. Precleaning with, for example, penetrant remover is often recommended. Surfaces and components should be at an acceptable temperature before testing, according to the specification of the chemicals used. Too hot a surface can cause one of two main problems:

- the chemicals applied could rapidly evaporate and may come above the flash point of the solvents used, giving a risk of explosion;
- the chemicals used in the dye can lose their (colour) effectiveness, or the penetrant can lose some of its penetrating ability due to the surface temperature.

Surfaces that are too cold may prevent the penetrant from penetrating properly.

The penetrant is applied to the surface to be tested by suitable means, spraying, dipping, brushing etc. It should be allowed to remain in contact with the surface for a specified time, and must not dry out in this time. The soaking, or dwell time depends upon the type of penetrant and the required sensitivity, and can vary between 1 minute and 24 hours. For general applications, 10 minutes is a reasonable dwell time if not otherwise specified. For solvent-removed penetrants, excess penetrant is removed from the surface with dry cloths. The surface is further cleaned by wiping it with a cloth dampened in remover until all traces of the penetrant are removed from the surface. The surface should then be allowed to dry before applying the developer.

For water-washable penetrants, again excess penetrant is removed with dry cloths, and then an emulsifier is applied over the surface. This is allowed to remain in contact with the surface for a prescribed time, before washing the surface of the component with a brisk but not forceful spray of water. After this, the surface is allowed to dry, where necessary, before applying the developer. Some water-washable penetrants have a self-emulsification feature, and then this extra emulsifier is not necessary.

Caution: You can over-remove penetrant from surfaces, and this will also remove penetrant from the cracks that you are trying to detect, making them invisible.

When the surface is clean and dry, the developer may be applied according to specification. This should not be applied too heavily or else you may mask defect indications. The developer should be allowed to dry and then to stand to allow indications to develop. This time is called the development time and can vary from a few seconds to about an hour. Five minutes is a reasonable time if no time is specified. After this, the surface to be inspected can be viewed. With fluorescent penetrants, this should be under a suitable intensity of ultra-violet light with reduced ambient lighting. For coloured penetrants, a suitable white-light illumination is necessary for inspection.

The surfaces under test should be investigated with suitable illumination and magnification. Where very small defects are to be detected, a microscope can be used for inspection, and under the most sensitive testing conditions, defects as small as 25 μm can be detected. For lower sensitivities, no magnification is necessary. Defects to be detected **must** be significantly larger than the features found on the surfaces of the component being tested, otherwise they will not be detected. This means that detection of, for example, 25 μm defects cannot be achieved on a rough casting.

Generally, penetrant indications from cracks will show up as lines on the surface. These will generally be fine, but heavy application of either remover or developer can make these indications less distinct. Partially closed cracks will appear as dotted lines. Pores open to the surface will appear as spots, and inadequate removal of penetrant from the surface will appear as smears. Some of the chemicals used in penetrant inspection can be hazardous, both to humans and to some safety critical components. Accordingly the following should be noted.

- There is a risk of asphyxiation when using volatile solvents in confined spaces. Adequate precautions must be taken here to avoid this.
- The chemicals used can be aggressive to the skin. Suitable precautions to avoid excessive exposure to the chemicals must be taken.
- When applied to hot surfaces, some solvents that may be present in the chemicals of penetrant tests, could be heated above their flashpoint. This gives a risk of an explosion.
- Exposure to ultraviolet light can be dangerous, especially to the human eye. Normal ultraviolet lights when undamaged will have an insignificant amount of harmful ultraviolet radiation, but direct viewing of the light should be avoided. Damaged filters on the lights are dangerous and should not be used.
- Some components for use in liquid oxygen (LOX) systems should only be tested with halogen-free chemicals and solvents. A risk of explosion

is present if liquid oxygen becomes contaminated with halogens. Only LOX approved materials should be used for these applications.
● Some high temperature or exotic alloys used in the aerospace and nuclear industries can react with sulphur and halogens, initiating stress corrosion cracking. Suitable penetrant systems must be used here to avoid this potential hazard.

5.6   RADIOGRAPHIC TESTING

This is one of the earliest known NDT techniques, and was discovered by Wilhelm Röntgen on 8 November 1895, actually it was the forerunner of today's real time radiography or radioscopy. He found that with a mysterious machine he could 'see through solid objects'. He recognized that it was a form of radiation or rays, so he named it X rays. The radiation discovered by Röntgen was a part of the electromagnetic spectrum that we all see every day around us. This ranges from DC (direct current), for example a battery, right through the mains electricity frequency of 50 or 60 Hertz (Hz), radio and television broadcast frequencies, past light and onto X and gamma rays (frequencies of thousands of Terahertz).

Radiation can pass through solid objects, but is attenuated, or reduced in strength, in the process–the denser or thicker the material, the greater the attenuation. This means that if X or gamma rays pass through a solid, dense, object such as a metal, then holes and less dense material enclosed within the metal will attenuate less radiation than the metal itself, while more dense inclusions will attenuate more. If we could see the intensity of the radiation after it has passed through the metal, we could detect these inclusions of different density, as shown in Figure 5.14. Fortunately for us radiation affects a photographic film in the same way that light does. If we can project a radiographic image onto a film, and then develop it, we can see if there are any defects in the object inspected. Radiation also makes some materials fluoresce, i.e. to emit light, so we can also use this feature in the inspection process.

Different materials attenuate radiation by different amounts, and shorter wavelengths, or higher energies, of radiation can penetrate more through materials than longer wavelengths. By knowing the material density and thickness, also the equivalent energy of radiation, it is possible to calculate how much exposure is needed to produce a radiographic image of a component or part of a structure. An example is a 25 mm-thick steel plate and a 200 kV X-ray unit, where a typical exposure time of a few minutes is normal. With a thicker specimen, e.g. 100 mm, a higher energy is needed, for instance 350 kV, and a longer exposure time, several tens of minutes.

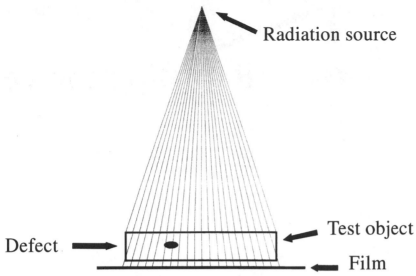

Fig. 5.14 Detecting radiations of different density.

An advantage of radiography over ultrasonics is that a remote surface can be inspected, as shown in Figure 5.15, while for ultrasonics, only the material of the scanned surface is normally inspected. This is particularly useful when inspecting tubes and pipes, when both the nearest and farthest wall can be inspected together, even through an insulation layer of several centimetres, although the image contains little or no depth information. With ultrasonics, the other internal defect detection technique, normally only the nearest wall of the tube can be inspected. One disadvantage of radiography over ultrasonics is that access to both sides of the work-piece is needed, one side for the radiation source and one side for the film.

The detection limits of radiography are about 1% to 2% of the total thickness of material that is being inspected. That means that if the example shown in Figure 5.15 had a wall thickness of 10 mm, then the best detectability would be about 1.5% of 20 mm, which is a defect of 0.3 mm in thickness. This method is known as the 'Double Wall, Double Image' technique. Normally the source of radiation would be offset axially along the tube by a small amount with this technique, producing an elliptical image of a circumferential weld in such a tube. Two separate radiographs separated by 90° would be used for the inspection of welding with this technique to ensure good definition of any possible defects.

If access is available to the inside of a tube, as shown in Figure 5.16, then the detection limit would be a defect of about 0.15 mm thick. Here the radiation source is shown at the centre of the tube. Wrapped around

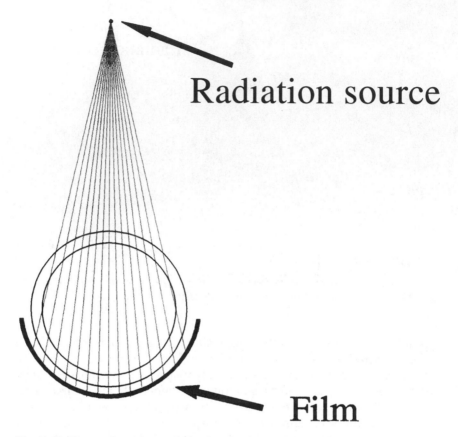

Radiation source

Film

Fig. 5.15 Inspection of remote surface by radiography.

the circumference are several films covering the complete 360°, but slightly overlapping each other. With this technique, the entire circumference can be inspected in one exposure. This method is known as the 'Single Wall, Single Image' technique. Another alternative, with the source placed close to the outside of the tube, and the film placed directly opposite the source, but outside the tube is called the 'Double Wall, Single Image' technique. There is no 'Single Wall Double Image' option.

Defect orientation plays an important aspect in detectability. Narrow defects oriented parallel to the radiation produce small images with good contrast. While those that have a very small dimension in the axis of the radiation, of less than 1% to 2%, produce images with larger dimensions and much lower contrasts, generally not being detectable. The sources of radiation normally used in radiography are X-rays that are generated electronically, and gamma radiation, 'generated' as a

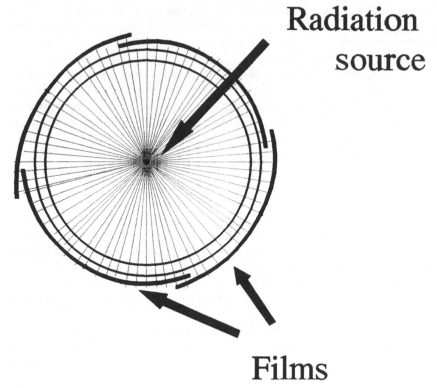

Radiation source

Films

Fig. 5.16 Radiation source at the centre of the tube.

natural process of decay of a radioactive isotope. With an X-ray source, electronic control of the source is easy to arrange, while for gamma radiography, normally a mechanical system is employed to transfer the radioisotope from a shielded container to the holder during the exposure.

Afterwards the same mechanical system is used to retract the source into the container. These points are important, as the radiation used in industrial radiography is extremely dangerous and cannot be 'seen' by any of the human senses. Instruments are needed to detect this radiation. Exposure to radiation can cause cancer, sterility and even death, and with the radiation sources employed in industrial radiography today, a lethal exposure could be reached within a matter of hours. To work safely with radiography, adequate protection **must** be available to the operators and to the public who may be near the exposure. The safest way of protecting personnel is to use a specially constructed exposure cabin with adequately lined walls, floor and ceiling.

This is not practical to apply on a construction site, or in most cases of condition monitoring. Here the next best protection available is distance. Radiation intensity decays with the 'inverse square' law. That is to say that if the distance from the source doubles, then the radiation intensity reduces by a factor of four. Four times the distance means an intensity reduction of 16 times etc. For practical purposes, it is often necessary to restrict access to radiographic sites over large areas, which can sometimes extend for 100 m diameter or more from the source. We must also consider areas above and below the level where radiography is being performed. Workers on other floors of a building or structure could also become exposed.

To produce clearly defined radiographs, the geometry of the inspection must be considered. Figure 5.17 shows the effect of geometric unsharpness, where the image of the defect becomes 'blurred' or unsharp. To reduce this effect, we must use a radiation source that is as small as possible, and at the same time have the maximum practical distance between the source and the test object. The film must also be as close to the test object as possible. These requirements are not always consistent with obtaining rapid radiographs. To reduce the exposure time, the source must be as close as possible to the test object, remember that doubling the distance means an increase in exposure time of four times.

With gamma radiography, a larger physical size of source means generally that there is a potential for a larger amount of activity, and radiation, but also greater de-focusing, while for an X-ray tube, a larger target will permit a greater current to be used, and therefore reduce the exposure time. Also because of the geometry involved, a thicker test piece will require that the source–film distance is increased, while in practice the thicker test piece will anyway require a longer exposure time.

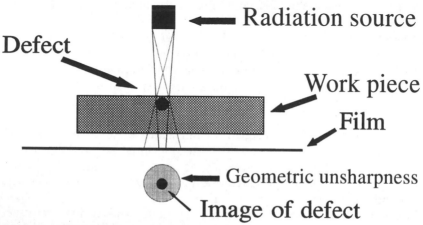

Fig. 5.17   Effect of geometric unsharpness.

What we have described above is called film radiography. This produces a film that can be remotely interpreted, and can be stored for recording purposes. A variant of radiography is called radioscopy or real-time radiography. With radioscopy, a screen of fluorescing material is placed behind the test object where the film would normally be placed. Any radiation falling onto this screen emits light according to the intensity of radiation that falls upon each grain of the fluorescent material. We can inspect this screen either by eye, after reflecting the image by 90° with a mirror to take it outside of the radiation hazard zone, or else with an image intensifier and television camera.

Neither of these techniques produce a storable film for record purposes, although with the TV camera, one can record a video image of the tested object. In practice, radioscopy is very difficult to apply 'on-site' or in a condition monitoring environment because one has to carry much more equipment to the radiography site. However, a recent development, suitable for inspecting insulated pipes for outside corrosion, uses radioscopy with low energies and tangential inspection (Figure 5.18). Here, the 'good' material of the pipe provides an almost total attenuation, while corrosion will show an irregularity in the 'horizon' of the picture displayed. Because of the low energies involved, the radiation hazards are considerably reduced. Typical inspection speeds achievable here are about 10 cm/sec, allowing a practical inspection rate of up to 100 m of pipework per day with a two-man crew (Enters, 1995).

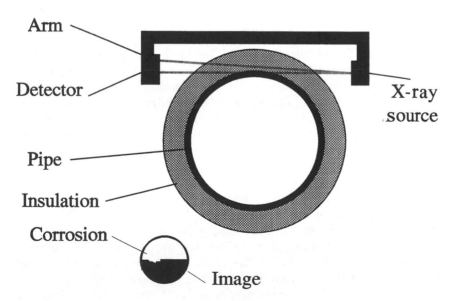

Fig. 5.18 Radioscopy with low energies and tangential inspection.

5.7   RESONANT SPECTROSCOPIC TESTING

This is a new technology that has been available in its present form only since late 1993. It is not really a new technique, but an old technique, probably one of the oldest known in the NDT world. The principle of the technology is whole body resonance of structures and components. In simpler terms, wheel-tapping. The technique was used until the 1950s or 1960s by railways worldwide. The wheel of a railway locomotive or wagon is tapped with a hammer, and an 'expert' listens to the ring of the wheel. If a defect is present in the wheel, then the expert can detect this in the tone that he hears.

The REsonant Spectroscopic Testing (RES) applied today can analyse the tone or ringing of the component or structure, and pronounces if there are any defects present. There are two forms of RES, active and passive. Active RES means that the component or structure is excited with a slow sweep of sine-wave form, 'narrow' band, vibrations, and the amplitude of the vibrations in the component recorded against the exciting frequency. Passive RES means that some external, natural or 'broad' band, excitation is used to excite the component or structure, and the amplitudes and frequencies of the resultant vibrations are analysed.

Active RES is more suited for monitoring components, and passive RES more suited for monitoring structures. Resonant frequencies of components and structures are dependent on many factors. They can be listed as dimensions, shape, defects, material type and the elastic constants of the material type. RES technology was developed at Los Alamos National Laboratory in New Mexico USA, as a tool to detect changes in elastic properties of sugar-grain sized super-conductors while they were subjected to different temperatures. In this way, the superconducting critical temperature could be determined (Migliori *et al.*, 1993).

Resonances can exist in four main forms or modes, and harmonics of any of the modes can be present, and used for diagnostics, in evaluating with RES. The four forms of resonance are **bending**, like a ruler held by one end on a desk, and the other end excited, **torsional**, twisting both ends of a component in opposite directions, **extensional**, growing and shrinking along one axis of the component, and **surface acoustic**, similar to propagation of surface waves in ultrasound, or like the ripples on a pond after a stone is thrown into it.

In active RES, resonances of components can be calculated, and then the component can be tested, using active RES, by sweeping the excitation frequency over a little more than the predicted range. This is to see if the resonances of the component measured coincide with the predicted ones, or alternatively if they differ from the spectra of known 'good' components. Deviations from the predicted, or 'good' condition indicate a component that is defective, and spectra of components coinciding with the prediction indicate a component that is good. Fingerprints of individual components can be recorded for comparison with later tests. If the fingerprint changes, this indicates a change in the condition of the component.

Active RES excitation can be with a piezo transducer, shaker, ElectroMagnetic Acoustic Transducer (EMAT), magnetostrictive or any similar device. Detection of the amplitude of vibrations can be achieved also with piezo, EMAT, microphone or accelerometer. Computers record the amplitudes of vibrations, and can plot them in graphic form with frequency on the X axis and amplitude on the Y axis as the fingerprint for the component or structure.

In passive RES, bridges, refinery columns, chimneys etc. can be scanned remotely to detect the resonances of the structures while they are subjected to natural, or 'broad' band, manmade excitation. This natural excitation can be in the form of, for example, a train or goods vehicle passing over a bridge, or a chimney or column subjected to a strong wind. With passive RES used on structures, predictions of resonances are almost impossible to make, cf. the Tacoma Narrows bridge disaster. Normal test procedures require that the structure is tested and a fingerprint taken, then subsequent fingerprints are compared with the first. Changes in the fingerprint indicate a change in condition of the structure.

Passive RES, as defined above, relies mostly on natural excitation of structures, although a 'broad' band mechanical 'shock' can be generated with a laser in a similar way that laser ultrasound is generated, or by hitting the component/structure with a hammer. Detection of vibrations in passive RES can be achieved with laser interferometry, accelerometers, EMAT, piezo transducers or microwave interferometry. Powerful computers are required to analyse the frequencies present in the recorded vibrations.

Active RES can produce much more sensitive spectra than passive RES, because all the energy in the excitation can be at one specific frequency during the sweep, typically with single numbers of Hertz steps in the sweep, while passive RES must analyse the entire spectrum of the vibrations. This means that the signal to noise ratio (S/N) is much better in active RES. Active RES is possible on structures such as bridges, chimneys etc., but practicalities normally require that passive RES be used.

5.8 ULTRASONIC TESTING

Ultrasound is the use of sonic energy at frequencies exceeding the human audible range. Typical test frequencies start at 250 kiloHertz (kHz), 250 000 cycles per second, up to 25 MegaHertz (MHz), 25 000 000 cycles per second, with ongoing research by various organizations into uses of even higher frequencies, exceeding 100 MHz. Ultrasonic energy is a form of mechanical energy excited by either piezoelectric, magnetostrictive methods, EMAT or laser. Mechanical shock waves are transmitted from probes into materials, producing a similar effect to striking an object with a hammer. Simple analogies can be drawn between the hammer and ultrasonic probe.

A small hammer will only excite weak shock-waves when striking an object; so too will a small diameter ultrasonic probe. Conversely large diameter probes and large hammers give strong shock waves which travel for greater distances through materials. For optimum penetration power, low-frequency large-diameter probes are used, enabling several meters of steel to be tested. However, this causes a problem in defining small flaws, and is termed 'Resolution'. High resolution or clarity is obtained by utilizing high-frequency probes with medium to high crystal damping. This means that the emitted pulse is very condensed and of short duration as shown in Figure 5.19.

Ultrasonic testing relies upon the measurement of time and amplitude or strength of a signal between emission and reception. Due to a mismatch of acoustic properties between materials, the sound will partly reflect at interfaces. The quantity of reflected energy is dependent upon the acoustic impedance ratio between two materials. For example, sound transmitted through steel reaching a steel/air boundary will cause 99.996% internal reflection, while a steel/water boundary would reflect only 88% within the material and transmit 12% into the water.

Bearing this principle in mind let us now consider detection of defects. If impedance ratios are widely different like an open crack with a steel/air interface, then adequate reflection will occur for detection of the

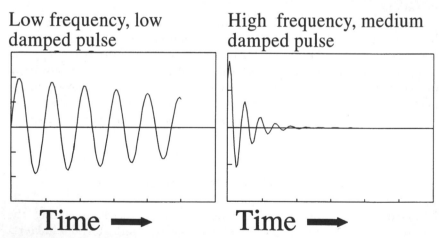

Low frequency, low damped pulse

High frequency, medium damped pulse

Time ⟶

Time ⟶

Fig. 5.19 Emitted pulse is condensed and of short duration.

Plate lamination

Direction of sound beam required for detection

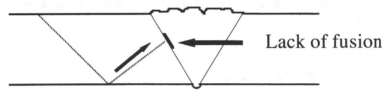

Lack of fusion

Fig. 5.20 Prior knowledge of weld geometry, preferential grain flow or positioning of casting chills promoting directional solidification aids the choice of probe to optimize detectability.

flaw; conversely, a small crack in a compressive stress field that does not have oxidized faces will yield a steel/steel boundary, and be transparent to ultrasound. Orientation of flaws with respect to the sound beam also affects reflected signal energy, and therefore signal amplitude on the flaw detector display. Prior knowledge of weld geometry, preferential grain flow or positioning of casting chills promoting directional solidification aids the choice of probe to optimize flaw detectability, as shown in Figure 5.20.

Maximum detection will occur when the ultrasonic beam is 90° to the major plane of a flaw. As with light, the sound energy will reflect at an angle equal to the incidence angle. This is shown in Figure 5.21. Several probes are often used with differing beam angles to detect flaws within an item. Choice of such is usually determined by an NDT expert, who will compile a detailed test procedure for a specific application. Figures 5.22 and 5.23 show some probe design types. Ultrasonic sensitivity is defined as the size of the smallest flaw detectable. This is like saying 'How long is a piece of string?' But quantifying sensitivity is often practically performed utilizing manufactured reference reflectors.

The theoretical smallest detectable flaw is considered to be $\frac{1}{4}\lambda$, where λ is one ultrasonic wavelength, and the following formula applies:

$$\lambda = \frac{c}{f} \tag{5.1}$$

where λ = wavelength; c = characteristic velocity of the material in millimetres per second (mm/sec); f = probe frequency in Hertz.

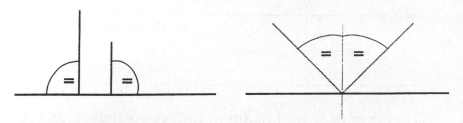

Fig. 5.21 Maximum detection occurs when the ultrasonic beam is 90° to the major plane of a flaw.

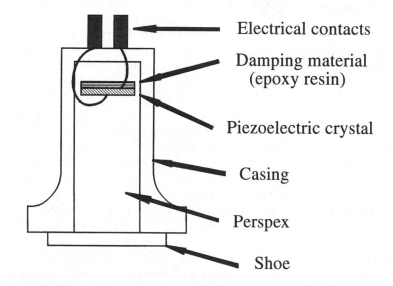

Electrical contacts

Damping material (epoxy resin)

Piezoelectric crystal

Casing

Perspex

Shoe

Fig. 5.22 Section through a 0° longitudinal wave probe.

For example, if a 4 MHz longitudinal probe is used to test steel, then:

$$\frac{\text{velocity of sound in steel}}{\text{probe frequency}} = \frac{5.960 \times 10^6}{4 \times 10^6\,\text{Hz}}\ \text{mm/sec} = \lambda$$

$5.96/4.00 = 1.49\,\text{mm} = \lambda$, and therefore the smallest detectable flaw will be $\lambda/4 = 0.372\,\text{mm}$.

From this we can see that high frequencies will improve sensitivity, and that is true. However, there is one major problem called 'attenuation'. Each material has a grain structure and the density of grain dislocation, together with the physical size of the grains, will govern attenuation. This strange word attenuation means the loss of sound as it

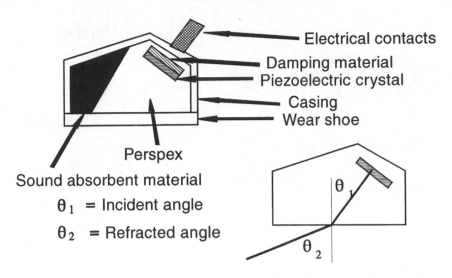

Electrical contacts
Damping material
Piezoelectric crystal
Casing
Wear shoe

Perspex

Sound absorbent material

θ_1 = Incident angle

θ_2 = Refracted angle

Fig. 5.23 Section through an angle probe: these generally produce refracted transverse waves; however, special applications may dictate longitudinal waves.

travels through a material. Coarse-grained materials cause large amounts of sound loss, and in extreme cases can be untestable. Heat treatment is often used to refine grain structures where high attenuation loss is a problem. The absence of fine grain structures necessitate utilization of low-frequency probes. The long wavelengths derived from low frequencies, 1 to 2 MHz, are less attenuated and penetrate greater distances.

The surface finish of the test component or structure is also an important factor. Rough surfaces scatter sound as it enters or returns through an interface. In general, surface finishes at 6.3 μm Ra or better are required for ultrasonic contact tests with ultrasound of 4 MHz, but high sensitivity examinations using high-frequency probes will require 3.2 μm Ra finish or better. The diagrams in Figure 5.23 compare attenuation and transfer loss mechanisms demonstrating the surface and internal scattering effects.

For sound to be transmitted in and out of the material to the probe, a liquid film known as a couplant must be used. This film has the primary function of expelling air. As noted earlier, air has a very poor match of acoustic impedance to most solid materials, so an interface with air on one side acts as an almost perfect reflector. The couplant also conforms to slightly irregular surfaces, and allows the probe contact to be bedded down, smoothing out small geometric changes (Figure 5.24). In practice,

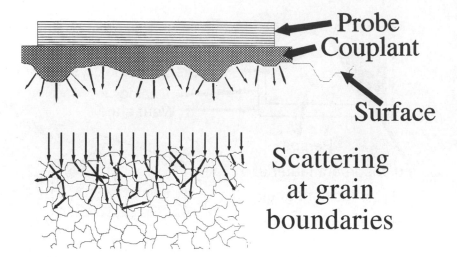

Fig. 5.24 Probe couplant conforms to slightly irregular surfaces, allowing the probe contact to be bedded down, smoothing out small geometric changes.

grease or a suspension of wallpaper paste in water are frequently used as couplants. These fluids have a much higher viscosity than water, and tend to 'fill' small irregularities better than plain water.

Ultrasonic testing can be conducted by either placing the probe on a test piece, known as contact testing, or by submerging an item in a water tank and transmitting the sound through water with the probe offset to a fixed distance, known as immersion testing. Most manual tests use contact methods, while automated systems can be either contact or immersion. With the improvement in microchip technology, the physical size of flaw detectors in general has been vastly reduced, aiding portability of both manual and some automatic equipment. Data presentation on manual flaw detectors is usually A scan. This system plots time or distance against signal strength or amplitude, as shown in Figure 5.25.

B scan, C scan and D scan are all derivatives plotting variations of depth and probe travel relative to a flaw. These presentations are predominantly automated techniques which can be stored digitally by computer-based systems. Digital technology has revolutionized data recording for both manual and automatic systems. Proforma reports can be downloaded from flaw detectors directly to printers or to computers, either directly or via modems from remote locations, thus allowing rapid data transfer of test measurements and parameters. Limited flaw location data can also be digitally recorded with manual flaw detectors, reducing operator time, potential errors, and saving time and money.

Applications associated with ultrasonic testing range from corrosion/erosion monitoring with thickness surveys, weld integrity testing,

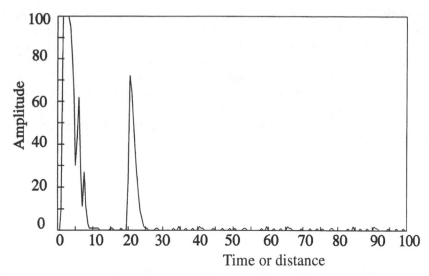

Fig. 5.25 Manual flaw detection system plotting time or distance against signal strength or amplitude.

wrought materials testing, cast material evaluation, composite material tests, bond testing, grain structure analysis, hardness depth analysis, fluid level monitoring. There are also many medical applications. Instrumentation can be varied, ranging from a calculator-sized wall thickness monitor up to a full 2 m high 19 in rack system for some automated installations, but the principle of ultrasound is common to all. One great advantage for condition monitoring purposes is that only one side of a material or component is required for access, while the test can evaluate internal and far-surface conditions. The method is generally portable, and becoming more so with today's micro-electronics, and no serious operating hazards are known.

5.9 VISUAL TESTING

Visual inspection is by far oldest method of non-destructive testing but is frequently viewed as a subordinate discipline. Basic techniques can be directed to achieve both simple or complex methods of inspection using the human eye as the principal instrument. Naturally the majority of visual tests are restricted to surface examinations, but translucent materials including some composite structures may enable subsurface flaw detection.

An observer's visual acuity essentially becomes the prime calibration requirement for a visual test, whereas optical tests in general can include automatic or semi-automatic machine vision systems. The standard visual tests for NDT personnel involves a near vision acuity test and a colour discrimination test. Certain applications warrant far vision and colour blindness examinations. Optical machine vision as a minimum requirement should be at least equal to direct human vision. While some video systems suffer from poor resolution, selective purchasing will often identify higher quality equipment.

Visual tasks can be affected by environmental, physiological and psychological factors which include:

- test illumination
- surface contrast
- surface reflectance
- perspective
- distance
- mental attitude
- angle of vision
- cleanliness
- temperature
- atmospheric condition
- physical health

Perception is the vital ingredient of a visual test. The eye allows light to focus on its retina, but the way the brain processes this information produces the image and the sensation of perception. What you see and what you think you see can be very different. Simple graphic illustrations like Escher's never-ending staircase or more contemporary illusions, illustrate how visual concepts can be manipulated. A person's preconceived thoughts will bias their perception during a visual test. This will not necessarily invalidate a test but may lead to variable results within a group of individuals (Juran, 1979).

Most NDT methods require some form of visual judgement. Magnetic particle testing, liquid penetrant testing and radiography are very common disciplines that enhance contrast to aid visual testing. Many other techniques electronically process signals before they are visually observed, helping to reduce the information that the human computer, the brain, has to handle. Chapters 3 and 6 of this book outline contemporary visual and optical equipment available, but remember, visual inspection must never be treated with complacency. In its own right visual testing is acknowledged as an individual discipline and simple techniques together with personal vigilance can prove an effective test strategy.

5.10   CONCLUSIONS

The above chapter has presented an overview of the techniques available in the area of surface and internal defect detection. It is important to note that the material above will hopefully act as a guide to the different technologies available in this area, their advantages, disadvantages, and the limitations associated with their application in industry. The information contained in this chapter is intended to act as an appreciation for the non-expert, which may stimulate the reader's interest in this area, and also provoke a desire to find out more about non-destructive testing techniques.

5.11 REFERENCES

Alferink, R.G. (1995) Bepaling restlevensduur van een warmtewisselaar op basis van inspectieresultaten, *Proceedings of KINT/BANT Biennale '95*, Breda, NL, 26 and 27 April.

Anon (1989) Skin effect, *ASM Metals Handbook*, Vol. 17, American Society for Metals, Metals Park, Ohio, USA.

Betz, C.E. (1967) *Principles of Magnetic Particle Testing*, 1st edn, Magnaflux Corporation, Chicago, Illinois, USA, 208.

Enters, A.C. (1995) Corrosion detection of insulated pipelines, *Proceedings of KINT/BANT Biennale '95*, Breda, NL, 26 and 27 April.

Juran, J.M. (1979) *Quality Control Handbook*, 3rd edn, McGraw-Hill, New York, USA.

Migliori, A. *et al.* (1993) Resonant ultrasound spectroscopic techniques for measurement of the elastic moduli of solids, *Physica 'B'*, **183**, 1–24.

CHAPTER 6

Commercial applications of visual monitoring

A. Davies

Systems Division, School of Engineering, University of Wales Cardiff (UWC), PO Box 688, Queens Buildings, Newport Road, Cardiff, CF2 3TE, UK

6.1 INTRODUCTION

The inspection of industrial and other types of machinery as an element of maintenance activity is, as outlined in many of the chapters contained in this book, heavily dependent on our visual sense. As human beings we have five senses, which arguably we use not only in everyday life, but also in the specific case of the condition monitoring of machines. We may do this quite unconsciously, as for example when we drive a car. Most experienced drivers do not consciously think about monitoring their vehicle as they are driving it, yet at the first sign of trouble our senses alert us to the fact that something is wrong. Any out of the ordinary noise is picked up by our sense of hearing, undue vibration by our sense of touch, fumes by our senses of smell, taste and vision etc. So although we are not consciously aware of it when driving, we act as a fully integrated condition monitoring system for the vehicle.

When you think about it, this is not that unusual. After all, any vehicle or other type of machine we use in everyday life or industry is a human concept, created by us for our use. Thus you would expect the monitoring and control systems to be designed for, and attuned to, our physiology. In the case of a car, this is evinced by the instrument panel, which is designed for rapid interpretation by our sense of sight. Just a glance reassures us that all is OK. The fact that this is less so in other forms of human transport, not designed by ourselves, is fairly obvious. For example, although we can tame and control a horse for riding, in the case of an average person, our senses are arguably less efficient in diagnosing

Handbook of Condition Monitoring
Edited by A. Davies
Published in 1998 by Chapman & Hall, London. ISBN 0 412 61320 4.

the horse's well-being compared to that of our car. A horse does not of course have an instrument panel!

In either case, however, our sense of sight for both monitoring and control is paramount, with the same being true of industrial machinery. Vision in human beings should therefore be recognized as our prime sense and may be defined as:

A sense of perception, within a defined range of the electromagnetic spectrum, which permits humankind to perceive, monitor, operate within, and control a portion of its environment.

The final few words of this definition contain the rub. For although sight is unbelievably important to us as individuals, so that we may operate efficiently within our world, in terms of range and flexibility, it has its disadvantages.

Two of these limitations, which are important from the point of view of monitoring, are of course our inability to:

- resolve objects at long or short ranges with sufficient definition for intelligent interpretation, and without some form of artificial assistance;
- perceive the complete electromagnetic spectrum, without some form of artificial assistance.

Fortunately, our curiosity, intelligence and manufacturing skill have allowed us to develop suitable instrumentation to overcome in large measure both these disabilities, thus permitting us to perceive our environment more completely. Indeed, it is this perception of the environment, or more specifically aspects of it pertaining to machinery well-being, which is the subject of this book. Although here we are mainly concerned with our sense of vision and the means by which we have enhanced nature's gift, it should be noted that the same argument can be made for all our other senses in respect of condition monitoring equipment.

Accordingly, in this chapter we will concern ourselves with the following aspects of commercially available condition monitoring equipment for visual inspection. These are:

The sensing and transforming of a physical feature or characteristic of the machine or equipment under surveillance, into a signal which can be visually interpreted, for clues, symptoms or a direct indication of the system's condition.

In addition, an outline is given in respect of the possible future for visual inspection, in terms of information acquisition, monitoring and display for maintenance.

6.2   BORESCOPES, FIBRESCOPES AND ENDOSCOPES

These particular instruments are quite well known within both the medical and maintenance environments, being used in both cases for diagnostic work. Thus, over very many years, all these instruments, in one form or another, have been used to enhance our visual perception of what is going on, either within an inaccessible region of the human body, or in an equivalent area of some industrial machine. In the industrial sense, they are:

Essentially optical systems for inspecting machinery spaces which are normally inaccessible without major plant disassembly.

Although performing a similar task, it is important to note the difference between them. These instruments are available in a rigid (borescope) or flexible (fibrescope and endoscope) form, with the amount of flexibility in each case being dependent on the design and attachments used. It should also be noted that they are available with or without magnification, and/or lighting attachments, depending on a particular user's need/application, purchase cost or multi-purpose requirement. In the case of the borescope, the image being inspected is transferred from the objective end of the instrument to the eye by means of a precisely aligned series of lenses. These normally include a prism at the tip of the instrument to allow more than simple forward vision (Cunningham, 1991).

The fibrescope on the other hand uses a coherent bundle of optical fibres to view the object, with each fibre being in exactly the same position at each end of the bundle. This results in an image built up from a pattern of dots similar to that on a television screen. The tip of the fibrescope can also be steered from its control body allowing the flexible shaft to be manoeuvred into hard-to-access areas. Modern systems have features which allow them to be focused to both eyesight and object, plus the ability to view and scan a wide area. Typical fibrescopes are available in diameters down to 1 mm, with adapters for television and still camera use. Some of these systems as with borescopes and endoscopes, can be made oil, water, heat, pressure and radiation resistant, for use in special applications or particular fluid environments such as the nuclear and other hazardous process industries.

Borescopes can, for example, be very useful in the inspection of machinery where the only access is along narrow bores, such as fluid circuits, or where small access holes have been included in the equipment design specifically for their use. Accordingly, they are used quite extensively to inspect various types of vehicle engines, and are also used widely throughout industry as a relatively low-cost aid for all manner of plant and equipment visual surveillance. Some of these instruments have

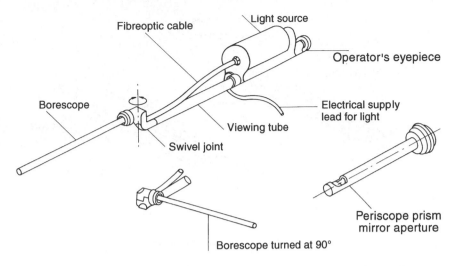

Fig. 6.1 Borescope with integral light source (Harley and Towch, 1993).

their own light source, as shown in Figure 6.1 (Harley and Towch, 1993), while in others, the light travels along a fibre-optic cable from a remote unit, and this is then connected to a short length of cable at the borescope itself. The integral light source types are claimed to be easier to use, with the light passing from the illuminated object along the borescope tube, via a mirror arrangement inside a pivot connector, and then through lenses in the viewing tube to the operator's eyepiece.

A variety of attachments and different borescope tubes can be fitted to a socket at the pivot connector, and thus the operator can rotate the viewing tube about the pivot, without bending the fibreoptic cable. This imparts a degree of flexibility to the borescope system. For example, to look in a direction at right angles to the borescope tube, the operator can use a tube fitted with a periscope type mirror, with the tube still being capable of rotation about the pivot as shown in Figure 6.1. A typical use of this sort of equipment is the inspection of aircraft engine components during pre-flight or routine maintenance. Some systems may also have the capability for use with photographic or television equipment, thereby allowing component measurement or trend monitoring to take place, plus providing a means by which workpart condition records may be kept after each inspection.

Endoscopes use controls wires, which extend along the inside of the bending tube to control the movement or flexibility of the device. Normally, these wires are adjusted using a control mechanism located at the operator's end of the endoscope. In other designs however, the wires may be pushed or pulled using actuators that are located inside the

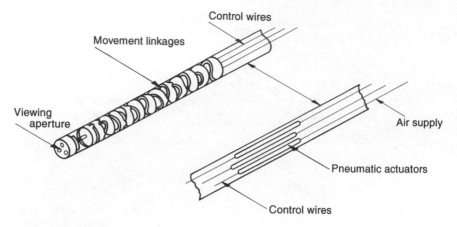

Fig. 6.2 Pneumatic actuation system for an endoscope.

endoscope tube itself. This is shown in Figure 6.2, where three rather than four actuators are used in this particular design, and which has the advantage of allowing the tube to be smaller in diameter. There are various designs of endoscopes in which the actuators are operated by hydraulic or pneumatic pressure, plus others which use wires made from shape memory alloys that expand or contract when they are heated or cooled.

As with borescopes, there are many attachments or specialist items of equipment which can be used with endoscopes. For example in one design, part of the endoscope is electically isolated. This is necessary because if an endoscope is used to inspect the engine in a motor vehicle, it may act as an electrical conductor and permit a short circuit. Should this happen, the endoscope may well suffer damage, and it is to prevent this situation occurring that a short section of the tube is made out of plastic or rubber. Inside the tube, the control wires may also be covered with a flexible insulating material such as nylon, to prevent damage or any conductive pathway being available. Alternative designs completely coat the tube in an insulating material which may be expensive and could possibly be open to damage in use. The in-situ repair of turbine blade scratches has also been carried out by using a special endoscope attachment, whereby a diamond can be unfolded from the instrument and used to file away the marks.

Other equipment which may be used with endoscopes include an attachment via which the instrument can be made to move along a pipe by virtue of pneumatically expandable rings (Uenishi *et al.*, 1993). The air is supplied via miniature hoses incorporated between the endoscope tube and an outer sheath. As shown in Figure 6.3, the ring 'a' is fixed to

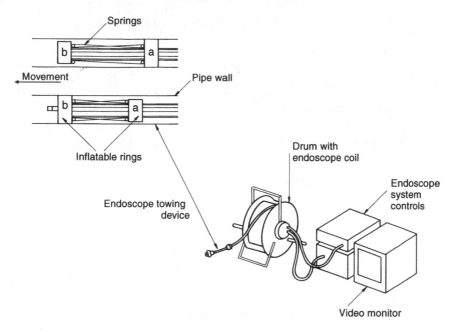

Fig. 6.3 Endoscope towing arrangement with video imaging (Uenishi *et al.*, 1993).

the endoscope tube and is inflated so that it grips the inside wall of the pipe. The deflated ring 'b' is then pushed along the tube by the action of a spring. When fully extended, ring 'b' is then inflated along with the flexible tubes shown, which shorten, and pull the deflated ring 'a' and the endoscope tube along the pipe. A reasonable length of pipework can be inspected in this manner, with the endoscope uncoiling itself from a drum to the desired length, and fitted with suitable controls plus a video monitor.

In serious applications, endoscopic systems now make extensive use of monochromatic or colour video equipment, when conducting the internal inspection of high-value plant and machinery. Not only is the image quality outstanding, but by use of 'state-of-the-art' electronics, low light conditions, freeze frame and memory recording, these features considerably ease the inspection process. Modern equipment tends to utilize a lightweight 12 volt direct current power source, which increases the system's mobility and its use in difficult application areas. Usually the cameras attached to endoscopic systems have a fixed focus. This means that to keep as much as possible within view, the camera has to have a deep field, a small aperture, and a good light source. In some designs, however, it is possible to have a variable focus while retaining the use of a standard light source (Krauter, 1993).

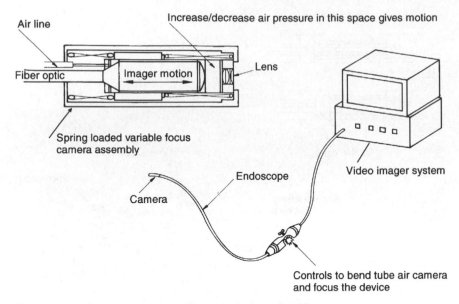

Fig. 6.4 Variable focus camera in endoscope assembly.

One such design is shown in Figure 6.4, where light from the object passes through the lens at the front of the device and falls onto the imager. This slides within the housing shown, along with a flexible tube, which is surrounded by a coiled wire sheath and fits onto the imager with an air-tight seal. Air under pressure is fed along a supply line into the space between the imager and the lens. Thus, by increasing or decreasing the air pressure, the imager can move either towards, or away from the object being viewed. The range of movement of the imager is normally quite small, around 2.5 mm; however, this is adequate for most applications and the equipment is compatible with existing video technology.

6.3   VIDEO IMAGING

As may be seen from the above section, video imaging offers many significant advantages over the conventional 'mark one eyeball' in rigid, or flexible endoscopy. During inspection, it not only reduces eye fatigue and the subsequent risk of operator error, but also allows a simultaneous group viewing of an internal machine examination. Small diameter, around 3–6 mm, cost-effective, and versatile systems are now available in either quartz or glass fibres, with interchangeable tips that provide variable direction and appropriate fields of view. One commercial system has an individual micro lens assigned to each pixel. These focus more light onto the Charged Coupled Device (CCD) chip, giving a considerable increase in sensitivity, quality and brightness of image, equal to that produced with larger diameter models.

Endoscopy, however, is not the only application to make use of video-imaging technology to improve visual inspection capability in condition monitoring. Microscopes have been used extensively over the years for a variety of inspection tasks, most notably within the condition monitoring area for wear debris analysis. However, other visual inspection tasks have also been undertaken using microscopes, and with the addition of colour video imaging they have become a very useful maintenance tool. The latest systems are relatively low-cost, portable, monitoring and measuring instruments, which can provide a sharp, deep field, true colour, and three-dimensional image on an appropriate monitor.

Normally, as the equipment is portable, the object being inspected is illuminated by a suitable integral light source, possibly via a fibre-optic cable, for the hand-held camera probe to examine. The image is then magnified and transmitted back to the monitor. Focus, brightness and colour adjustments to the images can be made, with a variety of lenses being available to give contact or non-contact operation, plus magnification factors from $\times 1$ to $\times 1000$, and a zoom option. The hand-held scanner is particularly useful for non-laboratory work, and providing the equipment is suitably 'hardened', this allows the video-imaging system to be used in 'on-site' conditions.

As many wear debris analysis systems are based on optical principles, in general they lend themselves quite well to enhancement via video imaging, although, providing a sample is available for inspection, the method by which it was obtained does not necessarily have to be integrated with the image analysis equipment. It should be noted, however, that debris particles have a complex response to illumination, in that they may well produce a combination of various effects. The major aspects which are likely to make up this response are reflection, absorption, scatter, refraction, diffraction and transmission. All these effects, aside from absorption, tend to deflect the illumination in perhaps an inconsistent manner, given the nature of the fluid, light, particle and air mix.

Thus, the available commercial wear debris instrumentation based on optical principles is quite varied, although generally falling into four areas. These are as follows.

- Those that provide a particle size count, such as 'light obscuration', where the shadow cast by the particle is detected in front of the light beam, and 'forward reflectance', in which the illumination present is reflected forward by the debris. The detected light intensity in the latter method depends on particle surface area, which in turn is dictated by size. The 'phase/doppler scatter' technique is another method used in this area, where two or more detectors are used to measure this value from moving particles.

- Those that provide a particle size distribution such as 'Fraunhofer diffraction', where a pattern of dark/light concentric rings are detected, or 'photon correlation spectrometry', in which dynamic light scatter due to particle motion is obtained.
- Those that provide a distribution for the particles such as 'light scatter', which is usually detected orthogonal to the illuminating beam, or 'time of transition', where either laser light blockage or back scattering is used, together with the time of scanning across the particle.
- Those that provide a general level of particles in the fluid, such as 'nephelometry', basically a measure of the cloudy nature of the fluid via light passing straight through, 90° scatter, or 180° reflection. Alternatively 'photometric dispersion' can be used, which analyses the fluctuations in the light levels around the average value.

The implementation of the above principles in commercial instrumentation is discussed in greater detail elsewhere, and the interested reader is referred to (Hunt, 1993). However, it should be noted that several of these systems have been integrated with video-imaging equipment such as the 'Galai Time of Transition' particle analyser.

This particular system is fully computerized, with the operator being able to see what is happening and what is being measured as it occurs. The system displays imagery concurrent with the transitional analysis. Particle illumination is effected by use of a synchronized strobing arrangement, and up to 30 images per second may be acquired for analysis. Special software allows for the separation/rejection of overlapping or out-of-focus debris particles, and a fully automatic lighting correction is also incorporated in the equipment. The video imaging system allows particle detail to be observed down to around a micron in size, and a debris shape factor is also included in the anslysis. The system is primarily designed for laboratory use, although an on-line version is available.

6.4 THERMAL AND ULTRASONIC IMAGING

Over recent years, video-imaging systems have become an increasingly popular addition to a wide range of visual inspection equipment. In their basic form, they are of course limited to the enhancement of what we can see as human beings within our optical spectrum. Accordingly, it was to overcome this difficulty, and because many industrial systems exhibit a characteristic rise in temperature prior to failure, that thermal imaging systems have been developed for condition monitoring. In general, two systems are readily available in the form of commercial equipment. These are, as outlined in a previous chapter, thermal imaging cameras and infrared line scanners. It should also be noted, however, that lightweight, low-cost, handheld and portable infrared thermometers are also now available, which are ideal for locating hot spots in machinery and can be computer-linked for data analysis.

The latest versions of the infrared line scanners are personal computer (PC) based, and use sophisticated software to produce two- or three-dimensional graphical images on the system monitor. These can, of course, be precise colour images of temperature maps, historical temperature profiles or any other management data for on-line, real-time, process monitoring and control. The scanning systems can be used for discrete or continuous product monitoring, and tend to find applications in the paper, textile, glass, plastic film, and metal processing industries. A typical unit may incorporate a built-in infrared detector, plus a laser guided sighting system, together with signal processor, power supply and a PC.

Many different spectral versions are available depending on the application and material involved. Detection temperature can vary in range from around 40 to 1700°C Celsius, with the scanner being configured to measure temperature in up to ten alternative locations. These measurement zones can be of any size in modern systems, with overlaps or spaces between them as required. The zones can also be set up to the full width or length of the product. Some units have the ability to detect product edges and to adjust the measurement zone accordingly, plus scan moving continuous products at up to 50 times per second, with a variable data acquisition time.

Such PC-based systems can allow three-dimensional thermal imaging to be displayed, which gives the real-time temperature of the product as it passes the scanner. This makes the identification of temperature change on the product simple, either in spot locations or predefined areas. In a two-dimensional mode, the use of a split-screen feature also permits an easy comparison between real-time temperature measurements and reference images, or, for on-line data matching between two scanners. By virtue of their database facilities and sophisticated software, temperature profiles can be trended and displayed using these systems, which relate to defined areas on the product, either in real time or from archived data. Some systems also have an integrating facility, which allows data from other condition monitoring sources such as vibration analysis etc. to be correlated with the thermal information.

The main disadvantage of infrared scanners is a lack of portability, most PC-based systems being fixed installations for specific applications.

Thermal imaging cameras on the other hand are designed to be portable, use anywhere systems, which is a useful advantage in terms of cost effectiveness, as this allows them to be designated as general-purpose rather than special equipment. Most now include a digital data interfacing capability, which permits information interchange with standard PC systems and some are Windows™ compatible. Once resident in the PC's database, software manipulation of the information can take place for colour imaging or trending etc.

Modern thermal imaging cameras are normally fitted with high-quality colour Liquid Crystal Display (LCD) viewfinders, to provide good image contrast for on-site viewing. They also have a built-in memory capacity for instant capture and recall of images via the viewfinder. In addition, archived data from a PC can, with some systems, be replayed through the imager if necessary, with a post-processing facility that allows for calculation and variable adjustments, along with temperature measurement at any point on the image. Extra displays, such as temperature profiles or isotherms may also be added if required, via this post-processing facility. For most camera systems, as with infrared scanners, the information link to the PC allows for easy word-processing of reports, and the inclusion of statistical analysis, raw data, or images as required.

In terms of application, camera-based thermal imaging systems find both wide and cost-effective use in a number of different manufacturing industries. These include identifying blockages in the water cooling systems of welding equipment for car body manufacture, and checking the paint stoving oven insulation for cracks in the same industry. In addition, they have also been used to detect scale build-up in oil pipelines, together with blockage, restriction and leakage problems in pressure safety valves for the offshore oil and gas industry (Ingebrigtsen, 1993). Many other cost-effective application uses for these systems have been cited in various chapters of this book. Indeed, commercial thermal imaging systems have been so successful in recent years, that they can be looked upon as a complete vindication of the concept of condition monitoring, as applied to industrial plant and equipment.

Ultrasonic imaging, on the other hand is perhaps, not quite so well known. These systems have been used for both weld inspection and corrosion mapping applications, plus the high-speed on-line inspection of steel billets and bar. Modern systems can operate a large number of ultrasonic transducers to provide real-time colour graphic imaging, via the latest computer integrated hardware and software techniques. On- or off-line data processing, presentation, trend analysis and reporting are all available on laptop, desktop or rack mounted systems. The typical on-line data presentations include conventional A, B, C, D and P scans in colour graphic composite format. In addition, high-resolution grey scale imaging of time of flight diffraction data, for rapid/reliable detection and accurate sizing, echo dynamic profiling and geometric overlays, is available, together with orthogonal, radial and isometric displays.

6.5   LASER SYSTEMS

The word laser is really an acronym for Light Amplification by Stimulated Emission of Radiation (LASER). However, by virtue of its widespread use in many applications, from supermarket checkout tills to the precise measurement of the Earth to Moon distance, it has become a word in its own right within the English language. Laser systems tend to fall into two groups depending on the type of arrangement used to produce the laser beam. This is an intense, highly directional and coherent, monochromatic beam of light. The groups are:

- gas lasers – which commonly employ either carbon dioxide, to produce machining process laser systems, or helium-neon for use in measurement/alignment applications.
- solid state lasers – such as ruby, crystalline aluminum oxide, and 8% neodymium in glass, which are also finding many applications in industry and everyday life.

The principles of laser beam generation are not our concern here and have been explained in great detail elsewhere (Mohungoo, 1992). We should note its characteristics, however, and as mentioned above, these are that it is:

- a highly collimated beam, with a very small divergence over a considerable distance; this may be less than 0.001 radians in some applications;
- it is monochromatic, or a single colour beam, usually red in most laser systems;
- it is also a coherent or 'in-phase' beam, where all the light waves tend to reinforce one another to produce very high intensity.

These features combine to produce in the condition monitoring context many extremely useful commercial sensing systems.

In the case of wear debris analysis for example, several particle detection systems are now based on the use of lasers in their optical arrangements. Most of the Fraunhofer diffraction, or forward scatter instruments use a laser light source to produce the light/dark banding at the edge of the cast particle shadow. In this system, as shown in Figure 6.5, a concentric ring pattern is produced, with particle size dictating the amount of scatter. The array of detector diodes recording the particulate distribution size. Another optical instrument in this class which also uses a laser beam is the 90° reflection light scattering system. The arrangement is shown in Figure 6.6, where light scattered

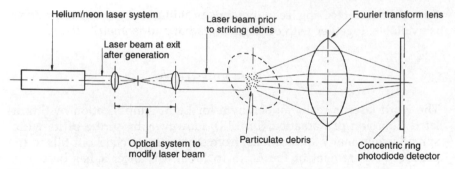

Fig. 6.5 Laser arrangement used in Fraunhofer diffraction (Hunt, 1993).

orthogonally to the source beam has an intensity proportional to the particle size. White or laser light may be used in the latter instrument, depending on the size of the debris to be examined.

The phase doppler scatter technique and the time of transition method also use a laser beam in practical instrumentation. In the first case, light is scattered from moving particles and sensed for phase difference by photodetectors set at different scatter angles, as shown in Figure 6.7. This phase difference, measured under the correct conditions, is directly related to particle size, and thus wear debris can be assessed in the range 0.5 microns to 10 mm with the technique. The fact that the time a particle

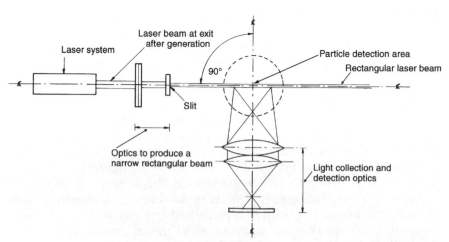

Fig. 6.6 Laser arrangement used in the light scattering method: 90° reflection (Hunt, 1993).

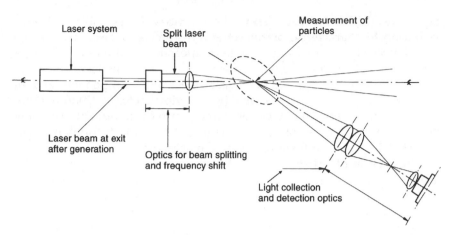

Fig. 6.7 Laser arrangement used in the phase/doppler method (Hunt, 1993).

interacts with a rotating laser beam scanner provides an indication of its size is the principle behind the time of transition technique. This method is independent of the liquid carrying the debris, does not require calibration, and is not affected by vibration, thermal convection, or molecular/particle collision. Most commercial instruments using this method, have a debris sizing/detection range of around a micron to about 4 mm.

In the wear debris analysis area, several other instruments also make use of the properties of laser beams. Accordingly, the interested reader is again referred to (Hunt, 1993), for a definitive outline of the commercial systems available in this field of condition monitoring. It should be noted, however, that laser systems also find use in alignment, displacement, speed and other inspection/measurement applications. A good example of their use in the alignment sense is given by the adoption of lasers to check the position of bearing pockets, seals and diaphragms in the overhaul of turbines, engines and compressors. Micron resolution has been claimed for these systems, over separation distances of up to 40 m, with measurement data being transmitted via an infrared link to a PC. The computer plots an alignment diagram of the datum point positions on the assembly, from a laser reference line as rebuilding proceeds. The combination of computer and laser in this system can accommodate offsets due to shaft sag, machine catenary, oil film or the effects of thermal growth (Stoneham, 1996).

In the displacement context, non-contact laser-based interferometric probes are available, which can be used to generate or detect subnanometre transient or permanent surface displacements. One system used heterodyne detection to give a high sensitivity across a bandwidth of 45 Megahertz. Acoustic waves are generated or detected by employing a pulsed BMI Nd:YAG laser. These probes are portable and have been used to characterise piezoelectric transducers. Vibrometers, of the laser doppler type, are also available for measuring vibration in bridges or buildings at ranges of over 40 metres. These systems have also been used at close range, on small delicate electronic components and in biomedical applications. A portable type of vibrometer, with no requirement for retro-reflective tape or target surface treatment, is being increasingly used as a convenient diagnostic tool in all forms of noise and vibration measurement on rotating machinery.

A final class of instruments in this area is the increasingly popular laser based tachometers which are used for the accurate measurement of rotational speed. They are claimed to have several advantages over other types of tachometer including a range of up to 15 m from the target. This ability reduces the risk of accident or injury when working in a hazardous environment with dangerous rotating machinery. They are also claimed to be effective in bright sunlight, via the use of an intense but safe laser beam, and not to require the use of reflective tape fixed to the target. Easy aiming is another virtue of these systems, which, via a highly visible beam, allows readings to be taken from areas of low contrast, poor reflectivity, keys, keyways or flats. Commercial instruments are rugged, waterproof, compact, portable and battery-powered with a digital readout giving the Revolutions Per Minute (RPM) value. They can also be used in permanent installations with a fixed power supply, or in conjunction with vibration equipment for process synchronization or balancing.

The application of lasers for on-line process inspection is well characterized by their use in steel strip and float glass manufacture. Laser scanner systems using a photo-multiplier tube and light collecting rod combination as a detector have been used for optical inspection in many diverse situations in the steel industry. These units, along with the more conventional CCD camera systems, have been used on line speeds of up to 600 metres/minute (Woodrough, 1995). In the float-glass application, such systems are capable of pin-pointing, measuring and classifying tiny bubbles and tin specks down to diameters of 0.1 mm. In addition, they can also locate distortions of 0.3 mm, in float glass up to 4 m wide, and travelling at speeds of up to 20 metres/minute. By using expert system software, this float-glass monitoring equipment can also differentiate between actual defects and normal process generated dust or glass chips.

6.6   SPECIAL APPLICATIONS

From the foregoing discussion, the future of commercial visual inspection systems can be seen to be linked to advances in computerized optical technology. Accordingly, such systems may be regarded as the combination and application of developments in several fields, which are then subsequently marketed as a single integrated unit for condition monitoring. In essence, they contain, among other items, improvements in optical sensor technology, electronics, computer hardware, image processing, pattern recognition and intelligent knowledge-based software, a trend which in commercial visual inspection systems, can be seen quite clearly by referring to the current special applications of this type of technology. Such applications in this day and age have a disconcerting habit of becoming commonplace quite quickly and are therefore instructive in what to expect in the next generation of visual inspection systems.

The use of fibreoptic sensors and cabling, for the secure acquisition and transmission of digitized data, is one such trend which may be readily identified in the marketplace. Fibreoptics are expected to be worth $935 million in 1996 via the increasing use of fibreoptic gyroscopes, displacement sensors and special visual inspection systems applied to condition monitoring (Anon, 1991). One such case in point, for example, is where on-line web inspection is carried out during papermaking (Nguyen, 1995). Here, crack detection in the paper web is

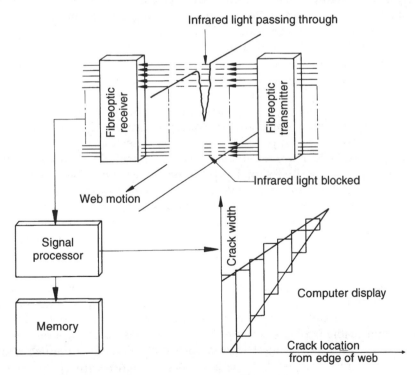

Fig. 6.8 Fibreoptic sensing of tears in papermaking (Nguyen, 1995).

being effected via the use of fibreoptic source and sensor cabling across the full width of the reel. Figure 6.8 shows the arrangement whereby the signal processor extracts and stores the digitized scanning data in a suitable computer system. Infrared light is sensed during the transition of a crack between the sensing unit, and its shape, dimensions and position from the edge of the paper can then be displayed on a suitable PC monitor or printer.

Barcoding is also a popular use of fibreoptic technology, which may not be immediately associated with automatic visual inspection systems, and yet is becoming increasingly vital to it. A barcode is a pattern of dark bars and spaces which can be read by passing a small spot of laser light over it. The light is absorbed by the bars and partially reflected by the spaces, so that a scanning device can convert it into a digitized electronic signal. This signal can then be transmitted via a fibreoptic cable to a computer system which decodes it and identifies the item to which the barcode is fixed. Barcoding is one of the fastest, most versatile and accurate methods of data entry into a computer system. The barcodes themselves can be produced in a wide range of materials which are proof against sunlight, moisture and scratching. In Printed Circuit Board (PCB) automatic inspection and test, they can be made small enough to fit, and therefore identify, miniature electronic components, having a special coating to resist the baking, soldering and cleaning processes involved in manufacture (Anon, 1990).

Another use of fibreoptic technology in a special application is given by its incorporation in a profile-measuring system for oil separation. The system is based on a technique that detects a temperature difference between the liquid or gas in a separator and a point of reference. This value varies according to the sensor's environment, so levels can be identified at various stages in the separation process. The system, which can be used in any application where the detection of strata between various fluids is required, comprises an optical time domain reflectometer and associated signal processing equipment. The arrangement can be used with up to six sensors, each with a maximum length of 10 km, and consisting of loops of standard telecommunications grade, multi-mode optical fibre. A typical installation, with a sensor cable of 4 km in length, can provide up to 16 000 data points, generating temperature versus distance measurements to a resolution better than 1°C. The data sampling frequency for the system can be set as required, via the software which runs on a standard PC in a Windows™ environment.

The monitoring of strain in a steel cable is another special application which can be quoted for the use of fibreoptics (Jouve, 1993). Here a fibreoptic cable is placed among the steel strands during the production of a rope. This carries light from a source at one end of the cable, to a detector at the other, so that when the steel stretches under load, so does the fibreoptic cable, thereby reducing the amount of light transmitted, and indicating the state of strain in the rope. The use of fibreoptics with laser systems can, as we have seen with barcoding, produce some interesting technology, which may have possible applications in the field of condition monitoring. One such item is an acousto-optical deflector which can produce a series of vibration hologram images of a vibrating object (Linet *et al.*, 1993). To produce a hologram image, a laser beam is divided in two. These are the reference and signal beams which are later re-combined to create an interference pattern on a holographic plate and hence the image. In this equipment a pulsed laser is used, the light beam passing through a vibrating acousto-optical deflector to produce several images in a short space of time.

In contrast, the automatic test and checkout of integrated circuits (ICs) on printed-circuit boards is possibly one of the classic applications of visual inspection systems (Zhang and Weston, 1994). With the increasing precision required in this application, and at the same time the dimensional reduction of the features to be inspected, automatic optical checkout systems have become essential for this task. Accordingly, a tremendous amount of computerized vision research has been done in this field over many years (Newman and Jain, 1995), with some of the more useful developments in this area having being reviewed in a previous chapter. These will not be reiterated here, except to mention one of the lastest techniques, which is the use of transistor luminescence as the inspection principle (Prasad, 1995). In this application, the performance of transistors in an integrated circuit are examined by detect-

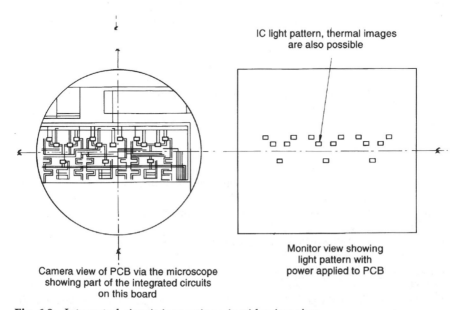

Fig. 6.9 Integrated circuit inspection via video imaging.

ing the light energy produced by the flow of electrons between layers of different materials used in their construction. The system is illustrated in Figure 6.9, and shows the activated transistors revealed by a pattern of lights on the display.

6.7 CONCLUSIONS

This chapter has attempted a review of the most popular and readily available items of commercial instrumentation and equipment suitable for use in the area of visual inspection or condition monitoring. Obviously, not all the available systems have been included, for to do so would take a book in its own right. In addition, simple tooling and instrumentation such as stroboscopes have been left out, as their use is well-known and requires no explanation. Hopefully, the discussion above has outlined those systems which are part of, or have adapted to, the current technological conditions which are to be found in modern maintenance practice. As industrial systems become increasingly more automated and sophisticated, however, so too must the equipment used to inspect them evolve and become highly efficient tools for the detection of defects or prediction of impending failure. Accordingly, this chapter has attempted to show not only the current state of technology in respect of this type of equipment, but also to indicate the direction of its improvement via the use of fibreoptic sensing and transmission technology.

6.8 REFERENCES

Anon (1990) Barcode monitoring can boost production and profitability, *Condition Monitor*, **42**, 6–10.
Anon (1991) Six-fold increase forecast for fibreoptic sensors, *Condition Monitor*, **50**, 4–5.
Cunningham, M. (1991) Remote visual inspection in the automotive industry, *Condition Monitor*, **59**, 6–7.
Harley, J.M. and Towch, A.W. (1993) Self-contained borescope has its own light source, *Condition Monitor*, **79**, 14–15.
Hunt, T.M. (1993) *Handbook of Wear Debris Analysis and Particle Detection in Liquids*, Elsevier, London, UK.
Ingebrigtsen, K. (1993) Infrared thermography targets offshore scaling problems, *Condition Monitor*, **81**, 6–8.
Jouve, P. (1993). Monitoring a steel cable using fibreoptics, *Condition Monitor*, **78**, 14.
Krauter, A.I. (1993) Variable-focus camera for endoscope, *Condition Monitor*, **80**, 14–15.

Linet, V. *et al.* (1993) Acousto-optical deflector produces vibration holograms, *Condition Monitor*, **78**, 14–15.

Mohungoo, P.S. (1992) *A Feasibility Study of Laser Measurement Systems for Determining the Spring-back on a Thread Rolling Machine*, ELYSM final year project report, UWCC, Cardiff, UK.

Newman, T.S. and Jain, A.K. (1995) A survey of automated visual inspection, *Computer Vision and Image Understanding*, **61**(2), 231–62.

Nguyen, D.D. (1995) Web inspection for papermaking machine, *Condition Monitor*, **105**, 14–15.

Prasad, J.S. (1995) Fault diagnosis of integrated circuit, *Condition Monitor*, **102**, 14–15.

Stoneham, D. (1996) Misalignment straightened out, *Condition Monitor*, **109**, 5–7.

Uenishi, N. *et al.* (1993) Device used for towing endoscope along pipe, *Condition Monitor*, **74**, 13–14.

Woodrough, R.E. (1995) Automatic inspection and the holy grail of production, *Steel Times International*, **19**(3), 34–5.

Zhang, J.B. and Weston, R.H. (1994) Reference architecture for open and integrated automatic optical inspection systems, *International Journal of Production Research*, **32**(7), 1521–43.

PART THREE

Techniques for Performance Monitoring

System quantity/quality assessment – the quasi-steady state monitoring of inputs and outputs

R.A. Heron

30 Church Lane, Stagsden, Bedfordshire, MK43 8SH, UK

7.1 INTRODUCTION

Most of the condition monitoring methods described in this book are targeted at the identification of developing faults, which could possibly cause a machine suddenly to stop functioning. The approach depends on seeking out and observing the symptoms of discrete faults, which may be expected to develop slowly before accelerating to failure; for example: bearings starting to break up, seal leakage, or fatigue cracks which, if undetected, could result in catastrophic failure with unscheduled downtime and consequential damage. Detection of this type of failure is the high-profile aspect of condition monitoring, where the costs and urgency associated with sudden machine failure and unscheduled shutdown are very obvious.

However, a machine is there to perform a specific function. How well it performs that function in the long term is vital to the economics of the operation. Although a machine continues to operate apparently reliably, there may well be a gradual degradation in performance which might not be detected by seeking localized symptoms, such as bearing vibration or hot spots. Steady-state monitoring of the machine's overall performance, particularly in terms of the key parameters which affect the economics of use, is both a valuable indication of the machine's condition and a check on the economics of operation. The method is not as powerful in identifying specific faults in a machine as, say, contamina-

Handbook of Condition Monitoring
Edited by A. Davies
Published in 1998 by Chapman & Hall, London. ISBN 0 412 61320 4.

tion analysis or vibration analysis, but it is capable of flagging up that an economically significant problem exists, or is developing, and that further detailed investigation is warranted.

Energy conversion efficiency in prime movers, pumps, compressors and drive trains which operate over extended time-scales is critical to the economics of ownership and is therefore an obvious parameter to monitor. Other examples where steady-state performance monitoring can be applied with benefit are observing variations in the quality of product, such as plastic mouldings or machined components. In the latter case, manufacturing technology has developed some very sophisticated condition monitoring methods, which depend on product sampling and trend detection to form the basis of product quality control.

This information can be related to machine condition, such as tool wear or drifting geometry, and is fed back to production engineers for corrective action. As such, this is an ideal form of steady-state monitoring, for the operation is repetitive and hopefully stable. This form of steady-state monitoring is widely practised and understood by the production engineering profession and, as a wealth of excellent reference material is available covering the subject, it is not examined in detail in this book. Accordingly, this chapter concentrates on the use of instrumentation to measure the rather less tangible machine products such as energy conversion.

All the usual problems of condition monitoring still apply and it should be noted that the underlying requirements of any form of condition monitoring are:

- to indicate the presence of a fault before it results in a substantial cost problem;
- the avoidance of false alarms.

The costs of stripping and servicing when it is unnecessary are compounded by the risk of introducing real faults in the repair procedure. 'If it ain't broke don't fix it' may be poor English, but it is very sound engineering. We should also note that a cardinal rule of condition monitoring is:

- that the instrumentation has to be more reliable than the equipment to be monitored.

This applies to instrument calibration stability as well as functionality. The validity of the calibration has to be maintained over extended time periods, or some economic form of calibration checking procedure against a reference standard should be an integral element of the process.

The steady-state monitoring approach is typically looking for a small percentage change in the level of the monitored value. This requires a high standard of measurement integrity which does not always readily translate to portable instrumentation. Permanent transducer installations need to be considered, particularly for 'difficult' parameters such as flow and drive torque. This tends to limit steady-state monitoring to high value plant, or where the equipment form lends itself to portable transducer installation for parameters such as speed and pressure. Dynamic monitoring of vibration, or looking for hot spots, does not need such a precise measurement, particularly where the dynamic form of the signal indicates the machine condition. Dynamic monitoring is therefore typically economic to apply to lower value plant.

7.2   ENERGY CONVERSION, FLOW MONITORING AND EFFICIENCY

Insidious forms of failure can be detected by close observation of the overall machine performance, by measuring the inputs, outputs and losses as summarized in Figure 7.1. As with other forms of monitoring, a trend may be apparent, or it may be possible simply to compare the measured values with a predetermined norm, such as the original manufacturer's test or commissioning data. A fundamental problem with the steady-state monitoring approach is that the machine's performance needs to be accurately measured under known load conditions, to allow direct comparisons and trend analysis. There are often problems in installing adequate instrumentation and in setting or holding an appropriate load for equipment in industrial installations. Accordingly, manufacturer's test beds are specifically designed to provide a known and controlled load condition, whereas a real installation is often subject to very variable loads, which may not be controllable within the operating envelope and where several parameters can vary independently.

If a test cannot be taken at a set loading condition on demand, some form of allowance should be made for the real variation in normal

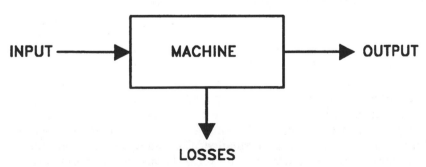

Fig. 7.1   Steady-state monitoring factors.

operation to allow long-term trends to be observed. It may be possible to devise a Taguchi-based test programme which will allow some load variables to vary, although this approach may be more expensive to set up initially. Alternatively it may be appropriate to monitor the machine over an extended time period and seek a particular load condition which is regularly applied and which is held sufficiently long for quasi-steady state readings to be acquired.

Monitoring a machine's energy balance appears at first sight to be relatively straightforward, mundane and often dismissed as a valuable condition indicator. However, a few global sums can highlight the real costs of allowing an 'off-colour' machine in service for extended periods. A liquid mass transfer pump will serve as an example. A relatively modest 30 kiloWatt (kW) machine running on continuous duty will consume about 263 000 units (kW hours) of electrical power per annum. Thus a 5% drop-off in efficiency will push the annual running charges up by more than the cost of replacing the pump's critical components.

Falls in efficiency are seldom caused by failures such as bearing wear. The resulting energy flows in the form of friction would finish off the bearing in a few minutes. These problems are more commonly associated with gross internal wear of major components. In the case of a centrifugal pump, impeller or seal ring wear affect the hydromechanics of the pumping process by allowing internal leakage. The energy lost to this form of inefficiency is carried away by the working fluid, and no hot spots or unusual vibration need occur. Therefore, to detect and evaluate the efficiency of a machine requires steady state measurement of the energy flows – the inputs, outputs and losses.

The general level of machine energy conversion efficiency affects whether it is better to compare input and output energy flows, or if as in the case of an efficient device, a change in the energy flow to losses may be more easily detected. For example, an oil hydraulic piston pump is typically an energy efficient conversion device, with an overall efficiency in the region of 94%. A drop in efficiency of a few percent due to internal leakage can be detected by comparing input energy, torque and speed with output energy, flow and pressure. However, we are making four separate measurements, and comparing two large and relatively close values, to detect a change in their difference. The difficulties of obtaining an accurate set of data for these parameters is discussed in more detail in the following sections.

It should also be noted that it is all too easy to end up with measurements showing that the pump is operating as a perpetual motion machine, rather than giving a realistic and reliable measure of its real performance. However, the change in energy flow to losses may double for a 5% drop in performance. The losses are therefore likely to be a more sensitive indicator if they can be accessed. In the case of a pump developing a high pressure and a relatively low flow, the change in the oil temperature increase across the pump, between inlet and outlet ports can be several degrees Celsius. Account has to be taken also of the temperature rise due to the adiabatic compression of the fluid, but the real value of this type of measurement is that, with a constant pressure system, the temperature rise is to some extent independent of flow.

This technique has been used with some success, in relatively low-pressure, high-flow pumps as operated by the water industry. The changes in temperature due to degradation of pump efficiency are only a few hundredths of a degree, but with suitable instrumentation this can be detected. The method only requires two temperature sensors placed in the inlet and outlet flows, and an approximate measure of the pump flowrate and pressure. Conversely, with low efficiency devices such as an internal combustion engine, the heat flows to losses are large, and therefore it is probably better to measure the input and output energies directly. The input energy in this case can be reduced to monitoring the fuel flow for a known fuel calorific value, and the shaft output power. Changes in the heat flow to exhaust, coolant and oil, are relatively small for substantial reductions in efficiency.

7.3   STEADY-STATE SIGNALS AND INSTRUMENT CALIBRATION

In precise terminology, the term 'quasi-state' monitoring applies to machines which are operated over the period of measurement with a fixed load condition. In normal operating conditions the load may vary, hence the term 'steady-state monitoring' is a loose use of terminology, but is used here to signify that:

> The machine should give the averaged set of signal levels over repeated tests under the same operating conditions.

All real signals will have a time-variant component superimposed on the true steady-state signal. This dynamic component may be a function of the machine, such as the pumping action, or it may be electrical noise from the measuring instrumentation. In order to arrive at the best estimate of the mean steady-state signal, the dynamic component must be averaged over a time period. With many forms of instrument, this averaging process is often hidden and unquantified, but as we need to compare values precisely over time, or between similar machines, we need to be aware of the potential pitfalls and effects of the averaging process.

Most forms of transducer are calibrated using reference standards such as dead-weight pressure testers, steady-state flow rings, mass and lever arms systems, constant frequency references and stabilized direct-current (DC) voltage and current sources. These calibration standards are truly steady-state devices. Under these artificial load conditions, the only noise on the signal should be that generated by the instrument's electronics, and it should be of a sufficiently low level not to affect the reading. Such noise is typically random in nature, and can easily be averaged out using a passive electrical low-pass filter. The form and nature of calibration inputs are therefore somewhat idealized compared to real signals, and provide little clue as to the instrument's dynamic response and true averaging characteristics.

Therefore it cannot be assumed that the dynamic performance of an instrument is the same as its steady-state response. While this statement is obvious, where the dominant proportion of the signal is dynamic, as with many of the other forms of condition monitoring approaches described elsewhere in this book, the dynamic component of a steady-state measurement can also introduce significant errors. A reasonably steady reading from an instrument with a known dynamic component on the physical input signal does not necessarily mean that the true average value is being displayed. The dynamic averaging characterictics of many instruments are unknown and are all too often simply assumed to be correct by default. The dynamic characteristics of the instrument and its detail installation should thus be assessed and either proven to be accurate or alternatively the dynamic component isolated and averaged in a controlled manner before the signal reaches the instrument. Three basic approaches to tackling this problem are employed. These are as follows.

1. To provide some form of mechanical or hydraulic filter in the raw input signal route, which will remove the dynamic component in a known manner to give a true mean steady-state value.
2. To incorporate an electronic 'low-pass' filter within the transducer, or its signal conditioning, before the signal is applied to the output indicator. This approach requires confidence in that the transducer and its interface to the filter is transferring the dynamic component unmodified, or at least symmetrically about the true mean value. The design of the electronic filter is straightforward, and can be readily shown to produce a true mean of the input signal, as long as the system is operating within its dynamic range.
3. To capture the signal digitally with an Analogue to Digital (A/D) converter, and calculate an average over a large number of samples. The sample rate and number of samples needs to be set to cope with the frequency range of the signal's dynamic component. This approach is potentially the most accurate method, as the averaging technique is well defined, and can easily be optimized for the particular application. Again confidence is required that the instrumentation chain is not introducing a skew into the dynamic signal component before it reaches the A/D converter, and that it is operat-

ing within its dynamic range.

Calibration methods are outlined in the following generic instrument sections, but general calibration principles apply to all instruments. Never fully trust an instrument, however expensive or sophisticated the presentation. If you do have an anomalous reading, do not disregard it, also do not take expensive actions based on uncorroborated evidence from a single instrument. Take stock of how the reading may have occurred. The laws of physics dictate that instruments have to go wrong at some time, and the laws of engineering dictate that they will go wrong at just the time when you need them most.

Equally do not dismiss unexpected readings as some undefined instrument or experimental error. It is very easy to end up with an anomaly which may well be due to human error, such as a logging error or a burst of electromagnetic interference from an unknown source, but just occasionally, the instrument may be indicating an intermittent real fault in the monitored equipment. Keep anomalies highlighted and on record, even if you suspect it was down to 'finger trouble'. A sizeable set of data completely free of anomalies has probably been 'educated', even with the best of intentions. Also be aware that the most stable set of data comes from a 'frozen' instrument.

A key method for combating instrument error is frequent calibration. Accordingly, calibration should be carried out regularly, and at a frequency related to the instrument type and stability characteristics, using properly maintained and traceable standards. Full records should be maintained of each instrument's test results, the test procedures used, the traceable test standards used and any adjustments carried out. Various calibration procedures for most types of instrument are set out in British Standards (BS) and BS 9000 sets out procedures for implementing an instrument calibration programme. Traceable standards can be expensive and it may be more cost-effective, to use a specialist instrument calibration service rather than maintain all but the most commonly used standards in-house.

However, a current calibration certificate is essentially a historical document, and accordingly should only be considered as a baseline control. It does not guarantee the gauge is accurate at the time of measurement. Before taking any potentially expensive decisions based on a transducer or instrument reading, it is wise to obtain a calibration check both before and after the critical measurement is made. This should avoid breaking down a machine in good condition on the basis of a sudden performance drop but conversely will support a decision to react if circumstances demand. A futher important calibration concept is 'through system calibration'. Individual transducers and readout devices are provided with separate calibrations, by the manufacturers or the in-house calibration service. However, better accuracy is assured if the whole instrument chain is calibrated, from the transducer input to the final readout and display. This not only avoids accumulating errors from multiplying each component's calibration together, but also takes into account any possible interaction between the components, such as poor impedance matching, excessive noise pick-up due to the signal transmission arrangements, differential earth potentials or some installation errors.

7.4 INSTRUMENT INSTALLATION – HYDRAULIC POWER

The interaction of the instrument and its installation can generate false readings in a variety of ways from simple orientation, such as Bourden tube pressures gauges, to electrical interference and mechanical vibration. Permanent instrument installations should always allow simple removal for calibration, or facilities incorporated to allow calibration in place and to allow re-installation without risking disturbance of any factors which could affect the readings. An example of the latter is re-establishing alignment of drive shafts linked by a torque transducer.

Hydraulic power for a liquid medium is a function of the product of volumetric flow rate and total pressure difference. Thus, Bernoulli's equation for the total head or pressure difference should ideally be used, as it takes account of the static head, dynamic head and differences in the height values at the points of measurement. The volumetric flow rate should include an allowance for the compressibility of the fluid. For high-pressure systems, this can amount to several percent of the original volume, due to the compressibility effect of the bulk modulus of the fluid.

Compressible flow energy requires a knowledge of the thermodynamics of the process, and is a much involved calculation. Although it is possible to calculate compressible flow energy transfers, it requires not only volumetric flows and pressures, but also absolute temperatures and a knowledge of the detailed fluid properties, including the moisture content for air. Any steady-state monitoring system requiring compressible flow measurement should be regarded as a non-trivial exercise and will need careful design. Accordingly, this is a measurement which falls outside the scope of this chapter. Specialist microprocessor based instruments are available for carrying out compressible flow measurements, combining several transducer outputs to compute mass flows. An excellent introduction to compressible flows is given in Miller (1990).

The typical good quality mechanical pressure gauge is capable of accuracies of the order of $\pm 1\%$ of the full-scale deflection value, while standard test gauges can give readings within $\pm 0.25\%$ of the full-scale value. Pressure gauges below 100 mm diameter are for general indication only and should not be relied on for taking accurate data. These general accuracies apply to gauges which are in the 'as calibrated' condition. Bourden tube gauges should always be calibrated and used with the gauge mounted vertically, as the weight of the mechanism will influence the reading. A temporary hose connection, with the gauge read while left flat on its back, is an all-too-common practice which can give several percent of error.

It should be noted that gauge calibrations are relatively easily compromised by pressure transients, or prolonged exposure to vibration. Corrosion of the mechanism can also give frictional hysteresis or even seizure, so beware of the absolutely steady reading. The pressure gauge can be one of the most reliable items of instrumentation if used within its operating limits, but any brief over-pressure excursion will upset the calibration and may still leave the gauge appearing undamaged. However, it is still worth checking that the needle is on the right side of the zero limit pin on the dial face when it is supposed to be reading zero pressure. It has been known to end up on the wrong side after a violent excursion! Many gauge indicating needles are held on a taper on the central shaft and, accordingly, they can come loose and will read anything you want.

Pressure limiting valves are available, and should be used for systems which are subject to overpressure transients. These close off and isolate the gauge if the pressure rises above a pre-selected level. They are remarkably effective but they can be beaten by rapid pressure transients. If you have reason to suspect a gauge has any over-pressure transients – check it. A further cautionary tale concerns gauges for monitoring slurry materials. Over a period of time the material can settle out inside the Bourden tube. As the extension of the tube involves a 'panting' action across its section, the material gradually compacts and prevents the tube recovering its shape, generating a false reading. If this situation is encountered, use a pre-filled gauge fitted with a small bladder which precludes the entry of debris. Also, avoid a pressure tapping form which could entrap particulates.

The blur of a vibrating needle on a simple pressure gauge is an all-too-common sight. The best any individual can do when faced with a vibrating needle is to take the mean of the two extremes of the needle vibration. This 'calibrated eyeball' approach is fraught with inaccuracy, as the wave form may be asymmetric and the dynamic response of the gauge complex, thus not necessarily providing an accurate average of the applied pressure. Two main approaches are used to smooth out any vibrations and hence obtain a mean pressure value from a mechanical gauge, these are as follows.

1. **Gauge restrictors**   The most common method is to place a restrictor or pulsation damper in the pressure line to the gauge. The proprietary versions are capable of giving a reasonably true mean pressure

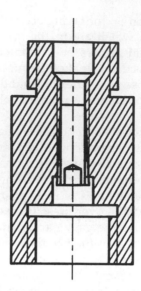

Fig. 7.2   Pressure gauge snubber.

reading. Such restrictors employ the long flow passage formed by a thread with relieved tips, which can be adjusted by being screwed deeper into engagement, as shown in Figure 7.2. These devices present an equal flow restriction in both directions and therefore give a smoothed mean pressure downstream.

Beware of the use of a needle or poppet valve employed as a damping restrictor, adjusted so that it is just cracked open. The construction of these valves can give rise to two separate effects which can bias the reading. The coefficients of the flow control elements within valves can be different in each flow direction. The result is a biased pressure reading. Such valves can also give a non-return action at very low openings, resulting in the gauge registering the peak pressures superimposed on the true mean level. This can easily give an error of several percent of the reading. I have seen errors of over 10% due to this effect on pulsile pressure signal readings. Some gauges are fitted with an isolating valve to facilitate removal for replacement or calibration with de-pressurizing the system. It is very tempting to use such isolation valves as a handy restrictor. They can provide a steady, but wrong reading.

It is usual practice to mount the damper directly on the gauge fitting. This gives a short time constant for the damping action, but does not cope with lower frequency variations, such as the output from a multi-cylinder reciprocating water pump. A length of hose placed between the restrictor and the gauge can substantially increase

the response time constant, as well as help to isolate the gauge mechanically from pipework vibration. Finally, beware of blockage of gauge dampers by particulates in the fluid.
2. **Viscous damping** Pressure gauges are available filled with a viscous clear liquid. This gives an accurate averaged reading of a signal with a fairly high-frequency dynamic component. It also avoids problems of restrictor blockage with dirty fluids. However, this approach does not protect the mechanical gauge mechanism from the vibrating loads which may result in fatigue failure and rack wear problems. This can affect the long-term calibration stability. The approach is more commonly employed with small indicator gauges rather than precision gauges. If gauge damping is essential, consider combining proprietary damping restrictors with a liquid-filled gauge.

Steady-state pressure transducers capable of accuracies of typically ±0.25% of full-scale reading are commonplace, with instruments capable of ±0.1% readily available. Beware the trap of using piezo-electric transducers for attempting to measure steady-state pressures. Piezoelectric transducers are suited for measuring high-frequency dynamic pressure signals, but are inherently unsuited to measuring the steady-state component. The similarly named but very differently constructed piezo-resistive transducers employing semi-conductor strain gauges are, however, ideal for steady-state pressure measurements. Apart from strain gauge diaphragm transducers, a wide range of other forms are available. They may have particular advantages, such as a differential pressure measurement capability, very low pressure ranges and exceptional calibration stability.

A particularly useful construction of diaphragm-based transducers uses non-contacting displacement measurement of the diaphragm, combined with full diaphragm support at maximum displacement. This allows the transducer to survive severe over-pressures without damage. However, the most common forms of pressure transducer are now all strain gauge types. When selecting gauges for low pressure measurement, determine whether they are referenced to an internal vacuum, or the reference is vented to atmosphere. If the transducer package includes a pre-amplifier, which enables the unit to transmit the signal over extended distances with reduced risk of degradation, it may be marketed as a 'pressure transmitter'. This is the preferred form for long-term industrial installation.

Unless the dynamic pressure disturbance is particularly severe, it is preferable to mount the pressure transducer directly to the pressure source, and to average the signal electrically. This can be achieved with a simple low pass passive filter in the signal conditioning chain. Pressure transducers are prone to high-frequency resonance effects if they are poorly installed. A commonly encountered configuration which can lead to mis-readings, is where a restricted tapping leads to a larger fluid filled chamber formed by the adapter fittings. This forms a Helmholtz resonator, which will have a strong response to a single frequency, in the order of typically a few kiloHertz (kHz) for liquid systems.

Broadband pressure disturbances in the pressure signal will excite this resonance, which may drive the transducer or its signal conditioning into asymmetrical saturation, and hence give a false mean reading. It also gives a false indication of an apparent high level of pressure ripple in the system if the dynamic component is monitored. The simplest approach is to place the transducer directly into the pipe, preferably flush with the pipewall. Check that the dynamic component of the pressure signal is within the transducer dynamic range when superimposed on the steady-state component. If the dynamic component is of such a magnitude that it is liable to saturate the transducer or cause physical damage such as fatigue failure of the diaphragm, it can be isolated using similar devices to those described for mechanical gauges. Be aware that cavitation can very rapidly damage many forms of pressure transducer. The high-speed transients can produce levels of negative pressure which will strip the diaphragm out of the transducer. If this statement seems biazarre (Trevena, 1987) explains the mechanism involved.

7.5 INSTRUMENT INSTALLATION – ELECTRICAL POWER

Electrical power input is a vital measure of the input and hence running cost of most industrial plant. The traditional methods of single-phase and three-phase power measurement are still perfectly valid techniques for many types of electrical load. Power inputs to resistive loads such as heaters and electrode boilers, as well as induction motors without speed control, can accurately be determined using electromechanical wattmeters. However, power semi-conductors have opened up many different methods of controlling electrical machines, and their application is now widespread. Speed control is also commonplace using multiphase variable frequency invertors, allowing far better matching of speed drive requirements, with huge energy savings and process improvements. The economics of employing variable speed drives, particularly with induction motors, has transformed power transmission technology, and the range of cost effective applications in the past decade looks set to continue apace.

A complication of this advance is the difficulty of accurate power measurement. Semi-conductor-based power control significantly distorts the current and voltage waveforms, and requires sophisticated electronic processing instrumentation to determine the real power consumption accurately. However, as manufacturers of control equipment are seeking technical differentiation in this highly competitive market, such power measurement capability is being built into the controller as a standard feature. The measurement of dangerous voltages and high current levels requires minimum standards of equipment safety and insulation. This involves the use of isolating transformers, current search coils and shunts, all of which need to be included in the instrument calibration procedures to ensure long-term validity of the data.

The power flowing in a single-phase AC supply is the instantaneous product of the voltage and current integrated over time. This takes account of the phase differences and any variation of the form profiles between the voltage and current waveforms. With a simple resistive or inductive load the waveforms will remain sinusoidal in form, and only the phase difference needs to be taken into account. We therefore have the familiar equation:

$$\text{power} = \text{voltage}_{\text{RMS}} \times \text{current}_{\text{RMS}} \times \cos\phi \qquad (7.1)$$

Where: power is measured in watts, the voltage in volts (RMS), the current is in amperes (RMS) and $\cos\phi$ is the power factor.

A dual coil wattmeter, set up as shown in Figure 7.3, monitoring both applied voltage and current, will take account of the phase, and for undistorted waveform can provide an accurate reading. Electromechanical wattmeters are calibrated for a specific frequency and waveform, typically 50–60 Hertz and a sinusoidal wave. Therefore such instruments will give an inaccurate reading of the power supply taken by semiconductor controlled equipment. Many electronic-based instruments are available for power measurement featuring true Root Mean Square (RMS) measurements of distorted waveforms. They are also available with 'clamp-on' type probes which facilitate simple application, but the user must carefully research the real accuracy limitations of any such device.

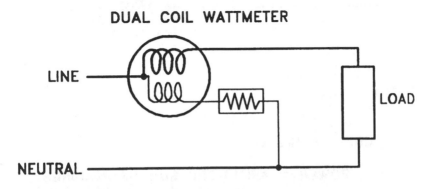

Fig. 7.3 Single-phase wattmeter.

Current and voltage transformers should be used with caution for all accurate power measurements, as they may introduce their own phase variations. As a safety note, current transformers must always be left in a safe condition, with the output leads connected together, when they are not effectively short-circuited by the current meter coil windings, or the current shunt of an electromechanical wattmeter. Open-circuit current transformers can overheat and cause a fire, and they may generate high voltages on the secondary coils.

For a balanced or unbalanced load connected in a three phase star or delta configuration, the two wattmeter method is required as shown in Figure 7.4. The arithmetic sum of the two wattmeter readings gives the average total power being transmitted through the three-phase supply. The dual wattmeter method takes account of phase imbalance, but three-phase power measurement is subject to the same limitations as single-phase measurements regarding the supply form factor, which is a measure of the waveform distortion from a sine wave.

For a balanced load, a single wattmeter is commonly used in one phase only. Three-phase power in a balanced star connected load is given by three times the wattmeter reading for a wattmeter placed in a single leg. For a delta-connected circuit the power is 1.73 times the wattmeter reading. The use of a single wattmeter may be acceptable for

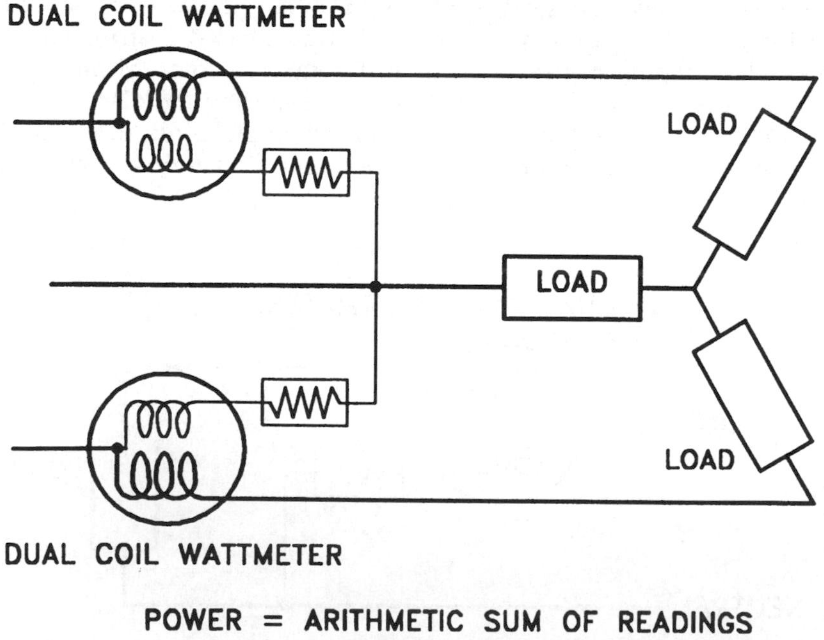

DUAL COIL WATTMETER

DUAL COIL WATTMETER

POWER = ARITHMETIC SUM OF READINGS

Fig. 7.4 Three-phase wattmeter: unbalanced load.

general power consumption evaluation, but is unlikely to be sufficiently accurate for condition monitoring measurements. Multiple 'clamp-on' probe instruments are available for three phase power measurements which will probably provide the best compromise. These employ 'Hall effect' transistors as the magnetic field sensing element. They do not have the same safety over-heating problems as current search coils, but you should not become blasé when placing them over live bus-bars!

7.6   INSTRUMENT INSTALLATION – TORQUE AND SPEED

Mechanical shaft power is proportional to the product of torque and speed. Torque measurement has therefore always been a critical parameter in mechanical efficiency measurements, and many ingenious methods have been devised for its determination. It is the most problematical of the common mechanical measurements, due to the inconvenience of the forces being transmitted by rotating shafts. A few years ago a 'round-robin' series of tests were carried out by British test-houses checking the performance of a series of hydraulic pumps. Significant, if not embarrassing, differences were found in the measured pump efficiencies, taken using the same pump but between different test houses. It was eventually established that the differences were almost entirely due to errors in torque measurement. In the main, two approaches can be used for torque measurement:

1. The first detects the reaction force of the drive force on the driving or driven device. This involves mounting the unit on gimbal bearings coincident with the drive line axis, and measuring the rotational reaction force. This method is most suited to a purpose-designed laboratory or test-bed measurements, rather than industrial installations. The method is also subject to several hidden sources of error, such as windage reaction forces from air-cooling blasts, magnetic torque coupling with stationary ironwork, supporting gimbal bearing friction, and changes in initial balance due to wind up of the support structure. It is an ideal method of torque measurement for dynamometers such as the 'Froude water brake', but the detailed implementation is critical if errors are to be avoided.
2. The second approach is to determine the angular torque via the angular strain in the spinning shaft. The advent of electronics has simplified the measurements required, rendering various elegant but complex mechanical and optical methods redundant. However, we are stuck with the problem of detecting and transmitting that data from the spinning shaft to a stationary instrument.

Thus, shaft torsion measurement is accomplished either by conventional strain gauges or by semi-conductor strain gauges. They can be applied directly to an existing drive shaft if the level of torsional shear stress and strain is adequate, that is if the shaft is not dramatically over-sized. Alternatively, they can be bonded to a dedicated torque transducer, which is mounted in the drive line. The latter device is much simpler to calibrate and can be bought as a fully operating package. Unfortunately, the torque transducer can very often be too bulky to retrofit to an existing machine. Space limitations may require that a prime mover or gearbox be physically moved back to allow the insertion of a torque transducer and its drive couplings. In this situation the fitting of strain gauges directly to the existing shaft may be the only realistic alternative.

Whether a purpose-built transducer or a shaft-fitted system, the strain gauges are usually arranged in a fully balanced bridge, detecting the torsional shear strains at an angle of 45° to the shaft axis. They are positioned to balance out any bending moments which may be present due to shaft misalignments, or unsupported shaft and coupling weights. The full bridge also provides good temperature compensation, and a relatively high level of output for a given torsional strain. The outer layers of a shaft dominate the torsion load transmission, giving the advantage that section changes in the shaft rapidly settle to the classical stress distribution. Therefore only a short length of shaft is required to give a valid measurement surface on which to apply strain gauges. Unfortunately, calculated levels of shear strain, and applying the nominal strain gauge sensitivity, are inadequate for determing the level of torque with sufficient accuracy for meaningful efficiency measurements. Calibration of each application is almost always required. This can readily be accomplished with a demountable torque transducer, although modern quality standards will require traceable reference standards of force and length.

The calibration of a set of gauges directly applied to a machine shaft is less straightforward. The drive line will almost certainly need to be broken to ensure freedom from residual torsional stresses, and to give access to apply a calibrating torque. A carefully applied lever arm, with low friction joint at the end carrying the weight hanger, is a valid calibration method, but the insensitivity of the gauges to bending moments applied to the shaft will have to be experimentally proven. The gauges should be finally calibrated with the adhesive fully cured and any protective layers applied.

A strain gauge amplifier, together with a frequency modulated signal transmission unit, is required to be mounted on the shaft to enable the signal to be transmitted to a stationary instrument. Power for the shaft-mounted electronics can be from a small battery pack or transmitted via an annular rotary transformer arrangement with a stationary primary and a rotating secondary coil. The use of battery packs is limited not only by their life but also by the tendency for the electrolyte to be centrifuged out of contact with the internal electrodes. This limits most battery applications to shafts running at speeds below 1500 revolutions per minute (RPM). Therefore battery powered systems are only applicable for slow-running temporary installations.

An alternative approach is to use silver slip rings to transmit the signals directly from the strain gauge bridge. The electrical noise inherent in the use of slip rings limits the rangeability of the instrument. A preferable alternative arrangement is to utilize a shaft-mounted preamplifier, with the slip rings providing the power supply and transmitting a high-level signal to the stationary instrumentation. This type of device forms the basis of many proprietary torque transducers which also employ tubular torsion tubes, sized to give optimum shear strain levels within the rated measurement range. More sophisticated transducers are also available, incorporating rotary transformers to transmit power and the frequency-encoded data signal. It is convenient to incorporate speed-monitoring transducers within a torque transducer to enable power transmission to be calculated directly.

The dynamic response of frequency-modulated signal transmission, utilized by torque transducers, limits their frequency response to a few hundred Hertz. The averaging characteristics of the frequency-modulated data transmission should also be reasonable, as long as the torque variations are within the dynamic instrument range settings. Torsional oscillations are commonly encountered in drive lines, and users should be aware that excessive levels can saturate the instruments, resulting in inaccurate readings. If excessive dynamic variations are suspected the user should intervene as far down the instrument data transmission chain as possible, observing the variations in the frequency-encoded signal using a digital storage scope or similar, to check that the signal remains within acceptable limits. A small passive search coil can often be effective in picking up the frequency-modulated signal without breaking into the instrument circuit.

Accurate speed measurement has become much simpler in the past decade, with the almost universal use of electronic crystal frequency references incorporated in the circuitry of tachometers. Non-contact methods using interrupted light beams on both permanent and hand-held instruments are ideal to avoid the danger of physical contact with moving machinery. As with all instruments, calibration checks are essential, although digital-based speed measurement instruments seldom drift or give small errors. They are either very precise or give an obviously wrong reading. The latter may be due to an inadequate input signal or some form of interference.

A simple method of cross-checking an optical RPM meter which registers one light reflection/revolution, is to direct it at a fluorescent light running from a mains electrical supply connected to the distribution grid. It should read 6000 RPM, corresponding to the 100 Hertz light wave form generated by a 50 Hertz supply. If the supply is 60 Hertz the RPM it should read is 7200. Beware checking a meter using a light not powered from the grid, as small generator supplies may not be so well frequency-regulated.

A more permanent and very accurate RPM meter can be constructed using a 60-hole disc or gear, mounted on the shaft with an optical or magnetic pick-up. The raw frequency signal will probably require conditioning by driving a 'Shmitt-trigger' circuit, to give a clean, bounce-free pulse to a frequency meter. This trigger circuit should be physically close to the pick-up, and supplied via a shielded cable to minimize the chances of electromagnetic pick-up. A frequency meter which records directly in Hertz will show a value in RPM with the 60 pulse/revolution disk.

Errors are possible from other light sources such as fluorescent lights which can produce stray input frequencies to an optical sensor. Beware any reading which corresponds precisely to the mains frequency. All squirrel cage induction motors will show a slight speed slip, even when running unloaded. A variable speed drive is required to run at the same speed as the mains, using a drive frequency slightly above the line frequency. Older types of thyristor-based variable-speed DC drives can give substantial levels of local electrical interference. Experience indicates that the level of electrical noise is inversely proportional to the power of the drive!

Avoid using mechanical contact-driven tachometers if possible, as a small angular offset can give a significant error in speed reading. Linear speeds are conveniently monitored by an optical sensor which is used to time the traverse time between two known points. Pick up wheels which drive a frequency generator are available as an accessory for digital and mechanical tachometers but beware errors due to wheel slip or poor alignment.

7.7   INSTRUMENT INSTALLATION–TEMPERATURE

Temperatures can be a valuable guide to machine condition, indicating excessive energy flows or losses. At one extreme it can be as simple as monitoring the temperature of a bearing or gearbox above ambient, or at the other, involve complex precision measurements of small temperature differences across a mass transfer pump. It is again all too easy to get a wrong reading. Installation factors and calibration drift can combine to mislead the unwary. Temperature monitoring is covered elsewhere in this book as a tool for detecting and identifying specific developing faults, so this discussion is limited to long-term steady-state monitoring of machine energy flows. Many excellent texts are available giving the theoretical background and details of the instrumentation

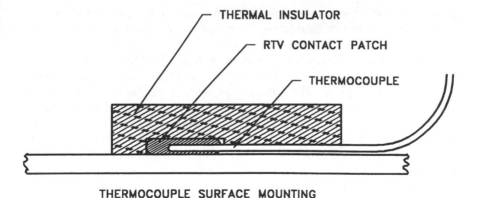

THERMOCOUPLE SURFACE MOUNTING

Fig. 7.5 Thermocouple surface mounting.

needed. Therefore the coverage of this topic here is limited to defining the common forms of temperature transducer, and to outlining the various practical techniques to assist with obtaining reliable readings.

The key requirement for accurate temperature measurement is to ensure the transducer sensing element follows the temperature of the target material. This requires good thermal contact with the target material combined with a minimum thermal gradient across the transducer. These requirements can be met providing a short thermal conduction path, with plenty of cross-sectional area between the material surface and the transducer. Always ensure the sensor is well-clamped or bonded to prevent the materials in the conduction path separating to form an insulating boundary layer. The target material, the transducer sensor, and about 50 mm of cable should be surrounded with a pad of thermal insulation, as shown in Figure 7.5.

For temperatures below 100°C, ideal heat-transfer materials can be formed from RTV compounds, which, although they have a relatively poor thermal conductivity compared with metal, can provide plenty of conduction area and excellent contact across material boundaries. If the temperature of the measured surface is more than a few degrees above ambient, lagging can be used to minimize the effects of thermal gradients through the target material, as long as this does not adversely affect the operating temperature of the monitored unit. With fluid flows through pipework, the lagged outside surface of the pipe is usually sufficiently close to the bulk fluid temperature for monitoring purposes.

If, however, you need to detect very small temperature differences, it is essential to mount the detecting element of the transducer within the pipe in a protective pocket filled with oil or grease, to ensure very good heat transfer to the transducer sensing element. Stainless-steel sheathed tubular sensors are ideal for mounting in fluid flows, passing through pipe walls with a ferrule sealed gland, but beware the possibility of fluid flow inducing vortex shedding from the probe which will result in transverse bending and fatigue failure.

The key factors determining the appropriate selection of a temperature transducer are the measured temperature range and the required accuracy, as well as price. The ubiquitous thermocouple is a very handy temperature transducer due to its apparent simplicity and small size. It can be packaged in a variety of very compact forms to facilitate mounting the thermocouple junction in good thermal contact with the target material. A range of material combinations is available depending on the maximum temperature, the output voltage generated and cost. Thermocouples are slightly nonlinear in performance, but calibration curves are available for couples manufactured to standard alloy specifications. They can be physically very small, and therefore have a low thermal inertia giving fast response rates to temperature transients.

Type 'K' thermocouples, chromal alumel, are commonly used for general engineering, and with appropriate mineral insulated sheathing, inconel, can be used up to 1100°C. Stainless-steel sheathed units will survive to 900°C. The main drawback with thermocouples is the very low voltages produced. This requires a sensitive detector circuit which can be susceptible to electromagnetic interference. The couple must also be protected from electrolytes, such as contaminated or salt water, which can result in spurious readings. Most commercially available couples are contained within a sealed stainless-steel enclosure. Other requirements are matching of the thermocouple type, that is the couple materials, to the signal conditioning, and ensuring that the sensor and the reference cold junction are the only thermally unbalanced alloy couples.

Thermocouple signal conditioning and read out devices should include provision for connecting the thermocouple without setting up additional unbalanced junctions. This is achieved by mounting all connection pairs in close proximity with good thermal conductivity between them. A reference junction, in the form of an electronically generated cold junction compensation, built into the thermocouple amplifier, is usually held at the equivalent of 0°C. This replaces the alternative reference of placing a thermocouple in crushed ice, produced from de-ionized water, floating in water, and contained within a vacuum flask as used in earlier times. Signal conditioning modules in the form of single integrated circuits are available which provide these functions at low cost. It is good practice to mount such a device close to the thermocouple, but in an area at a reasonable ambient temperature, and transmit the higher level output to a remote voltage display device.

A thin film platinum resistance temperature sensor can provide the most accurate and stable sensor commonly available. The sensor elements are typically trimmed to 100 ± 0.1 ohms. Sensors capable of surviving 500°C are available, but others are limited to 250°C by their packaging. The nonlinear relationship between resistance and temperature is well documented, so accurate calibration can be obtained by a single resistance measurement at a known temperature. A Wheatstone bridge can be used to sense the platinum film's change in resistance. A four-wire configuration is preferred, with the voltage tapped either side of the sensor, which avoids the excitation current losses in the conductors.

Encapsulated signal conditioning modules are available to provide a linearized output, typically covering the range -100 to $+500°C$, with an accuracy of better than $\pm0.2°C$ from -100 to $+300°C$. Specially matched pairs of platinum resistance detectors used in a bridge configuration can be used to detect very small temperature differences. This requires specially developed signal conditioning, with pulsed excitation to minimize possible self-heating effects, as well as very careful installation of the sensor to pick up representative fluid bulk temperatures. This equipment was developed for efficiency monitoring of mass-transfer pumps for the water supply industry.

Thermistors give a substantial but nonlinear change in resistance with temperature. They come in two basic forms, the first having a negative temperature coefficient and the second a positive coefficient. While they are valuable devices for temperature trips and alarms, or incorporating in low-cost temperature indicators, they require individual calibration and are therefore not ideal for long-term instrumentation applications. Other aspects to be aware of are self-heating characteristics with constant voltage excitation, and relatively slow thermal response compared to thermocouples.

Small integrated circuit temperature sensors are available in various standard transistor packages which produce an output voltage proportional to the case temperature. While these devices have a limited temperature range, typically between -40 and $110°C$, and are physically larger than a thermocouple, they are both cheap and simple to install. The output is suitable for interfacing directly to a monitoring computer or indicator. The usual calibration is 10 millivolts/$°C$. They only require an unregulated direct current (DC) excitation voltage between 4 and 30 volts, although a wary user would provide a regulated supply to minimize the possibility of spurious signal noise. Such devices can typically provide accuracies of $\pm0.25°C$ at $25°C$, and $\pm0.75°C$ over the whole range.

7.8   INSTRUMENT INSTALLATION – FLOW MEASUREMENT

Flow measurement is probably the most complex common industrial parameter to measure accurately. As hydraulic energy transmission is a function of the product of flow and pressure, it is necessary to measure

both parameters accurately for a meaningful estimate of machine efficiency. The following comments apply to both liquid and gas flow
although the latter is complicated by compressibility. This section does
not attempt to give a full description of flow-metering technology.
However, an excellent introduction to the subject is given in Baker
(1988), who presents the basic issues and techniques in a digestible form.

An absolute accuracy of 0.5% with a flowmeter is difficult to achieve.
While manufacturer's literature may claim better figures, these will be
only achieved under ideal conditions. Real installations are rarely ideal
and usually involve several compromises such as the position of valves
and bends upstream, fluid property variations and flow stability. Calibration of flowmeters requires an expensive calibration loop with traceable reference standards. This usually means a flowmeter has to be taken
out of the installation and sent away with both cost and time delay
implications. Some flowmeters, particularly ultrasonic types, claim to be
self-calibrating but will still need to be checked against a traceable
standard for this to be accepted.

The key factors defining flowmeter performance are:

- accuracy expressed as a percentage of maximum calibrated flow; this
  implies that the reading from a 1% instrument can be 10% out when
  measuring flows around 10% of its maximum rated capacity;
- rangeability, expressed as a measure of the 'turndown', or the ratio of
  the minimum flow the device can measure within the stated calibration accuracy, in relation to its maximum rated flow.

Typical turndowns are 3:1 for devices such as orifice plates, pitot tubes
and ventures, 10:1 for turbine flowmeters. Vortex shedding devices can
give 20:1 while some forms of fluidic flowmeter and positive displacement meters can give a turndown ratio as high as 250:1.

It should be noted that the majority of industrial flowmeters, with the
exception of positive displacement devices, detect the velocity of flow of
the fluid passing through a known area. Flow is the product of the flow
area and the mean flow velocity through that area. Therefore in order to
estimate the total flowrate, the flowmeter calibration has to assume a
stable and predictable flow velocity profile across the sensing element.
In turn, this velocity profile is very dependent on the geometry of the
upstream pipework as well as the flow conditions defined by the
'Reynold Number'.

This number is the ratio of viscous forces to dynamic forces and is a
useful dimensionless number which defines flow characteristics (Miller,
1978). A straight run of pipe of 20–40 diameters is required to allow
major flow velocity variations across the section of the pipe to even out
after a disturbance such as a bend or a valve. Flow straighteners are used
where space limitations preclude a long straight pipe run but these
should be regarded as a poor second best. They will generate small-scale

turbulence in their own right which again needs time and distance to settle down.

The main techniques of flow measurement are as follows (Baker, 1988; Cheremisinoff, 1979).

Differential flow techniques

Differential pressure sensing devices are sometimes called head or rate meters and measure the flow without sectioning the fluid into isolated quantities. They work by creating a differential head which is a function of the fluid velocity and density. Systems which incorporate this principle include venturi meters, flow nozzles, orifice meters and pitot tubes. It should be remembered that the dynamic head varies as the square of the fluid velocity. As the dynamic head increases the static head drops, resulting in a low pressure in the region of highest velocity. As most pressure measuring devices have a turndown range of about 10:1 the resulting flow-meter turndown is limited to $\sqrt{10}$, hence the limited flow range capacity of differential head devices.

The Venturi meter is the potentially most accurate form of differential head flowmeter as well as providing the least overall installed pressure loss. Its general form is shown in Figure 7.6. In the case of the Venturi meter as the fluid passes through the reduced area of the throat, its velocity head increases, and its static pressure decreases, resulting in a pressure differential between the inlet and throat regions measured by side tappings in the meter wall. This can be measured by use of differential pressure meters and capacity curves. The Venturi meter should be used whenever the loss of pressure across the meter installation is to be reduced to a minimum, or where the fluid is heavily contaminated with other material in suspension.

The Venturi meter can be installed in the horizontal, vertical, or inclined position preferably as far as possible downstream of any flow disturbance. In the vertical position, upstream and throat pressure taps

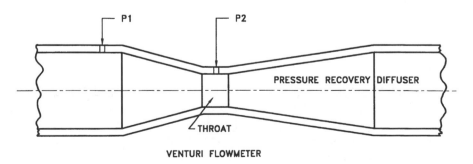

Fig. 7.6   Venturi flowmeter.

can be oriented at any angle around the tube, but in other installations, care should be taken to ensure these are in the proper locations to suit the fluid type or faulty measurements will result. In gas-flow applications or steam installations where the meter is above the line, pressure taps should be on top of the tube. While in liquid flow applications or steam installations where the meter is below the line, pressure taps should be at the side of the tube. Care should be taken to ensure the pressure tapping lines are free from trapped bubbles of gas or false readings will occur. A properly calibrated Venturi system should provide accurate measurements within ±0.5% over a wide size range, and maintain accuracy for a considerable period of time. They are also maintenance-free and self-cleaning.

The orifice plate is also a widely used device for both liquid and gas applications in this category. Its general form is shown in Figure 7.7. It has the advantages of simple design, circular, eccentric or segmental, a low-cost and ease of installation. Although it should be noted that they cannot be used for two-directional flow, and accurate positioning is essential to avoid incorrect differential pressure measurements. Other disadvantages, particularly of the concentric circular hole design can include, a poor ability to handle viscous or dirty liquids, inadequate condensate removal in steam and vapour applications, plus a higher pressure loss than a Venturi unit. As with the Venturi meter, the pressure tappings should allow gas or vapour to return to the line and avoid debris blockage.

Orifice plate calibration is sensitive to the quality of the orifice entry edge. This edge should be square, sharp and free from burrs. The quality of this edge can be easily affected by debris in the fluid which will quickly degrade the unit's calibration stability. Therefore the orifice plate

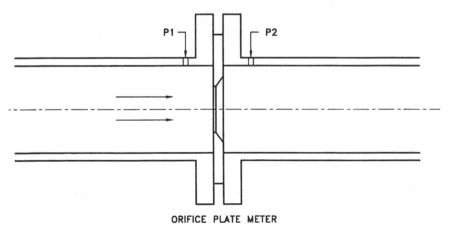

ORIFICE PLATE METER

Fig. 7.7 Orifice plate meter.

is not suitable for long-term installations with dirty fluids where calibration stability is important, as in the case of steady-state monitoring. Both the orifice plate and the Venturi meter can be calibrated by calculation using ISO 5167 and BS 1042, within defined limits of accuracy if the installation conditions are met. Higher accuracies can be achieved by physical calibration.

Flow nozzles are devices which embody most of the advantages of Venturi meters but which can be placed within existing pipework at a flange joint with very little additional space requirement. They do not incorporate a pressure recovery stage and therefore their installed pressure loss is rather worse than a Venturi meter. However, special calibration of these units is required, depending on both their design and finish. They are more efficient than orifice plates, being able to handle about 60% more fluid at the same pressure drop, and have been used for the metering of high-velocity fluids. Although capable of handling liquids with suspended solids, flow nozzles are not suitable for highly viscous fluids, or those containing sticky solids.

One of the oldest flow measuring devices is the Pitot tube as shown in Figure 7.8. It is essentially a device for taking a single point measurement of the velocity head of fluid flowing in a duct. The velocity profile has to be assumed from the flow conditions, or alternatively, the Pitot tube can be traversed across the pipe to measure the velocity profile. The device consists of two concentric tubes. The inner tube transmits the velocity head from an end tapping facing the incoming flow while the outer tube transmits the static pressure picked up by small cross drillings in the wall behind the velocity head tapping. Common designs have either sharp or rounded noses and static pressure taps well back

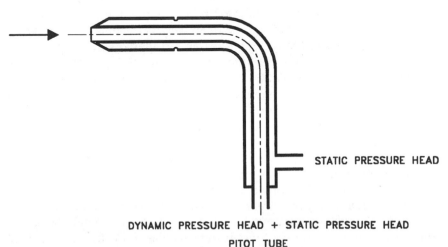

Fig. 7.8 Pitot tube.

from the front face. The flow rate range of these units is narrow, which has tended to limit their usage in industry plus the fact that they are also unsuitable for precise flow measurements in an industrial application. In addition, they can be blocked by debris quite easily and the pressure drops they create are rather small for accurate measurement for normal industrial velocities of flow.

Positive displacement flowmeters

These devices are used where high accuracy is needed. They are particularly useful for fiscal metering, mixing, blending or batch systems. The unit works by separating parts of the flow into discrete volumes, and then determining the flowrate by their summation. Normally, such systems are characterized by having one or more moving parts in the fluid, to separate it into the volume increments. The fluid motion supplies any energy required to effect this separation, and this gives rise to a pressure loss between inlet and outlet. As accuracy depends on achieving a moving seal between the moving and stationary parts, they are not suitable for metering fluids with high proportions of suspended debris.

Typical instruments in this class are the rotary or reciprocating piston types, nutating disk and rotary vane meters. The reciprocating piston meters are available in single or multi-piston versions with the application flowrate dictating which should be used. These instruments can be used to meter practically any fluid with an accuracy of up to $\pm 1\%$, or in some designs, down to $\pm 0.1\%$ should this be required. The nutating disk meter is essentially a movable disk mounted on a concentric sphere, and located in a chamber having spherical sidewalls intersected by inverted conical top and bottom surfaces. A radial partition, over which the disk is slotted, restricts rotational movement, but allows the disk to rock as the fluid passes through the chamber. As the disk rocks, gearing is activated to control the meter's register and thereby record the flow.

Rotary vane meters work on the basis of a revolving eccentric drum fitted with spring-loaded vanes. A known quantity of fluid is thus swept through each section of the meter, with its volume being measured by a register attached to the axis of the drum. These units are easily installed, lightweight, insensitive to viscosity changes, have good accuracy and are quite reliable in liquid or gas applications. Filters need to be used where there is a risk of heavy particulate contamination to extend service life.

The common domestic water meter is one form of positive-displacement flowmeter which is often overlooked. However, these devices are capable of providing the greatest degree of accuracy and turndown capability of any flowmeter outside of a National Calibration Laboratory. Extensive testing has shown these flowmeters last for extreme lengths of time with very little loss of accuracy. The fact they are typically one-tenth of the price of industrial flowmeters infers that they are in some way inferior, but for clean water or oil they are reliable precision devices. They are primarily flow totalizers, but they can be fitted with sensors to provide a low-frequency signal proportional to flow.

Variable area meters

These devices generate a fixed differential pressure across a restriction and allow the area to vary in response to the flow. This is the principle used in the float rotameter where a float is lifted by the fluid flowing vertically upwards through a tapered tube. The differential head to support the 'float' is fixed by its negative buoyancy in the fluid. Its position in the tapering tube, and hence its area, varies to expose sufficient flow area. The flow rate is therefore indicated by the height reached by the float, read from a linear scale. In general rotameters are used to measure small flowrates in clear fluids, and care is required in the design to avoid the problems of float oscillation or sticktion to the tube wall in an unstable liquid flow. A variant of the variable area meter uses a spring-loaded disk in the tapered tube to replace the float. An external sensor detects the disk position magnetically. This device is particularly useful for high-pressure applications such as oil hydraulic systems.

Turbine flowmeters

These devices are a commonly encountered form of flowmeter used throughout industry. The dominant form employs a small free-spinning axial turbine which turns in almost direct proportion to the fluid flow velocity through the meter throat. Radial flow designs are also common, particularly for smaller flows which use a vane rotor to pick up the velocity of a tangential fluid jet. Turbine rotation is detected by a magnetic sensor or similar device to generate a frequency signal proportional to the turbine speed. The turbine flowmeter typically has a 10:1 or 20:1 turndown capability. The life of a turbine flowmeter is limited by the bearing life. They are also susceptible to blockage by contaminant. Fibrous material is particularly unhelpful as it can accumulate on the blades and gradually degrade the calibration.

Turbine flowmeters are sensitive to fluid viscosity and density. Therefore the correct fluid should ideally be used for calibration, although manufacturers can provide correction curves to cope with minor deviations from the calibration fluid properties. Turbine meters provide accurate, low-cost metering for a variety of liquids but are generally unsuitable for those which are corrosive, abrasive or which contain an appreciable proportion of solids or dissolved gases. Miniaturized units are also available, to provide a traversing or point measurement capability which is useful for open channel work. They can be constructed for use in high-pressure applications such as fluid power systems.

Magnetic flowmeters

These instruments are based on Faraday's Law of Electromagnetic Induction, namely the fact that relative motion of a conductor and magnetic field induces a voltage within the conductor. In the case of an electromagnetic flowmeter, the conductor is the flowing fluid. The voltage produced is proportional to the relative velocity of the fluid through a magnetic field. In general they consist of two parts, firstly a magnetic flowtube mounted on the pipe and secondly a flow transmitter which can be remote from the flowtube. The flowtube consists of electrically insulated material, with an opposed pair of metal electrodes mounted in the tube wall. A pair of electromagnetic coils are mounted external to the metering tube which, when excited by an alternating electric current, generate the magnetic field orthogonal to the fluid flow.

Obviously, the fluid being monitored must be conductive to develop a voltage across the electrodes; this in turn is directly proportional to the volumetric flowrate. Normal tap water is usually sufficiently conductive to register the induced voltage, but oils and demineralized water have too high a resistivity for use with these flowmeters. These units can be used for all types of fluid monitoring, and have the advantages of minimal pressure drop, corrosion, and erosion resistance of the working parts. They are relatively unaffected by reasonable variations in fluid viscosity, temperature or pressure, but can be sensitive to changes in fluid conductivity or in some cases supply frequency/voltage variations. Modern flowmeters are equipped with electronic excitation drives which can accept reasonable supply variability. It should be noted that periodic cleaning and maintenance is required with these systems to ensure that accuracy is maintained.

Ultrasonic flowmeters

These units meter fluid flow via ultrasound beams traversing the fluid. They can operate in three basic forms. Firstly the 'time of flight' principle which depends on the change in transit time for a pulse to travel up-stream and down-stream over a known distance. Secondly the Doppler frequency shift of ultrasonic signals reflected from discontinuities, bubbles or suspended solids in the liquid. Thirdly correlation techniques which pick up random data due to fluid turbulence and calculate the time for such random variations to traverse between two sample points. This latter form is rather exotic and is not often encountered, being mainly developed to cope with particular problems such as multi-phase flows.

They have the advantages of being non-invasive, inducing no pressure loss, and of low maintenance by virtue of having no moving parts. Ultrasound transmitters/receivers are mounted on the outside of the pipe, usually transmitting the beam at a 45° angle to the flow axis, transmitting through the tube wall into the liquid flow. Reflected signals are detected by the units and compared with the transmission to give the time or frequency shift which is proportional to the flow velocity. It is claimed that on-site calibration can produce a measurement accuracy to within 1% of the actual flow.

However, it is all too easy to forget that the flow cross section must be accurately known and that the nominal bore of a pipe is not its real diameter. This is compounded by the area being proportional to the square of the diameter. The self-calibrating type automatically detect the internal pipe diameter but they assume the pipe is round. These systems are easy to install and align, minimizing plant downtime, and work well on most clean pipework materials or even open channel flow. The Doppler system has a requirement of around 2% suspended solids in the liquid being monitored. Full pipe flow is also a requirement, and the system does not work on solids, gases or some types of multiphase/multicomponent flows such as oil/water mixtures.

Mass flow measurement techniques are particularly useful for compressible flows as they automatically take account of gas temperature, pressure and composition variations. They may be of use for metering gaseous fuels so they are mentioned here in a condition monitoring context. However, the reader is referred to Baker (1988) for a full description if it is essential to investigate these devices.

7.9 CONCLUSIONS

This chapter has attempted to show how steady-state input/output monitoring can be useful in the condition monitoring of typical industrial plant and equipment. The technique can be used to good advantage to give an early indication of problems as they arise in machinery, and may then be supplemented by specific detection techniques to pinpoint the potential failure. In practice, technique implementation can be via standard instrumentation, which may already be used on the plant for measurement and control purposes. Although, as this chapter explains, care must be taken in the selection and calibration of such instrumentation, to ensure that satisfactory condition monitoring data is obtained.

7.10 REFERENCES

Baker, R.C. (1988) *Introductory Guide to Flow Measurement*, I. Mech. E, London, UK.

Cheremisinoff, N.P. (1979) *Applied Fluid Flow Measurement–Fundamentals and Technology*, Marcel Dekker Inc., New York, USA.

Miller, D.S. (1990) *Internal Flow Systems*, 2nd edn, Gulf Publishing Co.

Trevena, D.H. (1987) *Cavitation and Tension in Liquids*, Adam Hilger Ltd.

CHAPTER 8

System input/output monitoring

A. Davies and J.H. Williams

Systems Division, School of Engineering,

University of Wales Cardiff (UWC), PO Box 688,

Queens Buildings, Newport Road, Cardiff, CF2 3TE, UK.

8.1   INTRODUCTION

Input/output monitoring is a term which covers a wide range of different engineering techniques. It is also a term applicable to a variety of different condition monitoring (CM) methods and areas. As a consequence, it is important to explain initially what input/output monitoring is in a general sense, and subsequently in the context of the techniques to be explained in later sections of this chapter. Thus, input/output monitoring may be defined in general as:

> The dynamic measurement, recording and analysis of the input and output features of a System Under Test (SUT), with the objective of fault detection, identification, and ultimately correction.

This all embracing definition, encompasses the use of input/output monitoring and analysis in a wide variety of fields. For example, the use of input/output diagrams (Parnaby, 1987), is now routine in the case of manufacturing system design and analysis. Although perhaps not yet in a dynamic sense, as few companies have linked this technique to their performance monitoring in a 'real-time' input/output context. Another example, which can also be cited, is in the information technology area, where the system modelling and design tool SADT utilizes linked input/output diagrams to define the requirements of information flow (Williams *et al.*, 1994).

In a condition monitoring context, there are a raft of input/output

Handbook of Condition Monitoring
Edited by A. Davies
Published in 1998 by Chapman & Hall, London. ISBN 0 412 61320 4.

techniques available, many of which, as indicated above, are specific to particular applications and referred to in other chapters of this book. Accordingly, we need to tighten the definition given above, so that it accurately reflects the application of this methodology in the area of condition monitoring which concerns us here, and thus specifically relates to the techniques outlined below. So, for illustration purposes, this chapter will confine itself to the use of input/output analysis in the area of machine tool condition monitoring, and with this in mind, we may redefine the technique as:

> The dynamic monitoring and study of machine system input/output signals, to identify and correct a potential failure, prior to the event.

At this point, it is useful to remind ourselves of the types of failure we are trying to identify and correct via input/output monitoring, and to examine how diagnostic reasoning actually proceeds. In this regard we can identify two classes of possible failure, viz:

1. **hard faults**–defined as catastrophic failures which result in machine breakdown, together with a consequential and significant loss in production;
2. **soft faults**–defined as partial or degradation failures which occur over a period of time, and which are characterized by a loss of optimum performance in the machine; ultimately this type of failure may become a hard fault which results in machine breakdown.

To illustrate further the distinction between these two classes of failure, in the context of input/output modelling, consider the case of an ordinary household 100 watt lightbulb. Typically, if we switch on the light and the bulb is about to fail, it does so without any warning. It goes pop, and we change the bulb. This is an example of a hard failure, which is catastrophic, and which can only be repaired by replacement. However, if we had linked the bulb to some form of monitoring system, and looked at either the trend of input/output energy consumption, or some other measurable feature which could be presented as a dynamic signal, it is possible that we might have had a warning of impending failure.

Occasionally lightbulbs signal their degradation by dimming or flickering while still producing adequate but not optimum light. This is an example of a soft failure which ultimately becomes a hard failure as time passes. In the case of a lightbulb, repairs are not possible, so the detection of soft faults is obviously not worthwhile. However, for production machinery such as machine tools, soft-fault detection via some form of trend analysis is economically justifiable, and as a consequence, extremely desirable, to permit suitable repair actions to be undertaken at a convenient time and thereby to prevent catastrophic failure (Davies, 1990).

With the increasing use of automation in manufacturing industry, and the consequential cost associated with each failure/repair, an additional stimulus has been provided, to encourage the development of techniques such as input/output monitoring for use in automated failure diagnosis. Thus, we need to consider how diagnostic reasoning works, especially if we wish to automate the process. Indeed it is important to note that the process of fault diagnosis and problem resolution can be divided into the following four stages:

1. **detection:** this is the awareness that something has, or is going wrong with the operation, process or machine under surveillance;
2. **direction:** effectively, the orientation and navigation of the search process to the fault area;
3. **location:** the isolation and identification of the specific fault within the area in question;
4. **solution:** the provision of a procedure by which the fault may be rectified and the operation, process or machine restored to a nominal functioning state.

Reasoning or inference spans 'direction' and 'location' only, with 'detection' being part of a human or sensor-based monitoring scheme, and 'solution' forming part of the knowledge base which provides problem resolution. Thus input/output monitoring, as will be shown in the sections which follow, is an ideal technique for the automatic implementation of this process. Especially when used in conjunction with modern sensors, data acquisition methods and computer software to form an automatic, intelligent knowledge-based expert system.

8.2 SYSTEM MONITORING

Input/output analysis in relation to whole system monitoring can be explained very easily by use of a simple example. Let us suppose that we are monitoring a turning operation on a lathe which has the following known parameters:

depth of cut = 7.5 mm;

surface cutting speed = 30 metres/minute;

feedrate = 0.12 mm/revolution;

drive motor input = 3 kiloWatts (kW);

drive motor efficiency = 85%.

Now if we monitor the vertical force impacting on the tool top during the operation via a strain gauge, we would obtain a dynamic signal indicating the value of this force at any point in time. As the machining operation proceeds, the value of this force would tend to fluctuate. This is typically due to a variation in the property of the material being cut, or wear on the cutting tool. If we assume that in this example, an average value of the vertical force is equal to 3.5 kiloNewtons (kN), then we can calculate the power consumed in cutting as:

$$\{(\text{vertical force} \times \text{cutting speed})/60\} \text{ watts} \tag{8.1}$$

For the example above, this gives 1750 watts as the power consumed, or output of the machining operation. The input is given as 3000 watts to a motor which is only 85% efficient. Therefore the input energy to the lathe is 85% of 3000, or 2550 watts. Overall machining efficiency can now be calculated as output/input, or 1750/2550 which gives 0.686 or 68.6%. The loss of input energy can be attributed to deficiencies in the mechanism of the machine itself, such as gearbox drive losses, and overcoming friction on the slideways.

So far so good, but this sort of calculation only provides a snapshot in time of the average machining efficiency (one value), and does not help to predict failures either of the machine or of the cutting tool. Monitoring the trend of this average value might help in respect of say cutting tool failures, but this is more directly and adequately provided for by techniques which analyse in 'real time' the cutting force signal itself. Thus the problem is:

> isolating the cause of a loss of machining efficiency, where a downward trend in this value over time is indicating that something is wrong.

One way of doing this, as mentioned above, is to monitor directly via additional sensors, those aspects of a machining system, such as the cutting tool, which are likely to fail or cause problems in respect of component quality. Unfortunately this adds cost, complexity and perhaps unreliability to an already costly and complex system. A more elegant methodology is therefore, in the case of Computer Numerically Controlled (CNC) machine tools, to monitor, as far as possible, existing machine signals which are required for its functional operation (Williams and Davies, 1992). The trick is then to minimize the number of signals used, only one if possible, and to provide a technique, or techniques, which uniquely identifies the cause of a problem, from a range of potential causes.

Of course this is not easy to do, especially if we are attempting to do it via a general high-level signal like energy consumption. For within a mechanism like a machine tool, there will be a hierarchy of signals controlling its operation, via various subsystems, down to the most detailed level. Obviously, it is easier to detect and diagnose a fault in such a mechanism, if we are using specific signals which relate to a subsystem or particular part within the machine. However, we would like to use the highest level and most general signal possible, because then, any successful diagnostic or fault location technique would be more widely applicable, and ideally capable of identifying many different failures within a machine.

Such a technique would hopefully also have the advantage of not being machine or component specific, and a good example of this approach, using a general high-level signal, is vibration monitoring. This may be used on a wide range of machines, and via many available techniques, to diagnose down to a specific component level. Thus, the task set for research into input/output monitoring, is to identify signals and/or diagnostic techniques of use within the general area of condition monitoring for precise fault location. Accordingly, the techniques outlined below provide some examples of recent research into this philosophy as applied to machine tools.

8.3 MACHINE HEALTH ANALYSIS

A major subsystem, normally found in any type of machine tool, is an axis servo. In modern computer-controlled machines, there may be a multiplicity of such subsystems which are used to provide correct workpart/tool positioning during machining operations. The axis drive of a typical machine would consist of a direct current (DC) servo motor, ballscrew, slideways and a transmission belt. In order to monitor the condition of these components, it is possible to instrument the ballscrew, and thus obtain information directly in respect of preload variation, plus wear on the screw, slideways and transmission belt (Harris *et al.*, 1987; Hoh *et al.*, 1988). This approach has, however, significant operational disadvantages, including cost, complexity and unreliability in service. It also provides only limited information in respect of the servo motor, which is arguably, the most important component in the subsystem.

By applying the principles of input/output monitoring a more effective method can be devised. This uses the 'built-in' servo motor current sensor to monitor the health of the axis drive based on its dynamic performance characteristics (Hoh *et al.*, 1990). A health index of the axis drive system is then established, and subsequently used as the monitored parameter. This index checks the condition of the axis drive under a 'no-load' condition and utilizes a fault dictionary to facilitate the application of a pattern-recognition method of fault diagnosis. The condition of the axis drive system is monitored by measuring the steady state and transient behaviour of the direct current drive motor current and by use of the signal behaviour characteristics, a health index is set up which can be used to monitor drift in the condition of the system.

Of the three signals which are readily available from the axis drive subsystem, that is the commanded position signal, the tachometer feedback and the motor current, the last named was found to be the most sensitive to any parameter changes. When the machine table is required to move, the motor current exhibited both transient and steady-state features. Accordingly, it was found that the transient response was useful in monitoring the gain and balance of the servo, while the steady-state response was sensitive to the mechanical aspects of the drive assembly. Under nominal operating conditions data was then acquired to provide an overall average response together with an associated maximum envelope. The 'health' value was then calculated using:

$$HV(NOM) = \sqrt{[(\Sigma_i(((NOM_j - Y_n)/Y_n)_i \times 0.100)^2)/(N - 1)]} \quad (8.2)$$

where: $i = 1 \ldots \ldots N$.

   Y_n = average response.

   NOM_j = jth sampled response.

Thus, HV(NOM) is the reference value for the output of a healthy system, and in service, samples of the motor current data are taken from time to time, so that the prevailing health value HV(v) may be calculated using:

$$HV(v) = \sqrt{[(\Sigma_i(((v - Y_n)/Y_n)_i \times 0.100)^2)/(N - 1)]} \quad (8.3)$$

This represents current system condition, or input, for the purpose of generating the health index value as defined below:

$$HI = HV(NOM)/HV(v) \text{ if } HV(NOM) < HV(v)$$

or   (8.4)

$$HI = 1 \text{ if } HV(NOM) \geqslant HV(v)$$

Obviously, it may be seen that the health index value will range between zero (poorly), and one (fine), thereby giving an indication of the relative health of the drive. Over time, the health index values can provide a trend record in respect of the condition of the drive system. A feature which readily allows a condition monitoring scheme to be followed, whereby maintenance action can be introduced at a predetermined value of decay. When maintenance action is triggered, the lastest recorded dynamic response can then be used in a pattern recognition sense to diagnose the most likely cause of the loss in performance (Hoh, 1992).

Figure 8.1 illustrates the scheme in practice, where during trials, it was found that the transient response of the servo unit is sensitive to the resultant force acting against the motion of the axis. Consequently, the servo will give different transient responses along the length of the axis, and therefore it was necessary to restrict the test to a predefined area of the slideway. The middle range of the axis was chosen with the transverse feedrate set at 1200 mm per minute. A relatively high sample

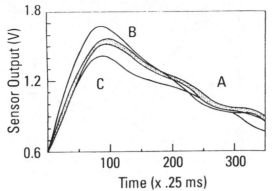

Fig. 8.1a A comparison of servo responses in time-domain.

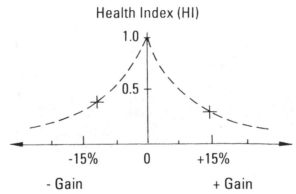

Fig. 8.1b A plot of HI values versus servo gain.

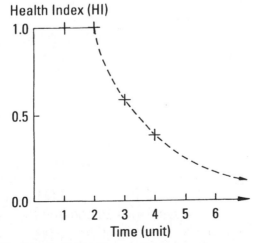

Fig. 8.1c An illustrated sketch of servo HI time series.

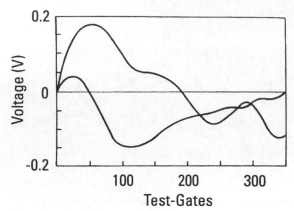

Fig. 8.1d Servo response deviation factors.

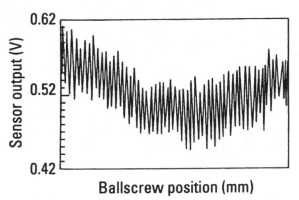

Fig. 8.1e A typical unsmoothed axis motor current profile.

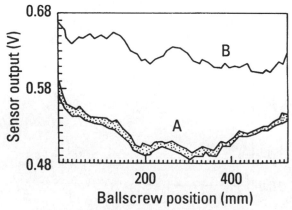

Fig. 8.1f A comparison of nominal and faulty axis profile.

rate is required to capture the servo's transient response, with typical response data being captured in under 0.1 seconds. This was smoothed to reduce noise and the value at each test gate noted.

The average 'spread' of the response was then known, and HV(NOM) was found to be 2.5, curve 'A' in Figure 8.1a showing the nominal response of the servo. Drift in the servo gain was simulated by adjusting the gain resistor in the servo unit. The response curves 'B' and 'C' in Figure 8.1a showing the effects of a 15% increase and a 15% decrease in the gain respectively; with the associated health index values, as shown in Figure 8.1b, being 0.304 and 0.422. Figure 8.1c illustrates the use of the health index as a time series, for predicting the next value, and also forecasting the time to reach an unacceptable limit.

The deviation vector of the fault response can be obtained by subtracting the nominal response from it, and Figure 8.1d outlines graphically the deviation vectors of the simulated faults. By storing these deviation vectors as a table, we can create a fault dictionary, which in turn can be used to diagnose all future occurrences of these faults by using pattern recognition techniques. Two such techniques are the Cross Product Method and the Nearest Neighbour Rule (Thorpe, 1988). These methods allow the faults to be ranked in order of likelihood, which is an important feature in the face of real-world uncertainty (Williams, 1985).

The nominal profile of the steady-state motor current across the axis was also obtained, by averaging a sample of profiles produced during 30 translations of the table along the ballscrew, at a feedrate of 800 mm per minute. Prior to averaging these profiles, each individual profile was smoothed. The ingress of foreign matter is simulated by increasing the ballscrew preload by approximately 7.5 microns (μm) or 3 kiloNewtons. A typical unsmoothed profile is shown in Figure 8.1e, while curve 'A' in Figure 8.1f illustrates the average steady state motor current profile along the machine's 'X' axis.

This profile shows indirectly the torque required to overcome the combination of forces against the motion of the table in the 'X' direction. These forces are thought to be the sum of the friction forces, table loading, the ballscrew preload, and the forces dependent upon the general condition of the ballscrew, carriage and the slideways. Loading due to machining forces is not considered. Curve 'B' in Figure 8.1f shows the profile produced under simulated fault conditions. The procedure for establishing HV(NOM) and HV(fault) is the same as for the servo condition monitoring. HV(NOM) for the axis motor current profile was found to be 2.0, while the calculated health value for the simulated fault was 20.56, and the calculated health index value was 0.097. This low value of the health index indicates that the fault is quite severe (Hoh, 1989). Similar health indices have also been successfully developed for machine tool coolant systems (Martin and Thorpe, 1990).

8.4 GRAPHIC PATTERN ANALYSIS

Thus a key subsystem, which may be present as one or more units in any computer-controlled machine tool, is the servo drive. These units vary in design and are used to control the movements of the tool/workpiece combination in the various axes present on a particular machine. They do so, such that the desired engineering component is manufactured to the required dimensions and tolerances as specified in the workpart drawing. Accordingly, the accuracy in operation of a machine's servo drives is fundamental to the performance of the machine tool, and by using an input/output modelling approach to the condition monitoring of these subsystems, the focus of the health of the machine is seen to present in the dynamic response relating to each servo drive.

Graphic pattern analysis is a manual input/output technique supported by computer graphics, and may be considered as an extension of some research work previously reported on the condition monitoring and diagnosis of machine-tool servo drives (Hatschek, 1982; Seeliger and Henneberger, 1980; Stoferle, 1976 and Shitov, 1980). The overall approach consists of four steps (Harris, 1987):

- on-line signal acquisition;
- use of a 'forgetting' memory;
- application of computer-implemented parallel models;
- off-line manual fault detection and condition monitoring using computer graphics.

A servo system on which this method is to be used must satisfy the following preliminary requirements:

- it should be divided into manageable subsections and suitable data-acquisition facilities provided;
- each subsection should be modelled so that a reference is available which is capable of reproducing its principle physical properties;
- each subsection should be digitally simulated as a digital transfer function model.

To illustrate the technique, two servo drive subassemblies are now considered, an operational amplifier and its associated power amplifier. The main section of the operational amplifier is shown in Figure 8.2a and comprises a differential amplifier (long tailed pair) and two emitter followers. This unit's input/outputs are:

- the limiter output, V1, (input to the op amp);
- the attenuated tacho output, V2, (input to the op amp);
- the operational amplifiers output, V3 fedback via Rf and Cf.

The overall system's output, Vo, is derived from V3, and smoothed by Ro and Co, with the points 'X' and 'O' in Figure 8.2a, being selected as the input and output monitoring points respectively. Because of the feedback action, point 'X' is a virtual earth, and hence the input signal at this point is exceedingly small. The two traces in Figure 8.2b show typical input and output waveforms at these two points and clearly the signal to noise ratio at point 'X' is very poor.

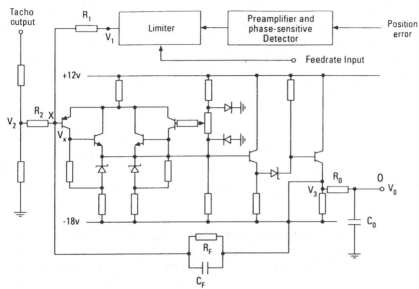

Fig. 8.2a   Operational amplifier circuit.

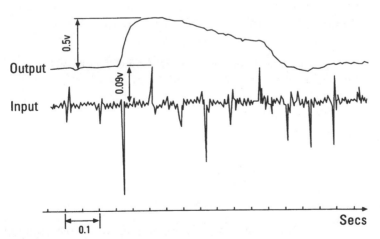

Fig. 8.2b   Input/output response of operational amplifier before use of access circuitry.

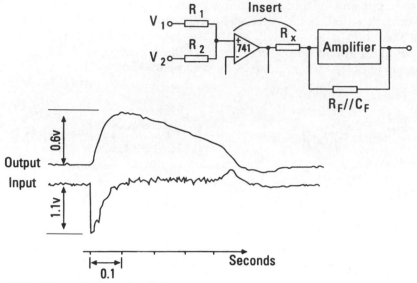

Fig. 8.2c   Input/output response of operational amplifier using access circuitry.

During tests on the system, it was not convenient to measure V1 and V2 simultaneously, so a voltage follower was used to extract the combined contributions of these voltages from the total error signal at 'X'. Figure 8.2c, shows the design of the voltage follower used, and also illustrates how the input signal has been considerably improved for monitoring purposes. Because of space constraints, it is difficult in this chapter to go into great detail on the method of modelling the operational amplifier, however, the basic stages are outlined below and a full explanation is given in (Harris, 1987). To model the operational amplifier:

- a theoretical analysis is required to yield an estimate of the structure.
- the systems frequency response needs to be measured;
- the actual response must be collated with the theoretical response and a transfer function postulated.

In this case the resulting transfer function was:

$$\frac{V_0(s)}{V_{in}(s)} = \frac{-K}{(1 + t_1 s)(1 + t_2 s)} \tag{8.5}$$

with $K = 6.31$; $t_1 = 10.2$; $t_2 = 0.00083$.

While the second time constant (t_2), is insignificant when compared to the first (t_1), it was included so that the software could be exercised with a second-order model.

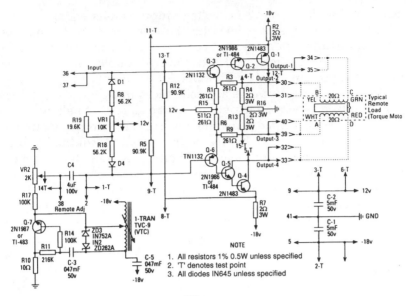

Fig. 8.3a Servo system power amplifier.

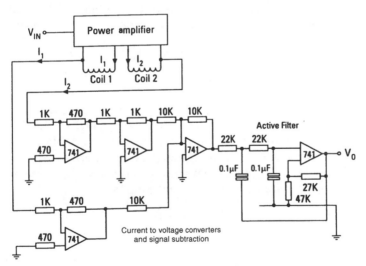

Fig. 8.3b Power amplifier signal access circuitry.

Figure 8.3a shows the circuit diagram of the servo system power amplifier. Positive or negative signals cause an imbalance between the currents flowing through the two sets of Darlington connected transistors Q1, Q2, Q4, Q5. This imbalance causes a torque motor arm to deflect and thereby permits oil to flow in one of two directions in this servo system's hydraulic circuit. Zero voltage input causes equal current to flow in the two transistor pairs, with the result that the torque motor remains undisturbed in its neutral position. Part of the circuit provides a 'dither' at approximately 250 Hertz and low amplitude, thus keeping the servo components 'live' and minimizing stiction.

The power amplifier input is an easily measured high signal-to-noise ratio DC voltage, whereas the output voltage signal is small, because of the low voltages developed across the low resistance torque motor coils, and this can be obscured by the dither noise. Accordingly current measurement is more appropriate and Figure 8.3b shows the circuit used to monitor the power amplifier output current. This circuit comprises two current/voltage converters, a subtraction circuit to measure the difference between currents, and a filter to suppress the dither noise. Naturally the frequency response of this monitoring circuit was obtained along with that of the power amplifier, and a transfer function fitted using a least squares method (Sanathanan and Koerner, 1963). This gave the following:

$$\frac{V_0(s)}{V_{in}(s)} = \frac{18.2 + 0.52s}{1 + 2.86 \times 10^{-3}s + 1.4 \times 10^{-5}s^2} \qquad (8.6)$$

Improved high-frequency matching was then obtained by adding an additional 'pole' by inspection, and this yielded:

$$\frac{V_0(s)}{V_{in}(s)} = \frac{18.2 + 0.52s}{(1 + 2.86 \times 10^{-3}s + 1.4 \times 10^{-5}s^2)(1 + 0.00067s)} \qquad (8.7)$$

Multiple order forms of equations (8.5), (8.6), (8.7) are not easily processed. These second- and third-order equations can however be separated into a series of first-order equations for which the solutions are well-known. Generally this is known as the state variable presentation and solution. These variables are the smallest number of variables which must be defined at $t = t_0$, so as to uniquely define the system's behaviour at any time $t > t_0$. The general representation is:

$$\dot{X} = AX \times BU \qquad (8.8)$$

This matrix equation must be solved to find the variation of the systems's states with respect to time given the input. Then the output $y(t)$ can be calculated. In general:

$$Y = CX + DU \qquad (8.9)$$

In practice, the discrete forms of equations (8.8) and (8.9) are used, with the stages in system simulation being shown in Figure 8.4.

This parallel modelling approach, can be used to construct a very straightforward diagnostic system using computer graphics. In a typical machine installation, the maintenance engineer is offered two adjust-

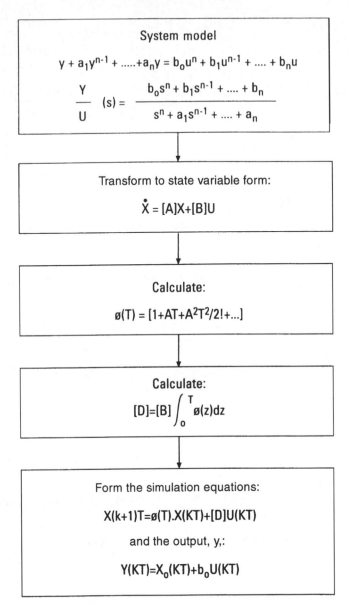

Fig. 8.4 Stages in system simulation.

ments, the gain and bias of the servo system. Both entities can be built into a computer model as adjustable parameters, and the technician invited to alter both as required, to match the model's response to the real system's output. A criticism of this type of parameter adjustment method is that they often require skill on the part of the technician to get

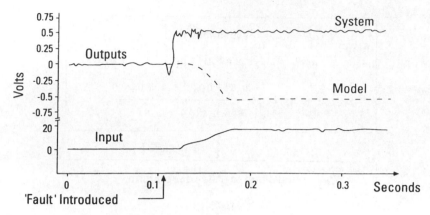

Fig. 8.5a   Operational amplifier hard failure simulation display.

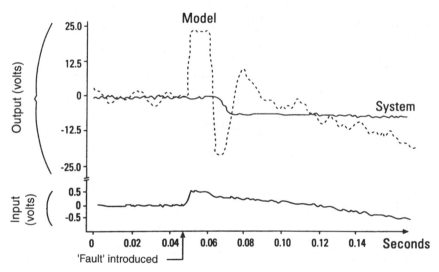

Fig. 8.5b   Power amplifier hard failure simulation display.

fast and consistent convergence. This is certainly true of complex models in which homing rules have to be devised (Towill, 1970), but in the present rather simpler case there are only two parameters and a technician can quickly learn the distinguishing features of each variable.

If a complex model is involved, there may be scope for using an expert system to accommodate the homing rules and assist in the diagnosis. In addition, when a servo system comprises a number of modelled sections, fault diagnosis can be made rapid if groups of components are considered together. Input/output tests made over two or more components, concatenated in this way, will alow a technician to home in quickly on a faulty region. Figures 8.5a and 8.5b show hard fault simulations for the operational and power amplifiers respectively. In the operational amplifier case the model and system outputs have diverged and saturated. In the case of the power amplifier, although the model output is slightly oscillatory before the fault is implanted, afterwards this state of affairs becomes rapidly worse.

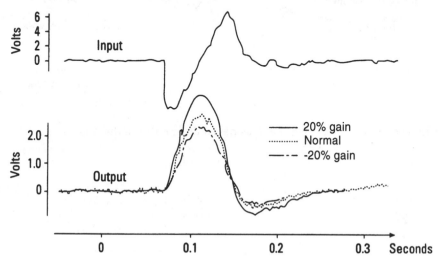

Fig. 8.6a Operational amplifier gain deviation simulation display.

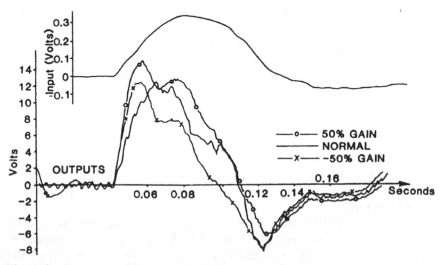

Fig. 8.6b Power amplifier gain deviation simulation display.

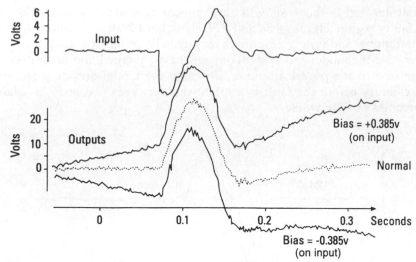

Fig. 8.6c Operational amplifier bias deviation simulation display.

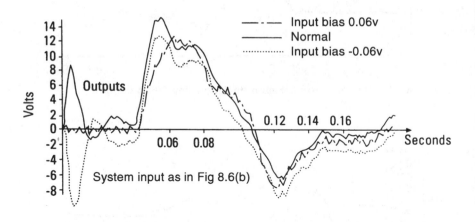

Fig. 8.6d Power amplifier bias deviation simulation display.

It is suggested that a technician examining these computer generated plots would be able to identify faulty behaviour readily, as would a suitable automatic expert monitoring system. Figures 8.6a–8.6d, show the results of soft fault simulations. The distinguishing features of each variation are summarized in Table 8.1. By using such information the technician can alter the appropriate model parameter, re-run the simulation and again compare the displayed outputs. By proceeding

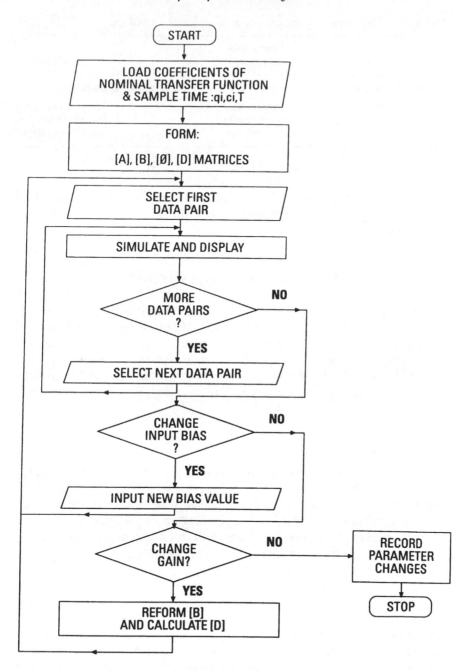

Fig. 8.7 Flowchart of simulation and adjustment process.

Table 8.1 Distinguishing features of gain and bias variations

	Operational amplifier	Power amplifier
± Gain deviation	± Deviation about nominal response	
± Bias deviation	Response superimposed on ramp	Initial deviation due to derivative action on step

iteratively in such a manner, he or she, or an automatic system, can effect a simple diagnosis. The simulation and adjustment process is summarized as a flow chart in Figure 8.7.

8.5   TRACKING RATE ANALYSIS

In this approach (Harris, 1987) the system under test is simulated in parallel with a digital model, but instead of letting the model have free run (i.e. letting it proceed from a set of initial conditions unaltered except by inputs), it is forced to track the actual system behaviour within a prescribed error limit. This is achieved by resetting the initial conditions of the model whenever the error limit is exceeded. Simulation errors will arise from two sources:

- model inaccuracy.
- system degradation.

Under these circumstances, the frequency with which the model is reset will rise. This frequency is called the tracking rate, and it is a useful diagnostic feature if the model is reasonably accurate. As the tracking rate is a continuous variable, and depending upon the chosen error limit, it will be seen that low, medium and high rates indicate healthy behaviour, degraded performance and catastrophic failure respectively. Figure 8.8a illustrates the basic approach.

The error at the ith simulation sample is defined as:

$$E_i = Y_i - O_i \tag{8.10}$$

To mininize false alarms due to signal noise or spurious modelling inaccuracies this error is not used directly. Instead account is taken of the error magnitude during the last n samples and an average A is formed from the weighted sequence of error values according to:

$$A1_i = \frac{E1}{2^{n-1}} + \sum_{i=2}^{n} \frac{E1}{2^{n+1-i}} \tag{8.11}$$

This error filter is in fact a particular case of the general class of exponential smoothing filters of the form:

$$\bar{E}_i = a E_i + b \bar{E}_{i-1} \qquad (8.12)$$

where $\bar{E}_i$ is the estimate of the smoothed error at the ith sample time taking account of the present error value E_i and the estimate of the smoothed error $\bar{E}_{i-1}$. These filters are forward sensitive since error values become less significant as they recede into history. Equations (8.11), and (8.12) are equivalent when $a = b = 0.5$. Computationally, these values are attractive as equation (8.11) can be implemented easily using additions and logical shifts.

The absolute value of $A1_i$ is compared with a selected limit and on exceeding it the model is reprimed with new initial conditions observed from the system output. The instantaneous tracking rate is proportional to the inverse of the number of simulation samples between the last and current reset points. Thus the greater the number of simulation samples completed between reset points the lower the tracking rate. Long uninterrupted runs indicate healthy system behaviour, given that the model is reasonably accurate. A flow chart of the algorithm is given in Figure 8.8b.

The number of simulation steps between the reset points is designated the Uninterrupted Run Length (URL). Tests have indicated that the URL varies randomly about a mean level. The URL is high if the system is healthy and vice versa. In practice a running group average was used,

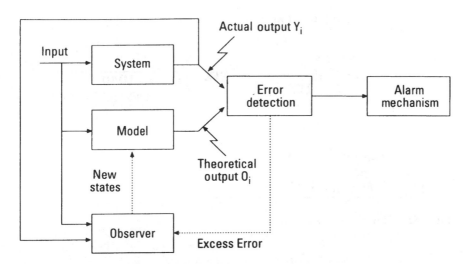

Fig. 8.8a Tracking rate analysis system.

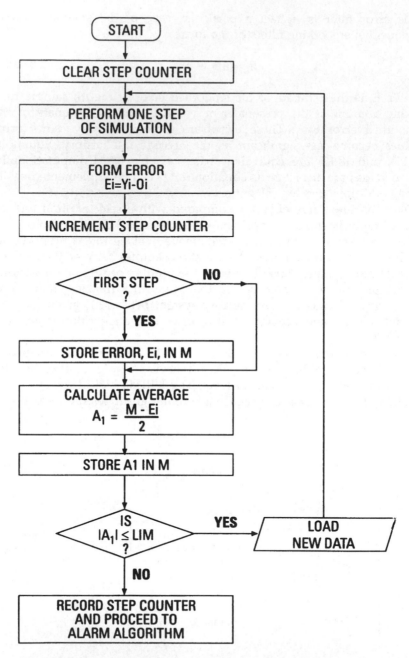

Fig. 8.8b Flowchart of algorithm.

defined as:

$$A2_j = \left(\sum_{i=0}^{N-1} S_{j-1} \right) \Big/ N \qquad \text{for } j \geqslant N \qquad (8.13)$$

where $A2_j$ is the running group average at the jth interruption, N is the sample size of the group and S_{j-1} is the number of steps occurring before interruption in the $(j-i)$th sample. Hard failures force this average to a consistently low value. The smaller the value of N, the faster the average reflects such system changes but at the same time it becomes more sensitive to URL noise. Therefore a compromise value must be chosen.

The basic tracking rate mechanism is illustrated in Figure 8.9a, and shows part of the arbitrary simulation of the operational amplifier detailed in section 8.4, together with the error detector output during the switching process. The size of the error limit chosen for the error detector has an important influence on not only the tracking rate but also on the diagnostic resolution. This can be clearly seen in Figure 8.9b, in which the minimum URL has been plotted against various error limits for a set of on-line signal distributions.

These distributions comprised three healthy data sets and two derived from the amplifier with injected hard failures. For both data sets it can be seen that the tracking rate decreases as the error limit is raised. It is also clear that two divergent distributions emerge corresponding to the two health states of the system. Bearing in mind the logarithmic scale used, the minimum URL value is strikingly different for the two health states at the higher error values, while the URL values tend to converge at the lower limits. This suggests that if the error limit is chosen at, say,

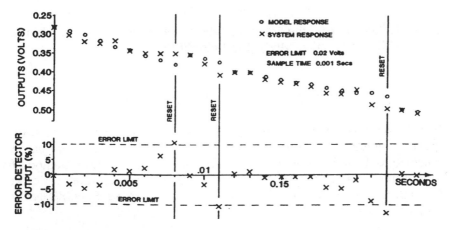

Fig. 8.9a Illustration of tracking process.

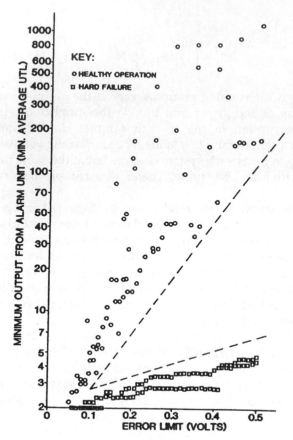

Fig. 8.9b Error limit versus minimum URL for healthy and hard failure operation.

0.05 for this system, the tracking rate as reflected by the minimum average URL is an effective diagnostic indicator.

8.6   NEURAL NETWORK APPLICATION

In the previous three sections, specific technique examples have been given to illustrate the use of input/output monitoring and its use in machine tool condition monitoring and diagnostics. Another method of monitoring and identifying signal features for failure diagnostics is through the use of Artificial Neural Networks or ANNs. Such networks may be defined as:

A model for parallel information processing which is nonalgorithmic, nondigital and intensely parallel (Caudill and Butler, 1992).

It is not a knowledge processor, although it can be regarded as a 'learning' mechanism via its ability to recognize from example, by changing the weight of 'synapses' that connect the 'neurons'. Thus, providing a signal can be monitored and stored as a simple numerical output data file, the information it contains may be analysed using neural net software to determine:

• if the equipment response is correct in respect of the input applied, or if not, to trigger the provision of diagnostic information regarding the reason for failure;
• if trends are present over a monitored period which indicate a potential failure is about to occur, and if so, to trigger the provision of diagnostic information by which it may be identified.

An example of such a monitoring scheme, as applied to a machining centre for signal response identification only, is illustrated in Figures 8.10, 8.11 and 8.12. The trend determination aspect is not shown, being the subject of further research work in this area. A Data Acquisition and Analysis System (DAAS) is used to collect the signal information, with interpretation being done by the ANN software. The DAAS used encompasses a fully integrated monitoring system based on the 68000 family of processors linked to a G96 bus, the system being capable of communication, signal acquisition, analysis and on-board diagnostics via a series of plug-in cards and software modules. In operation, the DAAS utilizes the OS9 operating system for programming and execution (Shaw, 1993).

As shown in Figure 8.10, the signals examined are from the axis drives of the machining centre and represent the tachometer feedback to the servo amplifier. To move to a specific point, the axis drives are simulated by a step input, which results in a response signal from the tachometer. This in turn is captured by the DAAS for subsequent interpretation by the ANN software, given that for a basic step input signal, four main types of system response curve can occur, viz ideal, overdamped, unstable or underdamped as illustrated in Figure 8.10. A feedforward, supervised learning network is used to perform the task of recognition as shown in Figure 8.11. The network being a multilayered arrangement of 'perceptrons', each of which is fully connected as shown in the diagram (Pao, 1989; Shaw, 1993).

In order to teach the network to recognize patterns, the DAAS files in this case, a series of sample files and their expected outputs were put to the network in a training mode. For instance, ten examples of the four types of response, each differing slightly, could be used for the learning process. Only one of each is shown in Figure 8.11. The network is then told what each data file represents, i.e. ideal, overdamping etc. and 'learning' is achieved by the network comparing the outputs with the inputs and computing an error. This error is then used to adjust the weights of the perceptrons, by back propagation, which changes the

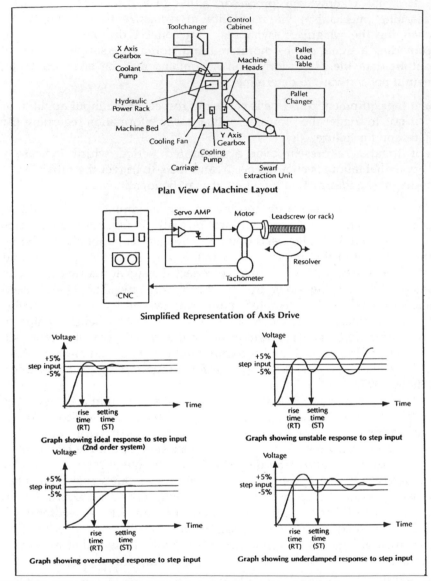

Fig. 8.10 Machine layout, axis drive and tacho signals used in neural network application.

interconnection patterns to reduce the error until the network converges. Once the network has converged it may be utilized as illustrated in Figure 8.11, where eight sample DAAS files are shown to be recognized, and classified correctly. The fault curves on recognition, triggering failure diagnostics as shown in Figure 8.12.

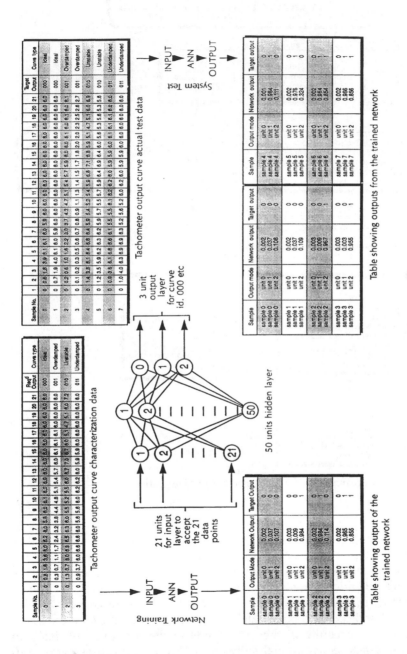

Fig. 8.11 ANN configuration for curve identification with training and test data.

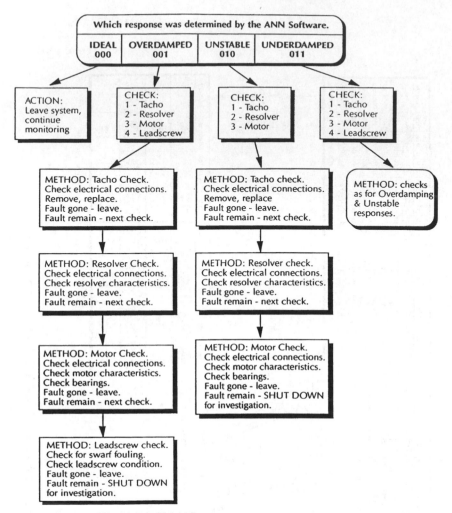

Fig. 8.12 Fault hierarchy for response curve diagnostics.

An advantage of this method over conventional algorithmic software is its robustness and its ability to deal with noisy, corrupt or incomplete data. For example, an incorrect data point would be ignored during curve identification, illustrating that the network is looking for trends rather than analysing individual data values. Conventional software on the other hand tends to be 'brittle', typical curve identification requiring it to take into account all the data points in the file. Corrupt data can therefore 'crash' conventional software, which also requires every conceivable type of curve to be accounted for, otherwise it may not recognize curves embedded in noisy data.

A neural network will, however, try to match the input file with that which it has learnt in the past. If it cannot find a perfect fit, then it will try to make an educated guess as to the type of curve the input file represents. Neural networks have been known to give satisfactory results even if a perceptron malfunctions, as each perceptron acts as an individual unit in a parallel environment (Pao, 1989). To enhance the scheme shown in Figure 8.11, automatic normalization of the data files is obviously necessary, for the envisaged response curves could take many forms and values. In addition, the network should be provided with the ability to handle any number of input data points generated as a DAAS file.

8.7 CONCLUSIONS

Hopefully this chapter has outlined how input/output modelling can be used in the area of condition monitoring and machine diagnostics, with specific examples being quoted, which show the application of this technique to machine-tool condition monitoring, fault location and diagnostics. The development of a performance-related health index, which can be used to monitor the condition of axis drive systems in machine tools has been described, along with the graphic pattern and tracking rate analysis techniques. All three of these techniques, use as a basis, the underlying philosophy and concept of input/output modelling, applied in this case to machine tool servo drives. In the final section of the chapter, the use of neural networks in this context is also outlined, along with the implementation of these techniques in practical systems, via the use of modern computing methods.

8.8 REFERENCES

Caudill, M. and Butler, C. (1992) *Understanding Neural Networks*, MIT Press, Cambridge, Massachusetts, USA.

Davies, A. (1990) *Management Guide to Condition Monitoring in Manufacture*, IEE, London, UK, 12–18.

Harris, C.G. (1987) Fault Diagnosis and Condition Monitoring for NC/CNC Machine Tools, UWIST Ph.D Thesis, Cardiff, UK.

Harris, C.G. Florio, F. and Davies, A. (1987) A sensor system for wear assessment in powerscrews, *Maintenance Management International*, 7, 53–64.

Hatschek, R.L. (1982) NC diagnostics–special report, *American Machinist*, 161–8.

Hoh, S.M., Thorpe, P. Johnston, K. and Martin, K.F. (1988) Sensor based machine tool condition monitoring system, *Proceedings of Reliability, Availability and Maintenance of Industrial Process Control Systems*, IFAC Workshop, Bruges, Belgium, 103–10.

Hoh, S.M. (1989) *Condition Monitoring of Axis Servo Unit Using Built-in Motor Current Sensor*, UWCC Technical Note EP 170, Cardiff, UK.

Hoh, S.M., Williams, J.H. and Drake, P.R. (1990) Condition monitoring and fault diagnostics of machine tool axis drives via axis motor current, *MSET 21 – The International Conference on Manufacturing Systems and Environment – Looking towards the 21st Century*, Tokyo, Japan, 93–7.

Hoh, S.M. (1992) Condition Monitoring and Fault Diagnosis for CNC Machine Tools, UWCC Ph.D Thesis, Cardiff, UK.

Martin, K.F. and Thorpe, P. (1990). Coolant system health monitoring and fault diagnosis via health parameters and fault dictionary, *The International Journal of Advanced Manufacturing Technology*, 5, 66–85.

Pao, Y.H. (1989) *Adaptive Pattern Recognition and Neural Networks*, Addison-Wesley, Wokingham, UK.

Parnaby, J. (1987) Competitiveness via total quality of performance, *Progress in Rubber and Plastics Technology*, 3(1), 42–9.

Sanathanan, C.K. and Koerner, J. (1963) Transfer function synthesis as a ratio of two polynomials, *IEEE Transactions on Automatic Control*, AC-8, 56–8.

Seeliger, C. and Henneberger, H. (1980) Uberwachen und Dignostiziern CNC-gesteuer Machinen, *Machinenmarkt*, 15, Wurzburg, Germany.

Shaw, M.W. (1993) FMMS – An Expert Data Acquisition and Diagnostic Analysis System Study, UWCC M.Sc dissertation, Cardiff, UK.

Shitov, A.M. (1980) Diagnosis of machine tool mechanisms and units by the control oscillograms method, *Machines and Tooling*, 50, 5–8.

Stoferle, T. (1976) Automatische Uberwachung und Fehelerdiagnose an Werkzeugmaschinen, *Annals of the CIRP*, 125, 369–74.

Thorpe, P. (1988) *Wadkin V4-6 FMC X-Axis Mechanical Drive System Description*, UWIST Technical Note EP 142, Cardiff, UK.

Towill, D.R. (1970) Development of analogue modelling strategies for dynamic production testing, *International Journal of Production Research*, 8, 285–91.

Williams, J.H. (1985) *Transfer Function Techniques and Fault Location*, Research Studies Press Ltd, Letchworth, UK.

Williams, J.H. and Davies, A. (1992) System Condition Monitoring – an Overview, *Noise and Vibration Worldwide*, 23(9), 25–9.

Williams, J.H., Davies, A. and Drake, P.R. (1994) *Condition Based Maintenance and Machine Diagnostics*, Chapman & Hall, London, UK, 29–34.

CHAPTER 9

System monitoring and the use of models

J.H. Williams

Systems Division, School of Engineering,

University of Wales Cardiff (UWC), PO Box 688,

Queens Buildings, Newport Road, Cardiff, CF23TE, UK

9.1 INTRODUCTION

Maintenance and associated condition monitoring schemes use templates of many diverse forms as references against which the manufacturing processes may be compared. When this 'check-out' indicates a divergence from normal operation then a diagnostic routine is invoked. The template may be described as a 'model' of the manufacturing system. One can classify these models into two broad types, statistical and time-based. The statistical models can be further split into discrete, continuous and distribution free groups. However, these classifications are not distinct, while the time-based class can be split into time series and frequency-based (spectra), models. This subdivision is shown in Figure 9.1.

A statistical model can be described as static, in that the data is considered as a collection of readings with no concern in respect of the order in which they were realized. On the other hand, time-based models are dynamic and show the behaviour of the process over time. To illustrate their different features consider the data set displayed in Figure 9.2. The two time series illustrated show quite different characteristics. Series 'a' shows a steadily increasing level while series 'b' is cyclic. However both series have the same histogram as they each have identical values in the data set. So statistically the two series have the same characteristics.

Thus, the best advice when evaluating numerical information is to always plot the data of interest as a time series. Subsequently, one can

Handbook of Condition Monitoring
Edited by A. Davies
Published in 1998 by Chapman & Hall, London. ISBN 0 412 61320 4.

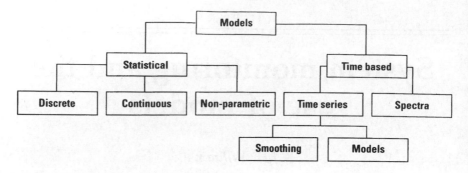

Fig. 9.1 Model classification.

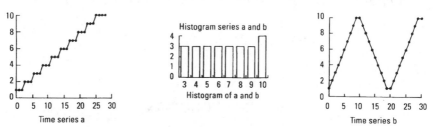

Fig. 9.2 Loss of dynamic characteristics.

decide whether a statistical analysis would be pertinent to any investigation. All the techniques considered in this chapter are presented at a level to enable the reader to understand their fundamental uses. However, the range of techniques which are available in this area is too wide to give fully detailed applications in this review chapter. Accordingly, the bibliography at the end of the chapter lists those useful texts covering both theory and practice, that the author has found stimulating and informative.

9.2   STATISTICAL MEASURES

There are many useful distributions available for discrete and continuous data. This is not the place to give a comprehensive treatment but the salient characteristics will be examined. It is difficult to draw precise boundaries around the particular techniques which are applicable in the general area of maintenance. So for completeness, useful practical distributions will be considered, whose direct applicability is not immediately obvious.

Naturally, discrete distributions are used on data that is discrete, such as the number of failures is a given time-period. Important distributions in this area are the Binomial, and the Poisson. The Binomial distribution is used in situations where the probability of a particular event, such as the number of defects found in a sample, is estimated when the percentage of defects in the population is known and does not change from sample to sample. Formally we have:

p = the probability of finding a defect or ('success');

q = the probability of not finding a defect or ('failure');

$(p + q = 1$, it is a binary situation);

n = number in samples or (number of independent trials);

r = number of defects found in the sample size n or (number of successes);

r_1 = a particular number of defects found in the sample.

Then the probability that the number of defects r found in the n trials is equal to r_1 is given by:

$$\text{prob} (r = r_1) = \binom{n}{r_1} p^{r_1} q^{n - r_1} \text{ for } r_1 = 0, 1, 2, \ldots . n \qquad (9.1)$$

where $\binom{n}{r_1} = \dfrac{n!}{r_1!(n - r_1)!}$ and $n!$ etc. represents factorial.

and also where $0! = 1$.

The probability distribution function given in equation (9.1) is the Binomial distribution with a mean equal to np, and a standard deviation equal to the square root of npq.

Example A population contains 10% defectives. What is the probability of finding 3, 2, 1 and 0 defects in a sample of 3 items?

Solution $n = 3, p = 0.1, q = 0.9$ and r_1 successively takes the values 3, 2, 1, 0.

$$\text{prob} (r = 3) = \binom{3}{3} (0.1)^3 (0.9)^{3-3} = \frac{3!}{3!0!} (0.1)^3 (0.9)^0 = 0.001$$

$$\text{prob} (r = 2) = \binom{3}{2} (0.1)^2 (0.9)^{3-2} = \frac{3!}{2!1!} (0.1)^2 (0.9)^1 = 0.027$$

$$\text{prob} (r = 1) = \binom{3}{1} (0.1)^1 (0.9)^{3-1} = \frac{3!}{1!2!} (0.1)^1 (0.9)^2 = 0.243$$

$$\text{prob} (r = 0) = \binom{3}{0} (0.1)^0 (0.9)^{3-0} = \frac{3!}{0!3!} (0.1)^0 (0.9)^3 = 0.729$$

Note that the sum of these probabilities containing all possible outcomes is equal to 1.

The interpretation of these results means that in a sample of 3 drawn from a population having 10% defectives, 72.9% of the time we would find no defectives, giving the impression that the population is defect-free, while 24.3% of the time we would infer that a third of the population are defective. This example illustrates the danger in using small samples.

The Binomial distribution is used when the percentage of defectives is constant throughout the tests. In practice, this can be considered to be a reasonable assumption when the sample size is small compared to the size of the population, or if the population size is not large, the sample is replaced before the next sample is taken. If these constraints are not met then the Hypergeometric distribution should be used. In this case it is common to use the term 'lot size' rather than population size.

Suppose the lot size is N, and the number of defectives is Np, and Nq are not defective. $(p + q = 1)$. If a sample size of n is withdrawn, and not replaced, the Hypergeometric distribution gives:

$$\text{prob } (r = r_1) = \frac{\binom{Np}{r_1} \binom{Nq}{n - r_1}}{\binom{N}{n}}$$

$$\text{Where again } \binom{a}{b} = \frac{a!}{b!(a - b)!}$$

Example   A wholesaler has a stock of components of lot size 20. He sells 10 of them. Later he is told by the manufacturer that 6 of the 20 components were defective. What is the probability that the customer received 3 defectives?

Solution $N = 20$, $n = 10$, $Np = 6$, $Nq = 14$, $r_1 = 3$.

$$\text{Prob (defects = 3)} = \frac{\binom{6}{3} \binom{14}{7}}{\binom{20}{10}} = \frac{6!}{3!(6 - 3)!} \frac{14!}{7!(14 - 7)!} \frac{10!(20 - 10)!}{20!}$$

$$= \frac{6!\,14!\,10!\,10!}{3!\,3!\,7!\,7!\,20!} = 0.372$$

Thus the probability of the customer receiving 3 defective components is 37.2%.

In the discrete distributions we have considered thus far, the probabilities of a 'success' and a 'failure' have been known. In the situation of isolated occurrences in a continuum this does not hold, and the Poisson distribution is used. Examples of this would be the number of faults in a length of cable or the number of breakdowns in a factory per year. In other words, isolated events occurring in a particular interval. The Poisson distribution is given by:

$$\text{prob } (r) = \frac{\lambda^r e^{-\lambda}}{r!} \tag{9.2}$$

Where lambda (λ) = average number of events in a specified interval.

r = the number of events occurring in the same interval.

Mean = lambda (λ) and variance = lambda (λ)

Example A maintenance department uses on average 16 spares of a particular machine component. To make up the stock the department orders on a weekly basis. What stock level must the department maintain so that the risk of running out of stock does not exceed 5%?

Solution The model is Poisson with lambda = 4/week. If the department stocks k components then the stock will run out when the demand exceeds k. This means:

$$\sum_{r=k+1}^{\infty} \frac{4^r e^{-r}}{r!} \leqslant 0.05$$

or

$$1 - \sum_{r=0}^{r} \frac{4^r e^{-r}}{r!} \leqslant 0.05$$

thus

$$\sum_{r=0}^{r} \frac{4^r e^{-r}}{r!} \geqslant 0.95$$

We successively accumulate the sum on the left hand side until it just exceeds 0.9. The calculation shows that $r = 8$ is the solution. This means that the department should order 8 components per week. Under the right conditions the Poisson distribution can be used to approximate the Binomial distribution. These conditions are that n is large and p small. This will now be illustrated by two examples.

Example (i)   A machine produces on average 0.5% defectives when
batches of 100 are tested. What is the probability that a
batch is free from defectives?

Solution   Binomial: $n = 100$, $p = 0.005$

$$p(0) = \frac{100!}{0!\,100!}\,(0.995)^{100} = 0.6058$$

Poisson: lambda $(\lambda) = np = 0.5$

$$p(0) = e^{-0.5} = 0.6055$$

So the percentage of defect-free batches is 61%.

Example (ii)   As example (i) but the batch size $= 10$ and the average
number of defects is 1 in 20. Then:

Binomial: $n = 10$, $p = 0.05$

$$p(0) = \frac{10!}{0!\,10!}\,(0.95)^{10} = 0.5987$$

Poisson: lambda $(\lambda) = np = 0.05$

$$p(0) = e^{-0.5} = 0.6055$$

For this example the percentage of batches free from defectives is 60%.
The Poisson approximation to the Binomial gives an error of about 20%.
So care must be taken when using the Poisson distribution as an
approximation to the Binomial distribution. The requirements are that
the batch size is large and the probability of a 'success' is small.
Mathematically, the Poisson distribution is a limiting case of the Binomial distribution.

Signals emanating from transducers are of a continuous nature. Naturally statistical analysis of such data will use continuous statistical
distributions. The dominant distribution is the Normal or Gaussian
distribution as shown in Figure 9.3. It is particularly applicable to
variations due to measurement errors or noise. Its dominant use in
analysis is due to the fact that the mean values of samples from any valid
distribution have a Normal distribution. It is fully determined by the
mean, 'mu (μ)', and variance, 'sigma^2 (σ^2)'. The probability distribution
is written as $N[mu\ (\mu),\ sigma^2\ (\sigma^2)]$ and is given by:

$$p(x) = (1/(\sigma\sqrt{(2\pi)}))e^{-[((x-\mu)^2)/(2\sigma^2)]} \tag{9.3}$$

In hypothesis testing one uses the standard Normal distribution. If we

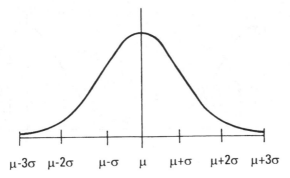

Fig. 9.3 Standard normal distribution.

change the variable, x to z, using the transformation:

$$z = \frac{x - \mu}{\sigma}$$

then z has a Normal distribution with a mean equal to zero and a variance equal to one, $N(0, 1)$.

$$p(z) = (1/(\sqrt{2\pi}))e^{-(z^2)/2} \qquad (9.4)$$

The integral of this expression is used in hypothesis testing and it is tabulated for values of z ranging from zero to three. As the distribution is symmetrical there is no need to tabulate values for z between minus three to zero.

Example   A machine produces components of mean diameter 15.35 mm with a standard deviation of 0.05 mm. The diameters are assumed to be normally distributed.

1. Find the probability that a component has a diameter between 15.37 mm and 15.44 mm.
2. If all the components that are outside the range 15.28 mm to 15.40 mm are rejected, what proportion of the components are rejected?

Solution (1)   The first task is to normalize the distribution to $N(0, 1)$. The limits of interest are shown in Figure 9.4 and calculated as:

$$z_1 = \frac{15.37 - 15.35}{0.05} \qquad z_2 = \frac{15.44 - 15.35}{0.05} = 1.8$$

The shaded area, is the area between 0 and 1.8, minus the area between 0 and 0.4. These are the areas read from the standard Normal distribution tables. The first of these areas is 0.4641 and the second is 0.1555. Therefore the required

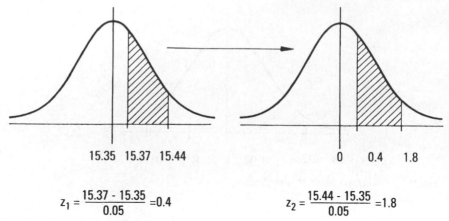

$$z_1 = \frac{15.37 - 15.35}{0.05} = 0.4 \qquad\qquad z_2 = \frac{15.44 - 15.35}{0.05} = 1.8$$

Fig. 9.4 Problem normalization.

area equals $0.4641 - 0.1555 = 0.3086$. Thus we can say that the probability of a component diameter lying between 15.37 mm and 15.44 mm is 30.9%.

Solution (2) In a similar manner, the area we require is shown in Figure 9.5 and consists of the area from -1.4 to 0.

$$z_1 = \frac{15.28 - 15.35}{0.05} = -1.4 \qquad z_2 = \frac{15.40 - 15.35}{0.05} = 1.0$$

which is the same as the area from 0 to 1.4, together with the area from 0 to 1. From the tables, these areas are 0.4192 and 0.3413 a total of 0.7605. So we can say that the percentage rejected is $100\% - 76.05\% = 24\%$.

While the Normal distribution holds the central position in statistical analysis, other distributions are used in particular situations. Two of the more important ones are the Exponential and Weibull distributions. The Exponential distribution is used as a statistical model of system failures. Its probability density function is:

$$p(t) = \frac{1}{\theta} e^{-t/\theta} \tag{9.5}$$

The Exponential only has one parameter, 'theta (θ)', and is shown in Figure 9.6(a). For convenience log/linear paper is used to plot the data. This results in a straight line plot, as shown in Figure 9.6(b). Its cumulative distribution is given by:

$$p(t) = 1 - e^{-t/\theta} \tag{9.6}$$

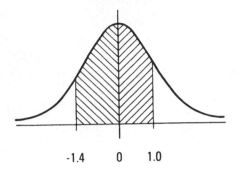

$$z_1 = \frac{15.28 - 15.35}{0.05} = -1.4 \qquad z_2 = \frac{15.40 - 15.35}{0.05} = 1.0$$

Fig. 9.5 Problem (ii) display.

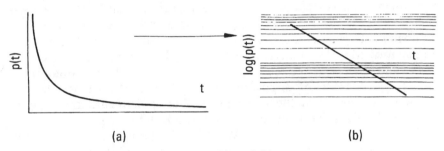

Fig. 9.6 Model linear transformation (a) and (b).

The Exponential distribution is a special case of the Weibull distribution. The Weibull can be a three- or two-parameter distribution and is used to model failures in components rather than assemblies, which use the Exponential. Its probability density function is given by:

$$p(x) = [(b/(\theta - x_0))(((x - x_0)/(\theta - x_0))^{b-1})]e^{-((x - x_0)/(\theta - x_0))^b} \qquad (9.7)$$

Where x_0 = expected minimum value of x – (this is set to zero for the two parameter model).

The cumulative Weibull distribution is:

$$P(x) = 1 - e^{-((x - x_0)/(\theta - x_0))^b} \qquad (9.8)$$

This distribution is shown in Figure 9.7(a) and the linearized cumulative plot is shown in Figure 9.7(b). The various available distributions, their applications, and typical practical cases are given in Table 9.1.

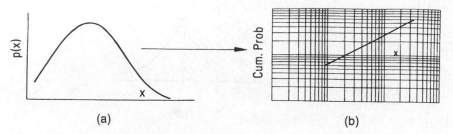

Fig. 9.7 Model linear transformation (a) and (b).

In situations where the underlying distribution is unknown, it is possible to use non-parametric statistics. We will briefly consider some of the popular techniques. These are the 'sign test', 'run test' and the 'Spearman rank correlation'. In the sign test $+$'s and $-$'s are used instead of quantitative data. It is a simple test to see if there is a difference between paired readings of two experiments. The test uses the Binomial distribution analyzing the sign of the differences of two data sets, x and y; $x_1 - y_1$, $x_2 - y_2$, $x_3 - y_3$, ... $x_n - y_n$, with the assumption that the probability of a positive or negative difference is 0.5. If a difference is zero then the usual practice is to disregard that pair of values and reduce the value of the sample size n by one.

The run test is often called the Wald-Wolfowitz run test. It is used to test the hypothesis that one set of data is consistently larger or smaller than another set. It is based on 'runs' where a run is defined as a succession of symbols followed by another symbol. Before any comparison is undertaken both sets of data are sorted in ascending order. For example, suppose we have a series of $+$'s and $-$'s from a comparison of two sets of data:

Like so: $+, +, +, -, +, -, -, -, +, +, -$

Here, we have a run of $+$'s (three), then a run of $-$'s (one), then a run of $+$'s (one), then a run of $-$'s (three), then a run of $+$'s (two), and finally a run of $-$'s (one). So in this example we have six runs. This value is then used in conjunction with tabulated values to test the hypothesis that there is no difference, statistically, between the two data sets.

The Spearman rank correlation is a very easy test of correlation between two variables. Both sets of data are ranked in increasing algebraic order, (not paired). Then the difference between the ranks is

Table 9.1 Distribution, features and applications

Distribution	Application	Cases
• Normal	Various physical properties	Rivet head diameters, tensile strength of alloys
• Log normal	Life phenomena where occurrences are concentrated in the tail end of range	Downtime of large number of electrical systems
• Weibull	Same as log normal. Also where failure rates vary over time	Wear-out life
• Exponential	The life systems. Failures in components that are independent of time (occur randomly)	Life to failure of machines
• Binominal	Number of defectives in a sample lot drawn from a large population. Go-nogo testing	Inspection for defectives in production of goods-in
• Hypergeometric	Same as the binomial but taking the sample may change the percentage defects left, i.e. small lot sizes	Probability of obtaining 10 good resistors from a lot of 100 that have 3% defectives
• Poisson	Events randomly occurring over time. The number of times an event occurs can be observed but the number of times it does not happen cannot be observed	Number of machine breakdowns in a plant. Dimensional errors in engineering drawings. Defects along a long wire

found, $d_i = y_i - x_i$. If n is the number of data points (pairs) the Spearman rank correlation is given by:

$$r_s = 1 - \frac{1}{n(n^2 - 1)} 6 \sum_{i=1}^{n} d_i^2 \qquad (9.9)$$

If $|r_s|$ tends to 1 then one can say there is a definite relationship between the two variables. Alternatively, if $|r_s|$ tends to 0, one can say that there is no relationship between the two variables. An important measure is r_s^2. For example if $r_s = 0.9$ say then $r_s^2 = 0.81$, and we can say that 81% of the variation of either variable is due to the correlation with the other. Also 19% of the variation cannot be explained.

9.3   TIME-BASED ANALYSIS

Signals from transducers or data taken over time are examples of data conveniently presented and analysed as a time series. There are two basic approaches in analysing time series, either in the time domain such as smoothing the data to reveal any underlying trends, and representing the series by a time-based model, or in the frequency domain by estimating the spectrum of the data.

Smoothing routines are evoked when one wishes to see the underlying trends in the data set rather than fitting a model. There are many situations where it is not sensible nor valid to represent data as a mathematical model. Smoothing reduces the variability of the data. The variability may be due to variations in the process being monitored or due to noise corrupting the underlying signal.

The moving average is probably the most widely used smoothing scheme. It is easily understood and applied. A window is wiped across the data and the data points revealed in the window are averaged. Alternatively, the data points can be considered to be unit masses and the distance from the independent axis of the centre of mass calculated. In Figure 9.8, if $x_i = 1, 2, \ldots n$ are the data set then successive values of a three-point moving average scheme are:

$$m_1 = \frac{x_1 + x_2 + x_3}{3}, \; m_2 = \frac{x_2 + x_3 + x_4}{3}, \; m_3 = \frac{x_3 + x_4 + x_5}{3}, \ldots$$

As the window width is increased so the number of points increases. Figure 9.9 shows examples of three- and five-point moving averages. It is best practice to place the successive average at the mid-point of the window. It is seen that the end points are lost. This loss becomes larger as the width of the window increases. Also the wider the window, the more the variability of the smoothed output reduces, as illustrated in

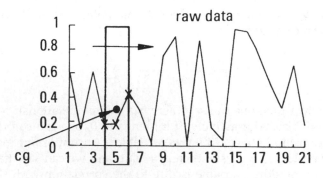

Fig. 9.8 Moving average window.

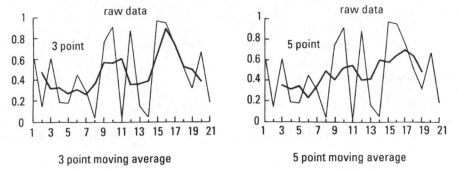

Fig. 9.9 3pt and 5pt moving average plots.

Figure 9.9. We can consider the scheme another way. The smoothed output can be considered as a weighted sum viz:

$$m_j = \frac{1}{3}x_{j-1} + \frac{1}{3}x_j + \frac{1}{3}x_{j+1}$$

where we have now centralized the output at the centre of the window as is done in practice. In general for the k point moving average we can write:

$$(m_k)_j = \sum_{i=j-(k/2)}^{i=j+(k/2)} w_j x_i \tag{9.10}$$

Where $w_j = 1/k$.

It can be shown that the reduction in variability is proportional to:

$$\sum 1/(k^2)$$

Also:

$$\sum 1/k = 1$$

Any smoothing scheme whose weights sum to unity is an averaging process. One can argue that the 'latest' data values should be given more weight than the 'older' data values. The moving average scheme gives equal weight to all the data points in the window. Alternative schemes have unequal weights that taper off giving less weight to the data as it ages. One such scheme is the exponential smoothing scheme. This can be represented as follows:

$$E_{n+1} = \alpha x_{n+1} + (1 - \alpha)E_n \tag{9.11}$$

where

$$E_{n+1} = (n + 1)\text{th smoothed output}$$

$$x_{n+1} = (n + 1)\text{th raw data point}$$

$$\text{alpha } (\alpha) = \text{smoothing constant}$$

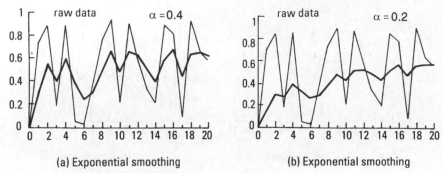

(a) Exponential smoothing   (b) Exponential smoothing

Fig. 9.10 Examples of exponential smoothing.

It can be shown that the data weights form a geometric progression (GP):

$$\alpha, \alpha(1 - \alpha), \alpha(1 - \alpha)^2, \alpha(1 - \alpha)^3, \alpha(1 - \alpha)^4 \ldots.$$

And as $\alpha < 1$ they taper off as the data ages.

Illustrated in Figures 9.10 (a) and (b) is the effect of α on the level of smoothing. The smaller α is, the greater the amount of smoothing. If we examine the weights we see that they taper off exponentially as shown in Figure 9.11. The exponential scheme uses all the data, unlike the moving average scheme, which 'loses' more and more end points as the window size increases. In the exponential scheme we can consider that the window stretches as the number of data points increase with time.

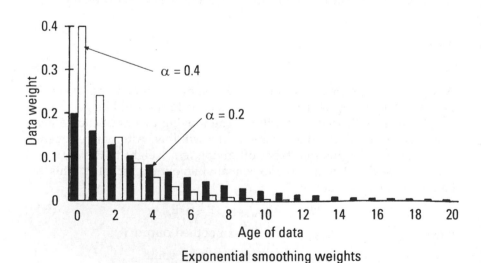

Exponential smoothing weights

Fig. 9.11 Data weight reduction as data ages.

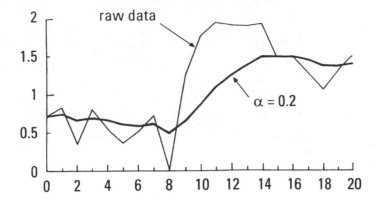

Exponential smoothing scheme response to step change

Fig. 9.12 Dynamic response of exponential smoothing.

Care must be taken in any smoothing scheme when the data set includes periods of changes in average level. Smoothing schemes introduce delays in response to such discontinuities. The plot in Figure 9.12 clearly shows the delay in the smoothed output when the data includes a step change.

9.4 MODELS

Models play an important role in the general field of maintenance. Process behaviour can be described using models which in turn can be used as templates against which performance can be judged. Mathematical models can take many forms. They can take the form of mathematical descriptions relating the independent and dependent variables, or dynamic models describing the fundamental dynamic behaviour of the system under consideration. Both these approaches assume there is an underlying law describing some particular characteristic behaviour.

Any measurements taken via transducers are likely to suffer some corruption due to inherent, or externally induced noise. In these circumstances, models are generally fitted to the data, by using a 'best fit' in the 'least squares' sense. The practitioner has to choose the model, then the least squares technique will produce the model's coefficients, so as to minimize the sum of the errors squared between the model and the data. An alternative to fitting a deterministic model is to fit a stochastic model. Stochastic models have the capability of producing a model of the system's behaviour, that is a typical realization, rather than an attempt to reproduce the original measured data. In other words, a stochastic

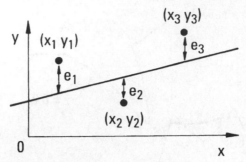

Fig. 9.13 Error display for linear least square modelling.

model produces different data each time, but the information produced would all have the same basic characteristics.

The fundamental curve fitting technique is the 'least squares' method. The technique will be illustrated by fitting a linear model (straight line), to a set of data, as shown in Figure 9.13. The equation of the line is:

$$y = mx + c \tag{9.12}$$

Where m = the slope of the line

and c = the intercept on the y-axis

Then the error between the actual data point and the model is:

$$e_i = y_i - (mx_i + c)$$

The sum of error squared is given by:

$$S = \sum_{i=1}^{n} [y_i - (mx_i + c)]^2$$

The values of m and c that minimize S are calculated by setting the partial derivatives of S, with respect to m and c, to zero. This procedure leads to the following expressions for m and c:

$$m = \frac{n[\sum_{i=1}^{n} x_i y_i] - [\sum_{i=1}^{n} x_i][\sum_{i=1}^{n} y_i]}{n[[\sum_{i=1}^{n} x_i^2]] - [\sum_{i=1}^{n} x_i]^2}$$

$$c = \frac{[\sum_{i=1}^{n} x_i^2][\sum_{i=1}^{n} y_i] - [\sum_{i=1}^{n} x_i y_i][\sum_{i=1}^{n} x_i]}{n[[\sum_{i=1}^{n} x_i^2]] - [\sum_{i=1}^{n} x_i]^2}$$

One important point has to be made. The method of least squares will not give you the best model but rather the fit for the model you choose. As the number of degrees of freedom of the model increases so the sum of error squared reduces. In fact if the data set has n values then an $(n-1)$th degree polynomial will fit exactly giving a sum of error squared of zero.

Table 9.2 Some nonlinear models and their transformations

Model	Linearised	Y axis Y	X axis X	Intercept	Slope
$y = Ae^{-bx}$	$\ln(y) = \ln(A) - b.x$	$\ln(y)$	x	$\ln(A)$	b
$y = Ax^b$	$\ln(y) = \ln(A) + b\ln(x)$	$\ln(y)$	$\ln(x)$	$\ln(A)$	b
$y = 1 - e^{-bx}$	$\ln\dfrac{1}{1-y} = b.x$	$\ln\dfrac{1}{1-y}$	x	0	b
$y = a + b\sqrt{x}$	$y = a + bX$	y	$\sqrt{x}$	a	b
$y = a + \dfrac{b}{x}$	$y = a + bX$	y	$\dfrac{1}{x}$	a	b

The linear model is one of the general set of polynomial fits which are easily calculated using the method illustrated above. For non-linear models the basic technique is the same but the data is transformed to give a linear model. For example if we decided that the appropriate model was:

$$y = Ae^{-ax} \qquad (9.13)$$

This model is linearized by taking logs (natural to the base e), of both sides giving:

$$\ln(y) = \ln(A) - \alpha x$$

If $\ln(y)$ is plotted against x we get a straight line with a slope equal to $-\alpha$ and the intercept on the vertical axis equal to $\ln(A)$. In general the nonlinear model is transformed to the linear model:

$$y = f(x) \text{ tending to } Y = mX + c$$

Typical nonlinear models and their transformations are shown in Table 9.2. This technique of linearising is commonly used in engineering. However, a work of caution is necessary. The natural variation about the curve that we get with practical data is also transformed, and this is illustrated in Figure 9.14.

Stochastic models are probability based models for time series. An observed time series is considered a realization, or one example of the infinite set of possible time series that could emanate from the process. This is shown in Figure 9.15. The infinite set of possible time series is called the ensemble. The two plots in Figure 9.15, clearly show different plots generated by the same stochastic model. There are two basic types of stochastic models – moving average and autoregressive. There are variations on these two such as a mixed moving average and autoregressive, and an integrated mixed model. The moving average model is

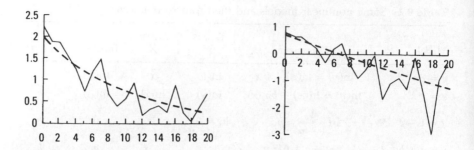

Fig. 9.14 Effect on error distribution due to transformation.

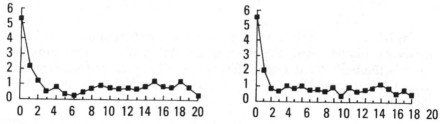

Fig. 9.15 Two realizations of the same stochastic model.

given below:

$$x_t = \beta_0 z_t + \beta_1 z_{t-1} + \beta_2 z_{t-2} + \cdots + \beta_q z_q \qquad (9.14)$$

Where

$$x_t = \text{the current model output;}$$

$z_t = $ the current variate drawn from a stationary random process;

$$\beta_j = \text{the } j\text{th model parameter.}$$

Identification of the moving average type model is difficult and beyond the scope of this review chapter.

The autoregressive model can be represented as follows:

$$x_t = \alpha_1 x_{t-1} + \alpha_2 x_{t-2} + \alpha_3 x_{t-3} + \cdots + \alpha_p x_{t-p} + z_t \qquad (9.15)$$

Where:

$x_{t-1} = $ the $(t-1)$th model output;

$z_t = $ the current variate drawn from a stationary random process;

$\alpha_j = j$th model parameter.

For the autoregressive model the time series' autocorrelation function is a useful vehicle for its initial identification. It is unlikely that stochastic models will ever become central in maintenance analysis. However, their probabilistic nature makes them likely candidates for future consideration in the move towards the use of fuzzy-based expert systems.

9.5   SPECTRUM ANALYSIS

Others chapters in this book cover this topic in great detail and expertise. Here we will attempt to show the basic mathematical characteristic of the Fourier representation of signals generated in numerous condition monitoring systems. Basically any continuous data or signal can be represented by an infinite series of sinusoids together and a constant. Consider the signal shown in Figure 9.16. Suppose we choose a polynomial model, say:

$$y = a_2 x^2 + a_1 x + a_0 \qquad (9.16)$$

If we decided that this was not a good fit we could try a higher order polynomial, say:

$$y = a_3 x^3 + a_2 x^2 + a_1 x + a_0$$

We would have to recalculate a_2, a_1, and a_0 because these would change. These polynomials are not orthogonal (at right angles), to each other. But sine and cosine functions are orthogonal. This means as we increase the order of the Fourier series model, we do not have to keep recalculating the previous terms. Consider the simple functions displayed in the left-hand column of Figure 9.17.

The first function is a constant, then $\sin(\omega_0 t)$, then $\sin(3\omega_0 t)$, then $\cos(\omega_0 t)$ and finally $\cos(2\omega_0 t)$. As we add or subtract each term in turn the result is shown in the right-hand column. We see that even with a few sine and cosine terms quite complicated signals can be generated. We have represented the function as:

$$f(t) = 0.5 + \sin(\omega_0 t) + \sin(3\omega_0 t) + \cos(\omega_0 t) - \cos(2\omega_0 t) \qquad (9.17)$$

Fig. 9.16   Arbitrary signal.

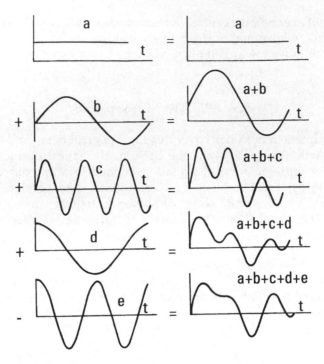

Fig. 9.17 Simple functions.

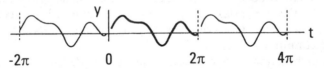

Fig. 9.18 Generated signal.

Due to the cyclic nature of the model the signal generated is as shown in Figure 9.18. Thus we can see how such a representation is useful and natural when dealing with rotating elements.

9.6 CONCLUSIONS

From the above discussion, the best advice is only analyse data if there is a real need. Too often, copious outputs from statistical or signal analysis are generated because computer packages can easily produce them. Simplicity is best. Often simple histograms or time plots are sufficient to answer the questions posed. All methods have their particular characteristics and remember that there are underlying assumptions for all models.

9.7 BIBLIOGRAPHY

Bowker, A.H. and Lieberman, G.J. (1963) *Engineering Statistics*, Prentice-Hall, Englewood Cliffs, New Jersey, USA.

Brown, R.G. (1963) *Smoothing, Forecasting and Prediction*, Prentice-Hall, Englewood Cliffs, New Jersey, USA.

Chatfield, C. (1989) *The Analysis of Time Series*, Chapman & Hall, London, UK.

Dixon, W.J. and Massey, F.J. (1969) *Introduction to Statistical Analysis*, McGraw-Hill, London, UK.

Draper, N.R. and Smith, H. (1966) *Applied Regression Analysis*, John Wiley & Sons, New York, USA.

Duncan, A.J. (1959) *Quality Control and Industrial Statistics*, Richard D. Irwin, Homewood, Illinois, USA.

Kraniauskas, P. (1992) *Transforms in Signals and Systems*, Addison-Wesley, New York, USA.

Moroney, M.J. (1965) *Facts from Figures*, Penguin Books, Harmondsworth, UK.

Walsh, J.E. (1962) *Handbook of Nonparametric Statistics*, Van Nostrand, New York, USA.

CHAPTER 10

Commercial applications of performance monitoring

A. Davies, K.F. Martin and P. Thorpe

Systems Division, School of Engineering,

University of Wales Cardiff (UWC), PO Box 688,

Queens Buildings, Newport Road, Cardiff, CF2 3TE, UK

10.1 INTRODUCTION

As outlined in the foregoing chapters of this section of the handbook, performance monitoring is normally undertaken to ensure that an item of equipment or machinery is operating correctly, to a required standard or known specification. As such, it is frequently undertaken immediately after manufacture, final assembly, installation or commissioning, and often under the guise of production, quality or safety testing.

In this context therefore, performance monitoring can be regarded as:

> The measurement, and subsequent assessment of, chosen system parameters which, if satisfactory, ensure that a machine or item of equipment is operating correctly against a known specification, or at a required level of in-service performance.

The initial data set, recorded during production testing, may be regarded as providing the 'as-new' characteristic equipment signature, against which all subsequent test information can, if required, be compared. Accordingly, should condition monitoring be carried out during an equipment's in-service life, a comparison of the data, made available by monitoring, to the 'as-new' signature, will provide not only a snapshot of plant performance at that point in time, but also extremely valuable trend information.

Handbook of Condition Monitoring
Edited by A. Davies
Published in 1998 by Chapman & Hall, London. ISBN 0 412 61320 4.

Thus performance monitoring may be considered as a two part technique, and the objective of in-service monitoring must be:

To assess the value of known performance parameters, in order to check firstly that the equipment is operating within its established performance specification, and secondly, to identify from the monitored data, trend information which may be used to predict when a loss of performance is likely to occur.

The technique implicitly requires two preconditions to ensure a successful application, i.e:

1. 'machinery or equipment which is stable in normal operation, and whose stability is reflected in the parameters under surveillance';
2. 'instrumentation whose measurements in respect of the machinery or equipment, may be recorded manually or automatically, on a periodic or continuous basis'.

Providing these conditions are met, any change in the normal operation of the system is easily detected, and through trend analysis may reveal the presence of a potential failure.

Critical machine components such as seals, bearings, ballscrews, pumps and motors etc. are, as shown in this book, typical application areas for performance monitoring. In addition, and as shown in earlier chapter examples, the technique can also use methods that are to some extent designed for a specific task. Accordingly, these applications may employ one or more of a whole spectrum of available sensing arrangements, as the operating principle to effect performance monitoring (Neale, 1979).

One prime advantage of the technique is its ability to provide during normal operation, an indication of the efficiency of the system or process under surveillance. As a consequence, if the sensor measurements are linked to an adaptive control unit, minor excursions from the optimal state may be detected and corrected by parameter adjustment, while a failure causing a deviation which defies rectification can readily be identified. Disadvantages normally associated with the technique are its possible insensitivity in attempting to relate output deviations to specific incipient component failures, and the aforesaid necessity to 'fingerprint' initially the machine under normal operating conditions, to provide the 'as-new' performance signature.

For complex machinery both these disadvantages give rise to problems in the design of practical condition monitoring units. The relationship between 'global' performance measures and a specific machine, component or software error is often obscure and awkward to identify accurately. Consequently, care is required in the selection of suitable parameters, as often a multiplicity of performance measures may be required to identify specific failures, and these can well have attendant disadvantages such as high cost, complexity and low reliability.

The definition and accurate specification of 'normal' operating conditions is also problematic in some cases. Accordingly, and as this is the 'baseline' for deviation and trend measurement from the initial 'as-new' signature, errors may occur which give rise to false alarms. This may well result in a consequent loss of confidence in both the condition monitoring system and the performance monitoring technique. However, as the following examples show, providing care is taken in the design of the system, performance monitoring can be successfully applied.

10.2 BEARING PERFORMANCE

In order to illustrate, in a commercial sense, the potential for the application of performance monitoring to industrial equipment, the following sections outline its use on a machine tool system and some of the common engineering components found therein.

Due to their widespread use in industrial machinery, rolling element bearings, for example, have received considerable attention in respect of condition monitoring. Various methods for monitoring these components being reviewed by Thorpe (1990), who confirmed that a combination of 'spectrum area' and the 'enveloping technique' was capable of detecting bearing faults. Other investigations by Connolly (1990) and others (Thorpe, 1990), have also indicated that a combination of 'Kurtosis measurement' and the 'Envelope Technique' will provide information on the condition of rolling element bearings.

In this example, the latter pairing of techniques are used to illustrate the performance monitoring of a machine tool's spindle bearings. As vibration is the measured parameter, it should be explained that the 'Kurtosis' value, is a measure of the impulsive nature of the signal (x) and may be defined as:

$$K = \int_{-\infty}^{\infty} \frac{[(x - \bar{x})^4 f(x)]}{\sigma^4} dx \qquad (10.1)$$

A less precise, yet easier to compute equation is:

$$K = (1/(n\sigma^4)) \sum_{i=1}^{n} (x_i - \bar{x})^4 \qquad (10.2)$$

Where: $f(x)$ is the probability density function of x;

sigma (σ) is the standard deviation of x;

$\bar{x}$ is the average value of x;

n is the sample size.

The use of this technique for monitoring bearings was first proposed by (Stewart and Dyer, 1978). Their results suggested that the use of various frequency bands for 'Kurtosis' measurement would allow continuous fault tracking, thereby permitting surveillance from the initial, to the final stages of the bearing defect. They also found that there was some speed independence for 'Kurtosis' measurements, particularly for 'as-new' bearings, whose Kurtosis value is normally 3. This offered the possibility of dispensing with the necessity of 'fingerprinting' each new bearing in a machinery performance monitoring scheme, as the standard 'Kurtosis' value of every new bearing would be 3.

The 'Envelope' analysis of a band-pass filtered signal is also a well-established condition monitoring technique (McFadden and Smith, 1984); one, in fact, which is currently incorporated into various commercially available frequency analysers. The technique involves initially passing an amplified accelerometer signal through a band pass filter which normally has a centre frequency corresponding to one of the resonant frequencies of the System Under Test (SUT). The output of this filter is then rectified and processed via a low-pass filter, which essentially demodulates any amplitude variation.

Subsequent frequency analysis of the demodulated signal then enables the detection of problems on any of the characteristic bearing defect frequencies. A particular advantage of the 'Enveloping' technique is its insensitivity to outside disturbing vibrations (noise), and to electrical interference (RFI). In addition, it is also capable of monitoring defects at low shaft speeds, thus making it suitable for monitoring bearings on slowly rotating industrial equipment.

The monitoring system used in this example is shown in Figure 10.1, where the signal from the accelerometer is initially passed through a charge preamplifier, and then used for both the 'Kurtosis' and 'Enveloping' analysis. For 'Kurtosis', the signal is passed through a filter box which contains both a low-pass, (0–5 kiloHertz (kHz)) filter and a high-pass, (>5 kHz) filter. The output from these, together with the signals from the spindle servo, are then fed into a switch box.

As shown in the figure, the purpose of the switch box is to allow the two channel Analogue to Digital converter (A/D) to receive data from any combination of the five inputs. For 'Enveloping', a second signal is obtained from the accelerometer preamplifier, and fed via the envelope detector to a spectrum analyser. The frequency analysis of the envelope signal is then fed to the controlling personal computer (PC), which utilizes the commercial data acquisition software (ASYST).

The PC controls the switch box, A/D converter and the spectrum analyser to effect fully automatic data acquisition and analysis. In the preliminary tests detailed here, the accelerometer was attached to the front of the spindle head casting and level with the lower bearing set. A more detailed explanation of the monitoring system and investigation

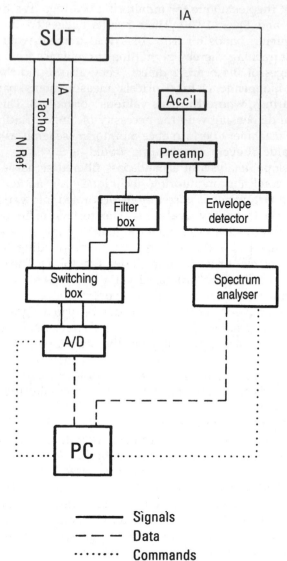

Fig. 10.1 Spindle bearing monitoring setup.

into the best position for the accelerometer can be found in Thorpe (1990).

Preliminary tests were performed to determine the optimum values for both 'Kurtosis' sample time and sample rate. The sample time being the number of revolutions of the spindle during which the vibration data is collected. For the in-process acquisition of data this time should be minimized and, accordingly, only 50 revolutions of vibration data were collected, at a sample rate of 83 kHz and a programmed spindle speed of 2000 Revolutions Per Minute (RPM). This gave a data sample of 128 000 points.

A proximity sensor was utilized to detect shaft RPM, Figure 10.2 illustrating a portion of its output and the corresponding vibration signal. The detailed analysis of this data gave the variation in 'Kurtosis' values for different sample sizes (Thorpe, 1990) and, based on this information, subsequent tests adopted a sample size of 10 spindle revolutions. In the initial test the sampling rate was set at 200 kHz; however, to reduce computing time, 4096 points were extracted from the original 128 000 and the 'Kurtosis' value calculated. This reduced the sampling rate to 6.25 kHz.

The 'Enveloping', as previously outlined, involved passing the amplified accelerometer signal through a band pass filter, and envelope detector before being analysed in the frequency domain. The centre frequency of the band pass filter was selected as one of the resonant freqencies of the spindle. Figure 10.3 shows the full vibration spectrum for spindle speeds of 1000 and 2000 RPM. Using this information, 20 kHz was selected as the centre of the bandpass filter, with the spindle speed for the envelope test set at 2000 RPM and the bandwidth of the filter as 4.6 kHz.

To establish the 'as-new' signatures for the spindle bearings, the following test procedures were adopted: for the 'Kurtosis' value, spindle

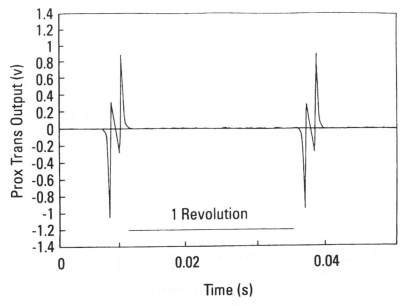

Fig. 10.2a Proximity sensor output.

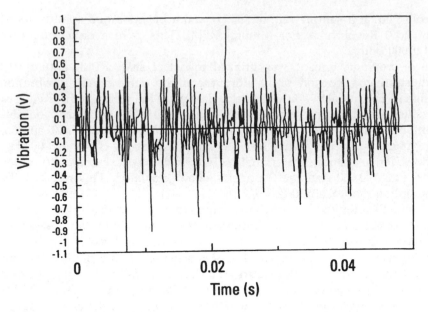

Fig. 10.2b   Spindle vibration signal.

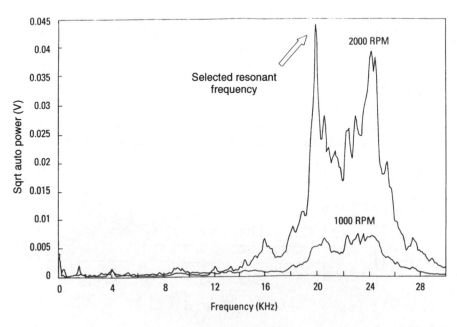

Fig. 10.3   Full-range spectrum for 'as new' spindle system.

speed constant in the range 1000–2000 RPM with clockwise rotation and no cutting; for 'enveloping', spindle speed constant at 2000 RPM with clockwise rotation and no cutting. Defects were introduced on the bearing components to simulate faults via the process of spark engraving, and tests were subsequently conducted, utilizing the same procedures, to try and identify these failures.

In addition, reference tests were carried out to investigate the effects of variations in machine temperature, speed and stripdown. These confirmed that no significant effects were experienced in respect of these parameters, on the 'Kurtosis' value or 'Envelope' spectrum generated from the monitored signals. The 'Kurtosis' value for 'as-new' bearings were confirmed as having a mean of 3 with a range of approximately ±0.5 (Thorpe, 1990).

Figure 10.4 illustrates the movement in 'Kurtosis' values found for three conditions of bearing life. Stage A corresponds with an 'as new' bearing having a 'Kurtosis' value of around 3. Stage B shows the effect on this value of a slight defect on the inner race of one of the lower spindle bearings; and stage C a deeper defect on the same bearing.

Table 10.1 details the characteristic frequencies associated with the bearings and the spindle shaft. These values are calculated using the standard equations for bearing defect frequencies as given in Collacot (1979). Figure 10.5 summarizes the 'Envelope' spectra for each stage of

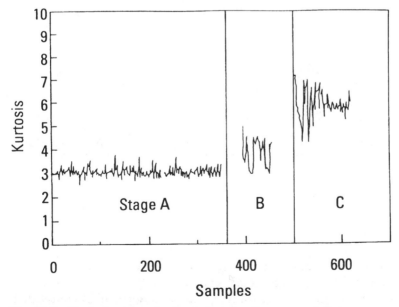

Fig. 10.4 Kurtosis values for lower bearing inner race defect.

Table 10.1 Discrete spindle and bearing frequencies

Defect		Defect number	Discrete freq. (*w)	Freq-Hz @ 2040 RPM
Spindle	1/2	1	0.5	17
	1	2	1	34
	2	3	2	68
Lower	R.E.	4	13.90	473
bearing	I.R.	5	10.23	348
	O.R.	6	7.77	264
Upper	R.E.	7	13.67	465
bearing	I.R.	8	9.68	329
	O.R.	9	7.32	249

R.E. = Rolling element; I.R. = Inner race; O.R. = Outer race; w = Spindle shaft speed (Hz).

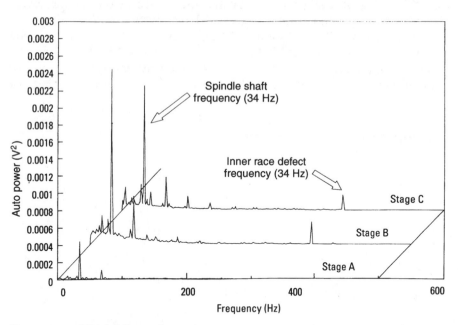

Fig. 10.5   Waterfall envelope spectrum of the three stages of lower bearing inner race defects.

bearing condition as above, with stage A being an 'as-new' bearing, stage B corresponding to the slight defect and stage C the deeper defect.

Each spectrum shown in Figure 10.5 is the average of all the individual 'Envelope' spectrums captured during each stage. The spectrum representing stage B, for example, is the average of six separate spectra

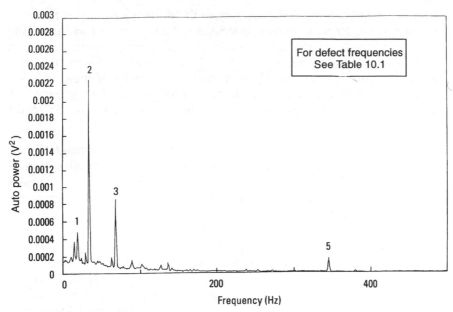

Fig. 10.6 Single envelope spectrum for lower bearing inner race defects.

captured during this test stage. Figure 10.6 illustrates one such individual spectrum for stage B. These figures show the prominent frequencies associated with a defect on the inner race of the bearing, which are:

- 0.5 × the spindle shaft frequency − 17 Hz;
- 1.0 × the spindle shaft frequency − 34 Hz;
- 2.0 × the spindle shaft frequency − 68 Hz;
- lower bearing inner race defect frequency − 348 Hz.

Due to the constant speed of 2000 RPM used for all the 'Envelope' tests in this example, it was only necessary to analyse the discrete frequencies outlined in Table 10.1, for the majority of faults to be recognized. When similar defects to those outlined above were introduced to the outer race of one of the lower set of spindle bearings, and equivalent tests performed, similar results to those given above were found; thus illustrating the effectiveness of performance monitoring in this application (Thorpe, 1990).

10.3   BALLSCREW PERFORMANCE

The specification of a positioning mechanism usually dictates a requirement in its design for the use of accurate and mechanically stiff components. These components are required to ensure a satisfactory level of performance during the service life of the system. It is therefore not surprising to observe that recirculating ballscrews, which have both these attributes, are extensively used in positioning system design. In CNC machine tool applications where high accuracy is essential during component machining, the reliability and performance of these units are crucial, and thus they are ideal candidates for performance monitoring; a suggestion which may be further justified by consideration of the following.

- Ballscrews are critical components in the design of a machine tool system for they translate rotary motor motion into linear table motion. They are high value items and are not used by designers unless accuracy, stiffness and efficiency are an important requirement in the positioning system design.
- Because of the task they perform, ballscrews are often embedded in carefully set up assemblies. Replacement on failure, therefore, involves a major maintenance strip down and can, on a typical CNC machining centre, take in excess of three days. This is not the sort of activity to embark upon at short notice or during a busy production schedule. Prior notice of impending failure is therefore desirable in respect of these components.
- The prospect of totally unmanned machining in 'flexible' factories makes the comprehensive diagnostic monitoring of machine tool systems an urgent requirement (Davies, 1990). In this respect ballscrews may offer a useful source of information relating to the operating conditions in their immediate environment. This may include data on the state of the slides, guideways and the lubrication systems on the machine.

Ballscrews are manufactured to provide very high standards of accuracy in a positioning system and they can be preloaded, in some cases adjustably, to provide good stiffness characteristics and backlash-free operation. They are extremely efficient linear to rotary motion converters via the rolling contact of ball bearings on hardened steel tracks. The preloading facility is of major importance, for it can be used to set the screw loading point, as shown in Figure 10.7, to the top end of its non-linear force deflection characteristic on final assembly.

As a consequence, the preloaded ballscrew is always suffering a degree of elastic deflection, and tends to act as a compressed spring in which the screw nuts force the balls against the screw track. This takes up any slack, and allows for moderate amounts of wear to occur in the screw prior to any major deterioration in stiffness, or the onset of backlash. Wear in operation will reduce the screw's performance by gradually degrading its preload, initially to a point where the unit's stiffness falls outside its specification, and ultimately, when the preload is entirely lost, to the point where backlash sets in. Thus the preload

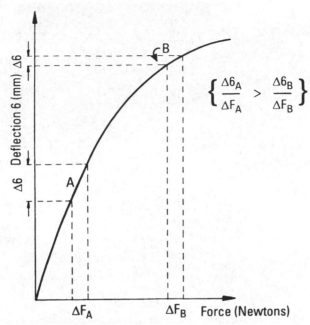

Fig. 10.7 Typical ballscrew force/deflection characteristic.

value makes an ideal parameter for assessing screw condition via a performance monitoring system.

In this example the preload, which is determined by the axial angular displacement between the nuts, is set up by means of an annulus vernier. In other ballscrew designs, different arrangements for preloading the screw are possible and these have been comprehensively reviewed by Hildebrant (1981). Several different transducers have been used to measure the variation in preload force. These include piezoelectric types (Goldie, 1982), strain gauges (Harris, Florio and Davies, 1987) and accelerometers (Drake, 1988). Of these, the most successful in practice have been the strain gauge transducers which can be internally or externally mounted. To outline their use in performance monitoring, the internal type is now used as a vehicle to explain the scheme.

The internal strain gauge transducer which is used to measure the variation in preload, is shown in Figure 10.8. Four strain gauges are evenly distributed around the external circumference of a steel ring which is mounted between the pressure faces of the ballscrew nuts. The ring's positional stability is assured by locating it over the annulus vernier. A stout rubber coating on the inside surface is intended to allow it to grip the vernier and so prevent rotation or transverse movement while at the same time being sufficiently flexible as to not disturb the

Fig. 10.8 Preload transducer.

ring's stress pattern. Figure 10.9 shows the unit in position on the ballscrew with its protective sleeve removed prior to installation.

The strain gauges are aligned with the screw's longitudinal axis and measure the strains induced into the ring when it is compressed by the preloaded nuts during their translation along the screw. Four gauges are used so that when opposite pairs are aligned with the ballscrew's polar axis, information is readily available about unbalanced couples acting along the screw. These couples could arise for example in a machine tool, by assembly misalignment, unevenly worn slideways, foreign body ingress or lubrication failure. For basic preload measurement, when the effects of unbalanced couples are negligible, the circuit shown in Figure 10.10 has been used to process the strain data.

In this arrangement, any transverse forces causing approximately equal and opposite strain changes in the series connected gauges will produce a zero net bias shift. Axial force variations will, however, similarly affect all four gauges and consequently the bias will shift appropriately. The transducer arrangement has been calibrated using a test rig, which allows the relationship between axial force and sensor output to be investigated, together with the effects of orthogonal moments via the use of a rotating load arm. A typical calibration plot for the transducer is shown in Figure 10.11.

Fig. 10.9 Ballscrew transducer and nut assembly.

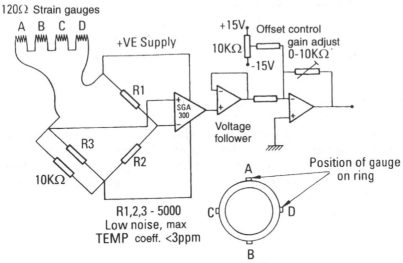

Fig. 10.10 Interface circuitry to obtain strain signals.

Unfortunately, strain gauges will not only respond to the strain in the surface to which they are bonded, but also to temperature variation. The gauge resistance changes with temperature and this alters the strain reading. At small actual strain levels as in this application, less than 20 microstrain, the effect is overwhelming, and thus requires pretest trans-

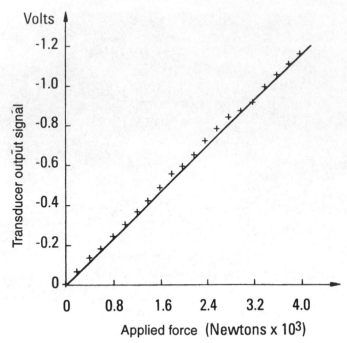

Fig. 10.11   Transducer calibration curve.

ducer calibration to fix the floating preload datum. This may be accomplished by externally loading the ballscrew in such a way that one of the nuts and hence the transducer ring is relieved of the preload.

Having established a zero preload datum, the wear test on the ballscrew is completed by simply moving the nut assembly along the screw length and recording the transducer signal. This signal can then be used in conjunction with the transducer calibration for comparison with the zero datum, or the 'as-new' ballscrew signature, thus evaluating how much preload is left in any position along the ballscrew, or how much wear has taken place since installation.

Given that the ballscrew nut and sensor assembly has a reasonably high thermal mass, small ambient temperature changes are unlikely to affect the validity of the datum providing the whole test is completed within a short time interval. This is quite feasible under computer control and a typical ballscrew preload signature is shown in Figure 10.12. The four strain gauges utilized in this sensor design can also be diametrically paired, to give information about any unbalanced couples acting on the ballscrew axis. The configuration which has been used to exploit this feature is shown in Figure 10.13.

It may be expected that a moment applied at different angular positions about the ring's central axis will cause uneven but symmetrical

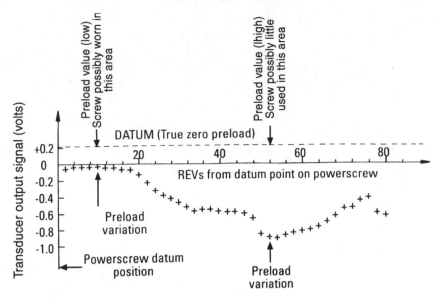

Fig. 10.12 Ballscrew preload signature (single clockwise traverse).

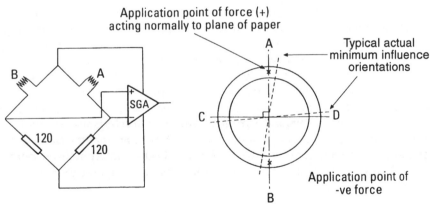

Fig. 10.13 Strain gauge configuration to detect unbalanced couples.

strain changes at the gauge positions. This is illustrated in Figure 10.14 where a torque applied on the 'AB' orientation should, in theory, produce minimal or no strain changes on the 'CD' line and vice versa. However, gauge positioning inaccuracies and uneven axial loading on practical transducers, do cause deviations from the ideal orthogonal relationship of these 'minimum influence' orientations.

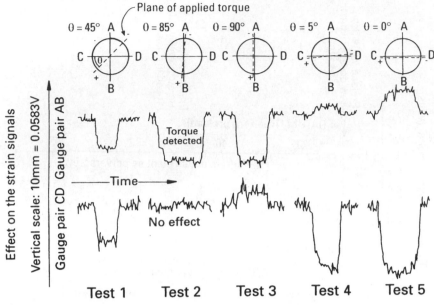

Fig. 10.14   Transducer response to external unbalancing torques.

The effects are illustrated in Figure 10.14, which shows the simultaneous responses of two diametrical gauge pairs on the prototype ring, to a 12 Newton metre (Nm) torque superimposed on a steady 2000 Newton (N) axial force at various angular positions. The discriminatory capability of the prototype system is clearly demonstrated and its sensitivity is quite remarkable. In tests the prototype ring exhibited a capability of detecting torques down to 0.68 Newton metres (Harris, Florio and Davies, 1987). Currently, these characteristics are under investigation for use in machine tool applications and may be capable of diagnosing uneven slideway wear, foreign body contamination, lubrication failure or loss of guideway stiffness in the yaw and pitch planes of a positioning system.

10.4   PUMP PERFORMANCE

Deciding what constitutes an item for inclusion in a condition monitoring system is an important consideration during the design process. An item could be a system, a component within a system, or a component within a component. In a machine-tool coolant system, for example, an item could be the whole coolant system, the pump unit, a bearing within the pump or a seal anywhere in the system. Obvious criteria for deciding what to monitor are, first, the critical components and secondly the high-frequency failure items.

Performance monitoring systems such as those described in this book do have the ability to predict failure via trend analysis. Unfortunately, the majority also have the disadvantage that a great deal of data needs to be collected and/or stored, for the system to monitor and predict failure accurately on a number of items within an overall surveillance scheme. Accordingly, methods by which the amount of data required can be reduced are an active research area, and such techniques are usually based upon simplifying procedures.

Not carrying out any diagnostic tests until the performance of the system has deteriorated to a measurable degree, or hit a tolerance threshold is one such procedure, which permits the minimization of data storage. In this example, to determine performance, a number of parameters are defined which are based on the steady-state characteristics of pump outlet pressure and motor temperature. Performance is then defined as the lowest value of any of these parameters at any particular time (Martin and Thorpe, 1990).

The performance information available from the monitoring system includes a prediction of the time to failure and the likely cause of the failure. When pump performance drops to a preset value, warnings are issued by the monitoring system and diagnostic measurements are taken. This consists of the measurement of the transient response of the pump outlet pressure when the flow valve is closed, the monitoring system being interconnected with the machine controller to ensure correct timing.

The transient response is then curve fitted to produce the coefficients of a polynomial by use of orthogonal polynomials (Bevington, 1969). These coefficients are then used in a cross-product fault location technique (Williams, 1985), to indicate the likely fault. Use of the technique requires a fault dictionary which is generated from known or simulated System Under Test (SUT) faults. In this example, tests on the machine indicated that the steady-state values of pump outlet pressure were sensitive to simulated faults such as:

- the partial opening of a control valve;
- the partial blockage of pump outlet pipes;
- the blockage of a filter;
- the partial blockage of the pump inlet pipe but not to drainage of coolant level.

Initially, it was therefore decided to monitor the pump outlet pressure with the flow valve in both the open (P1) and closed (P2) positions, plus the coolant reservoir level (L1).

The pump performance parameters were then defined as:

Open Flow Valve (OFV) = {[P1 (Actual)/P1 (As New)] × 100%} (10.3) where P1 (As New) = 307 kiloPascals (kPa) and

$$\text{Closed Flow Valve (CFV)} = \frac{[\text{P2 (Actual)} - \text{P1 (Actual)}]}{[\text{P2 (As New)} - \text{P1 (Actual)}]} \times 100\% \qquad (10.4)$$

where P2 (as new) = 561 kiloPascals.

In addition, the coolant reservoir level performance was defined as:

$$\text{Coolant Level (CL)} = \frac{[\text{L1 (Actual)} - \text{L1 (Failure)}]}{[\text{L1 (Good)} - \text{L1 (Failure)}]} \times 100\% \qquad (10.5)$$

OFV is sensitive to the pump condition and partial blockages in the pipe or filter, while CFV is a measure of the increase caused in the pump outlet pressure when the control valve is closed. By using the difference between P2 and P1, the effect of pump performance on CFV is reduced and makes it more sensitive to trends in the system relief valve. CL is a direct measure of the amount of coolant retained in the coolant reservoir.

These three parameters will not be sensitive to the relief valve seizing, the pump motor overheating or to the coolant quality. The first two could lead to catastrophic failure, and thus performance measures are desirable. Hence:

$$\text{Maximum Working Pressure (MWP)} = \frac{[\text{Pmax} - \text{P2 (Actual)}]}{[\text{Pmax} - \text{P2 (As New)}]} \times 100\%$$
$$(10.6)$$

Where Pmax is the maximum working pressure for the coolant system (2000 kiloPascals or 290 pounds per square inch).

The pump motor has a built-in thermometer cut out set at 45°C (Tmax), and although tests showed that at a given test point, the motor temperature (T) was affected by previous operating states, it was thought that motor temperature might provide some useful information.

Accordingly motor temperature performance is given by:

$$\text{Motor Temperature (MT)} = \frac{[\text{Tmax} - \text{T (Actual)}]}{[\text{Tmax} - \text{T (As New)}]} \times 100\% \qquad (10.7)$$

The overall coolant system performance at any point in time can now be defined as the lowest value of any of the above performance measures. Obviously, over time and after sufficient monitoring has taken place, trends in these performance parameters can be analysed and a time to failure for each determined. In this example, we are mainly concerned with illustrating their use in connection with pump performance, and for the interested reader, details of the full scheme are available in Martin and Thorpe (1990).

Additional diagnostic information in respect of the pump is also available via the pressure ripple (Silva, 1986), and as shown in Figure 10.15, this may be seen in the transducer response and the analysis of

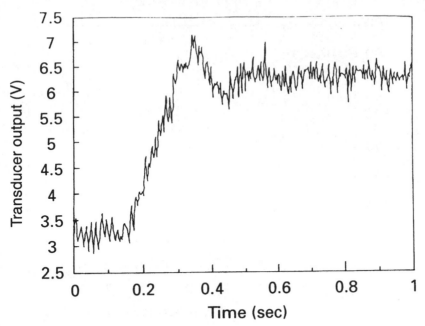

Fig. 10.15a Pressure transducer response.

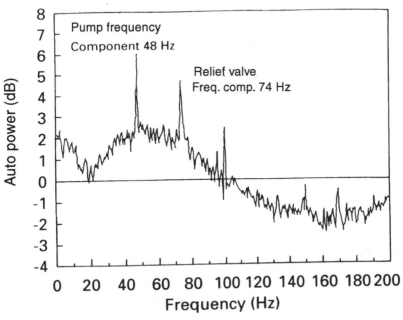

Fig. 10.15b Pump pressure ripple frequency analysis.

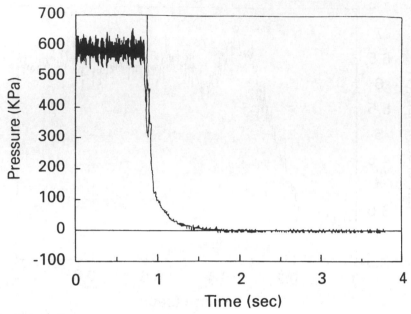

Fig. 10.16a Pump pressure as pump is switched off.

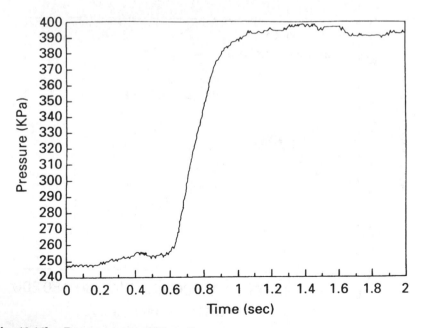

Fig. 10.16b Pump pressure (filtered) as valve is closed.

the associated frequency spectrum. The latter shows peaks at 48 and 74 Hertz. The former corresponds to rotational frequency, and monitoring its drift may yield information on pump performance. The latter frequency component is thought to be related to oscillations of the relief valve spring as this component is absent when the relief valve is closed.

Another source of information is the pump outlet pressure transient under various operating conditions, such as switching off the pump, and the closure of the flow valve. These are shown in Figure 10.16, with the latter signal resulting from the filtration of the original signal via a low pass filter. Because the pressure transients were easily available, it was decided to utilize these parameters within the overall coolant monitoring scheme, with the performance parameters defining when to collect fault location information.

Collecting such information at different health levels enables the relationship between fault level and diagnosability to be investigated. Three gates were selected at 75%, 50%, and 30% of performance, the width of each gate being ±2% as shown in Figure 10.17. The condition monitoring scheme is thus:

> To continually check the health of the overall coolant system, at hourly intervals, and when performance deteriorates to a value within any gate, then fault location information, the individual performance parameters and pressure transients, are stored for further analysis.

So, provided the coolant system performance remains above the upper limit of gate 1, at 77% no fault diagnosis is attempted. As the performance passes through the gate, the pressure response as the valve is closed is captured and stored. This continues while the performance is in the

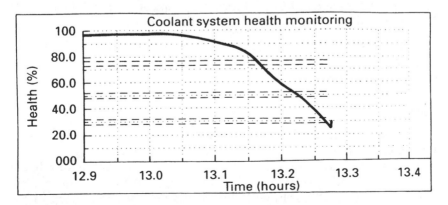

Fig. 10.17 Typical coolant system performance curve.

range 77% to 73%, and the average of all the pressure transients captured are stored for analysis. The cross products generated using this information, and data from the fault dictionary, are then compared, in order to identify the system fault (Martin and Thorpe, 1990).

10.5 MACHINE PERFORMANCE

In the foregoing sections, examples have been given of some performance monitoring techniques, which are available for both component and subsystem monitoring on machine tools. It is important to note, however, that in whole machine monitoring the above areas are not mutually exclusive, or in other words, many of the techniques available in condition monitoring, can be integrated and used in combination to monitor satisfactorily the equipment under surveillance.

This is an important point with relevance to many items of industrial equipment and especially to machine tools. These, in common with other items of equipment, are made up of an assembly of sub-systems and components, each of which contribute to the 'whole system'. This, in the case of a machine tool, has the primary task of efficiently machining workparts.

Hence no single surveillance technique or system is suitable to monitor all the aspects of such a complex assembly; rather several techniques may have to be employed, each dedicated to a specific area, and as shown in the examples above, integrated on the machine. In addition some components of the machine may require monitoring 'on-line' in 'real time' as the process proceeds. While others can be satisfactorily controlled 'off-line' out of 'real time' and on a periodic basis.

As we have seen in the examples above, performance monitoring is mainly aimed at using trend analysis to predict failure. It can, however, also provide useful information to assist in the location of actual or potential degradation failures. Accordingly, care should be taken to note the distinction between the two classes of failure, viz:

- **hard faults**–defined as sudden catastrophic failures which result in immediate machine breakdown and consequential lost production;
- **soft faults**–defined as partial or degradation failures in machinery, which progressively worsen over a period of time, leading to a loss in performance, and ultimately to a catastrophic or hard failure.

The fitting of complex data acquisition systems may also cause rather than reduce machinery reliability problems unless carefully designed, for the associated sensor requirements, harsh environment, poor reliability in use, and additional cost all mitigate against this approach (Williams and Davies, 1992). In addition, a complete understanding of the many component failure mechanisms involved is absent, as is the integration of suitable monitoring techniques. Thus a single, comprehensive, condition monitoring package for use on commercial machining systems is not yet available.

In order to give the reader an appreciation of the current situation in respect of whole machine monitoring, an example is now described. The complete system involves:

The automatic sensing and acquisition of data from a machine tool via a monitoring system, coupled with the subsequent signal/data processing, analysis of system condition, and finally the prediction/diagnosis of failure.

At present, the system can best be described as incomplete, although it does demonstrate in certain specific areas the capability of existing technology. It does not, however, demonstrate all the features necessary for an 'intelligent' machining system, as some aspects of this concept are still speculative in terms of hard engineering (Wright and Bourne, 1988).

The demonstrator, is a unit developed at the School of Engineering in Cardiff, and is based on a Flexible Manufacturing Cell (FMC) which consists of a vertical machining centre, automatic tool changer and magazine, plus an eight-station pallet pool. The unit features the three performance monitoring examples quoted above, being controlled via management software running on a PC and utilizing a Data Acquisition System (DAS) (Martin and Williams, 1990). The monitoring includes:

- on the axis drive sub-system via motor current and force sensing: ballscrew wear and friction, thrust bearing and drive belt wear, gain and bias of the controlling amplifier with associated diagnostics;
- on the main spindle drive: bearing surveillance via accelerometer, (vibration), measurements plus associated performance and diagnostic software; motor voltage and current transient response monitoring for amplifier/motor failure.
- on the coolant supply sub-system: steady-state pressure and reservoir level monitoring for pump faults; plus associated diagnostics from transient pressure measurements; hard fault diagnostics covering filter, pipework, loss of coolant and relief valve failure.

The main continuation of this research at Cardiff is the MIRAM project, 'Machine Management for Increasing Reliability, Availability and Maintainability'. The objective of this collaborative European research project is the development of a predictive maintenance methodology for industrial processes. Application of the methodology to machining systems is the chosen development area, with MIRAM ultimately being transferred to other industrial processes.

Figure 10.18 illustrates the practical implementation of the MIRAM DAS unit on the Wadkin V4-6 research demonstration system at Cardiff. Three industrial examples of this data acquisition system are currently undergoing trials on Heckler & Koch BA-25 turning centres. The DAS

Fig. 10.18   Experimental Data Acquisition System (DAS).

being fully integrated within the machine control unit, for the automatic collection of field failure data.

10.6   CONCLUSIONS

The performance monitoring techniques outlined above have attempted to show, in a machine tool context, the usefulness of this approach in assessing equipment condition. By employing illustrative examples based on common engineering components such as bearings, ballscrews and pumps, it has hopefully allowed the reader to appreciate the use of performance monitoring on other industrial systems.

10.7   REFERENCES

Bevington, P.R. (1969) *Data Reduction and Error Analysis for the Physical Sciences*, McGraw-Hill, New York, USA.
Collacot, R.A. (1979) *Vibration Monitoring and Diagnosis*, Goodwin, London, UK.
Connolly, P. (1990) *Evaluation of Vibration Based Condition Monitoring Techniques for Rolling Element Bearings*, UWCC Final Year Project Report, Cardiff, UK.

Davies, A. (1990) *Management Guide to Condition Monitoring in Manufacture*, IEE, London, UK.

Drake, P.R. (1988) *Monitoring Ballscrew Preload / Wear using Vibration Signal Analysis*, UWIST Technical Note EP141, Cardiff, UK.

Goldie, D.M. (1982) *Transducer Design using Piezoelectric Ceramic Crystals*, UWIST Final Year Project Report, Cardiff, UK.

Harris, C.G., Florio, F. and Davies, A. (1987) 'A sensor system for wear assessment in power screws, *Maintenance Management International*, **7**, 53–64.

Hildebrant, H.J. (1981) 'Ballscrew technology', *Engineering*, **7**, 543–47.

McFadden, P.D. and Smith, J.D. (1984) 'Vibration monitoring of rolling element bearings by the high frequency resonance technique – a review', *Tribology International*, **17**, 3–10.

Martin, K.F. and Thorpe, P. (1990) 'Coolant system health monitoring and fault diagnosis via health parameters and fault dictionary, *International Journal of Advanced Manufacturing Technology*, **5**, 66–85.

Martin, K.F. and Williams, J.H. (1990) *Sensor Based Machine Tool Condition Monitoring System*, Final Report SERC ACME grant GR/E 12818.

Neale, M. and Associates (1979) *A Guide to the Condition Monitoring of Machinery*, HMSO, London, UK.

Silva, G. (1986) *Pump Failure Mode Forecasting through the use of an Integrated Diagnostic Methodology*, Society of Automotive Engineers, Report No 861307.

Stewart, R.M. and Dyer, D. (1978) Detection of rolling element bearing damage by statistical vibration analysis, *Transactions of the ASME Journal of Mechanical Design*, **100**, 229–35.

Thorpe, P. (1990) *Condition Monitoring of a Machine Tool Spindle*, UWCC Technical Note EP173, Cardiff, UK.

Williams, J.H. (1985) *Transfer Function Techniques and Fault Location*, Research Studies Press, Letchworth, UK.

Williams, J.H. and Davies, A. (1992) System condition monitoring – an overview, *Noise and Vibration Worldwide*, **23**, 25–9.

Wright, P.K. and Bourne, D.A. (1988) *Manufacturing Intelligence*, Addison-Wesley, Massachusetts, USA.

PART FOUR

Techniques for Vibration Monitoring

CHAPTER 11

Review of fundamental vibration theory

D.W. Gardiner

Training Manager – ENTEK IRD International, Bumpers Lane, Sealand Industrial Estate, Chester, CH1 4LT, UK

11.1 INTRODUCTION

Vibration, simply stated is:

> The motion of a machine, or machine part, back and forth from its position of rest.

Whenever vibration occurs, there are actually four forces involved that determine the characteristics of the vibration. These forces are:

1. the exciting force, such as unbalance or misalignment;
2. the mass of the vibrating system;
3. the stiffness of the vibrating system;
4. the damping characteristics of the vibrating system.

The exciting force causes the vibration, whereas the stiffness, mass and damping forces oppose the exciting force to control or minimize the vibration.

The simplest way to illustrate such a vibration, is to follow the motion of a weight suspended on the end of a spring, as illustrated in Figure 11.1a. This is a valid analogy since all machines and their components have weight (mass), spring-like properties (stiffness) and damping. Until a force is applied to the weight to cause it to move, we have no vibration. However, by applying an upward force such as imbalance, the weight will move in an upward direction, compressing the spring. If we release the weight, it will then drop below its neutral position to some lower limit of travel, where the spring will bring the weight to rest. The weight

Handbook of Condition Monitoring
Edited by A. Davies
Published in 1998 by Chapman & Hall, London. ISBN 0 412 61320 4.

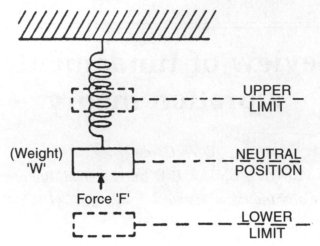

Fig. 11.1a   Vibration of a simple spring-mass system.

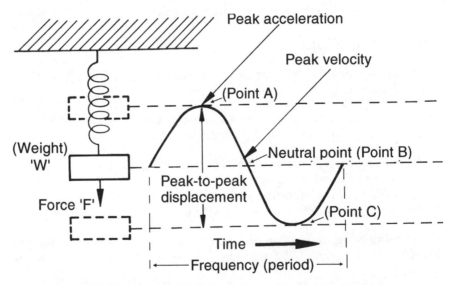

Fig. 11.1b   Vibration of a simple spring-mass system with respect to time.

would then return in an upward direction through the neutral position to the upper limit of motion, and then back again through the neutral position to the lower limit of motion. This movement will continue in exactly the same manner, for as long as the force is reapplied, and it is this motion which is called vibration.

With few exceptions, it is the mechanical problems associated with machinery operation which cause vibration. To list all the possible causes of machinery vibration would be quite a complex undertaking; however, some of the more common problems which are known to produce vibrations are as follows:

(a) imbalance of rotating parts;
(b) eccentric components;
(c) misalignment of couplings and bearings;
(d) bent shafts;
(e) component looseness;
(f) worn, or damaged gears;
(g) bad drive belts and drive chains;
(h) bad anti-friction bearings;
(j) torque variations;
(k) electromagnetic forces;
(l) aerodynamic forces;
(m) hydraulic forces;
(n) resonance;
(o) rubbing.

All of these causes can be reduced to one, or more, of five different types of problem. That is, parts will either be unbalanced, misaligned, loose, eccentric or reacting to some external force. Regardless of the cause or causes of the vibration, one basic thing must always be true, i.e.

The cause of vibration must be a force, which is changing in either its magnitude or in its direction.

It is the manner in which this force is generated which determines the resulting characteristics of the vibration. Each cause of vibration therefore has its own individual combination of vibration characteristics. Simply by recording these characteristics a machine's mechanical condition and/or operating problems can easily be defined.

Referring back to our weight suspended on a spring shown in Figure 11.1a. If we now plot the movement of the weight against time, the plot shown in Figure 11.1b, will be generated. The motion of the weight from its neutral position to the upper limit of travel, back through the neutral position to the lower limit of travel, and its return to the neutral position, represents one cycle or period of the motion. The waveform shown in Figure 11.1b contains all the characteristics required to define the vibration, and the continued motion of the weight will simply repeat these characteristics.

11.2   MEASURING VIBRATION

The three characteristics that are required to define any vibration are:

(a) amplitude,
(b) frequency,
(c) phase.

The term amplitude is now internationally recognized as the term used to describe 'how much' in whatever measurement parameter is being used. Vibration amplitude is measured in three main engineering units:

(a) displacement,
(b) velocity,
(c) acceleration.

These measurement parameters can be expressed in either metric or imperial units. Most modern vibration measurement instruments have the ability to measure either metric or imperial as a user defined option.
   We will now describe each of these measurement parameters in detail.

(a)   Displacement

As shown in Figure 11.1b, the total distance travelled by the vibration part from one extreme limit of travel to the other extreme limit of travel is referred to as the 'peak-to-peak' displacement. In metric units the peak-to-peak displacement is usually expressed in microns, where one micron equals one thousandth of a millimetre (0.001 mm). In imperial units it is usually expressed in mils, where one mil equals one thousandth of an inch.

(b)   Velocity

As the weight shown in Figure 11.1b has to travel a certain distance (the displacement) in a certain amount of time, it follows that it must be moving at some speed or velocity. At the limits of the motion, point A and C, the speed is zero as the weight must come to a stop before it can move in the opposite direction.
   At point B as the weight passes through the neutral position, the part is moving at its highest velocity. Hence, as the velocity or speed of the weight is constantly changing throughout the cycle, the highest or 'peak' velocity is selected for measurement. In metric units, vibration velocity is expressed in millimetres per second peak. In imperial it is expressed in inches per second peak. Velocity can also be expressed in units of velocity RMS (root mean square). Most modern measuring instruments can be set to read in terms of velocity peak or velocity RMS.

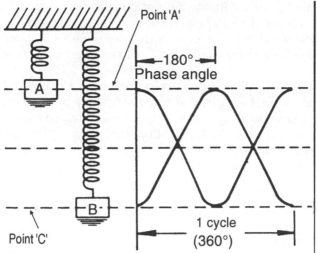

Fig. 11.2a Weights vibrating 180° out-of-phase.

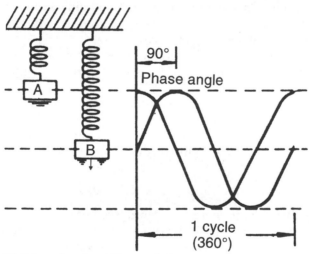

Fig. 11.2b Weights vibrating 90° out of phase.

(c) Acceleration

Acceleration is a measure of 'the rate of change of velocity'.

If we refer to the diagram of motion shown in Figure 11.1b, the acceleration of the part is maximum at the extreme limit of travel (point A), where the rate of change of velocity is the greatest, the acceleration is at its maximum. As the velocity of the part increases, the acceleration decreases, until at (point B), the neutral position, the velocity is constant and at its maximum and the acceleration is zero. As the part passes through the neutral point, it must now decelerate as it approaches the other extreme limit of travel (point C), where acceleration is again at its peak.

Vibration acceleration is normally expressed in g's peak, where 1 g is the acceleration produced by the force of gravity at the surface of the earth. By international agreement, the value 980.665 cm/sec/sec has been chosen as the standard acceleration due to gravity. In summary, the principle measurement parameters and units which are used to describe the amplitude of vibration are:

(a) displacement (peak-to-peak),
(b) velocity (peak),
(c) velocity (RMS),
(d) acceleration (peak).

In addition to the above, there are two other forms of expressing vibration measurements which though rarely used may still be encountered. These are displacement peak and velocity average. Displacement peak is the distance travelled from point A to point B in Figure 11.1b or equally from point B to point C in the diagram. It is in effect half the value of the displacement peak-to-peak.

Velocity average is a fixed value and represents two points on the waveform. It is no longer considered to be a representative value of vibration velocity.

The displacement, velocity and acceleration of a vibration are all mathematically related to one another. For example, if the peak to peak displacement and the frequency of a single frequency sinusoidal vibration are known, then the velocity of vibration can be calculated as follows:

$$V \text{ peak} = 52.3 \times D \times (F/1000) \times 0.001 \qquad (11.1)$$

where: V peak = vibration velocity (mm/sec peak);
$\quad\quad\quad D$ = vibration displacement (microns peak-to-peak);
$\quad\quad\quad F$ = vibration frequency (CPM–cycles/min).

When it is required to calculate vibration acceleration, the following formula can be used:

$$Va \text{ (g peak)} = 5.6 \times D \times (F/1000) \times 0.0001 \qquad (11.2)$$

where: g peak = acceleration due to gravity;
$\quad\quad\quad D$ = vibration displacement (microns peak-to-peak);
$\quad\quad\quad F$ = vibration frequency (CPM–cycles/min).

These formulae illustrate the important relationships between the amplitude parameters of displacement, velocity and acceleration.

11.3 FREQUENCY

As shown in Figure 11.1b, the amount of time required to complete one full cycle of the vibration is called the period of vibration. If, for example, the machine completes one full cycle of vibration in 1/60th of a second, the period of vibration is said to be 1/60th of a second.

Vibration frequency is simply a measure of the number of complete cycles that occur in a specified period of time such as 'cycles-per-second' (CPS) or 'cycles-per-minute' (CPM). Frequency is related to the period of vibration by this simple formula:

$$\text{Frequency} = 1/\text{Period} \tag{11.3}$$

In other words, the frequency of a vibration is simply the 'inverse' of the period of the vibration. Thus, if the period or time required to complete one cycle is 1/50th of a second, then the frequency of the vibration would be 50 cycles-per-second or 50 CPS. Although vibration frequency may be expressed in cycles-per-second, the common practice is to use the term Hertz (abbreviated Hz) in lieu of CPS. Thus, a vibration with a frequency of 50 CPS would actually be expressed as 50 Hz.

In the real world of vibration detection and analysis, it is not necessary to determine the frequency of vibration by observing the vibration time waveform, noting the period of the vibration and then taking and calculating the inverse of the period. Although this can be done, nearly all modern-day data collector instruments and vibration analysers provide a direct transform from the 'time domain' to the 'frequency domain' to give a direct readout of the vibration frequencies being generated by the machine in the form of an amplitude verses frequency plot. This is usually referred to as a 'signature' or a 'spectrum' or, when using an instrument with digital processing, a Fast Fourier Transform (FFT).

For most work on rotating machinery, the vibration frequency is usually measured in cycles-per-minute, rather than Hz. Expressing vibration frequency in terms of CPM makes it much easier to relate the indicated frequencies to the rotational speed of the machine components. Rotational speeds are normally expressed in revolutions-per-minute or RPM. Thus, if a machine operates at 3000 RPM, it is much more meaningful to know that a vibration occurs at 3000 CPM (1 × RPM) rather than 50 Hz.

Of course CPM and Hz can be easily converted to one another as follows:

Given a frequency expressed in Hz, you can convert it to CPM:

$$\text{CPM} = \text{Hertz} \times 60$$

Given a frequency expressed in CPM, you can convert it to Hz:

$$\text{Hertz} = \text{CPM}/60$$

If we only have a portable vibration meter (which just measures the overall vibration), we may under some circumstances, such as emergency 'no phase balance', require to determine whether the total vibration is mainly at one dominant frequency. This may be done using the following sequence. (Entek, 1996).

Ensure that you use exactly the same measurement position and direction.

(a) Measure and record the displacement (D) at a given point.
(b) Measure and record the velocity (V).
(c) The dominant frequency can now be calculated using the following equation:

Dominant frequency (CPM) = (Velocity/Displacement) × 19 120 　　(11.4)

This equation can be used for both metric and imperial units.

11.4　PHASE

The third characteristic required to fully describe any vibration is phase. Phase is defined as:

> The angle between the instantaneous position of a vibrating part and a fixed reference position, or the fractional part of the vibration cycle through which the part has advanced relative to the fixed reference.

or if observing two moving parts of a machine:

> The fractional part of a cycle through which one part has advanced, relative to the fractional part of the cycle through which the other part has advanced, expressed as an angular difference.

In a practical sense, phase measurements provide a convenient way to determine the relative motion between two or more parts of a machine moving at the same frequency. For example the two weights shown in Figure 11.2a are vibrating at the same frequency and displacement; however weight A is at the upper limit of travel at the instant weight B is at the lower limit. The points of peak displacement are thus separated by one half cycle, or 180°, therefore, the two weights are vibrating 180° out of phase with each other. In Figure 11.2b however, weight B is leading weight A by a quarter of a cycle. The phase difference between the two weights is therefore 90°.

Determining the relative motion of various parts of a machine through the measurement and comparison of phase, is also essential in the diagnosis of specific machinery defects. For example, if a machine and its base are vibrating out of phase, this would indicate mechanical looseness. Techniques have been developed for distinguishing between bent shafts, misaligned couplings and misaligned bearings using phase measurement and analysis techniques. The diagnosis of resonance problems is also aided by observing changes in phase versus rotating speed, and phase measurements are also used for quick and easy in-place balancing without trial and error techiques (Fox, 1977).

Now that we have an understanding of the three characteristics of vibration, the next obvious question is:

Since the amplitude of vibration can be measured in terms of displacement, velocity or acceleration, which of the three measurement parameters should we use?

Overall vibration amplitude readings indicate the severity of the vibration. But which is the best indicator of vibration severity: displacement, velocity or acceleration? To answer this question, consider what happens when a strip of sheet metal is bent repeatedly back and forth. Eventually, this repeated bending will cause the metal to fail by fatigue in the area of the bend. This is similar in many respects to the way in which a machine or machine component fails–from the repeated cycles of flexing caused by excessive vibration.

The amount of time required for the sheet metal to fail may be reduced by:

(a) increasing the amount of the bend or displacement; the further the metal is bent each time, the more likely it is to fail;
(b) increasing the rate of bending or its frequency; the more times per minute the metal is flexed, the quicker it will fail.

Thus, the severity of this bending action is a function of both how far the metal is bent or displaced, and how fast the metal is bent or the frequency of the bending action. Vibration severity is therefore a function of displacement and frequency, and since vibration velocity is also a function of these two parameters, it is reasonable to conclude that:

A measure of vibration velocity is a direct measure of vibration severity.

Through experience, this has been shown to be basically true, and vibration velocity is accepted as the best overall indicator of machinery condition. Displacement and acceleration readings are sometimes used to measure vibration severity. However, when displacement or acceleration is used, it is also necessary to know the frequency of the vibration.

General Severity Charts similar to those shown in Figures 11.3 and 11.4, are used to cross reference the displacement or acceleration with frequency, and thereby determine the level of severity. It will be seen from Figure 11.3 that a displacement of 25 microns occurring at a frequency of 1200 cycles per minute (CPM) is in the good range,

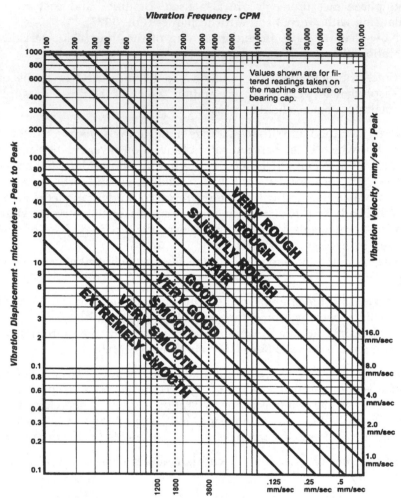

Fig. 11.3 General machinery vibration severity chart (metric units).

however, the same displacement at a frequency of 20 000 CPM is in the very rough range. Note also that the diagonal lines dividing the zones of severity are constant velocity lines. In other words, a velocity of 12.7 mm per second peak is in the rough regardless of the frequency of the vibration. In Figure 11.4 it can be seen that an acceleration of 1.0 g at a frequency of 100 000 CPM is in the good region of the chart, while 1.0 g at a frequency of 18 000 CPM is in the slightly rough region.

VIBRATION SEVERITY CHART — ACCELERATION

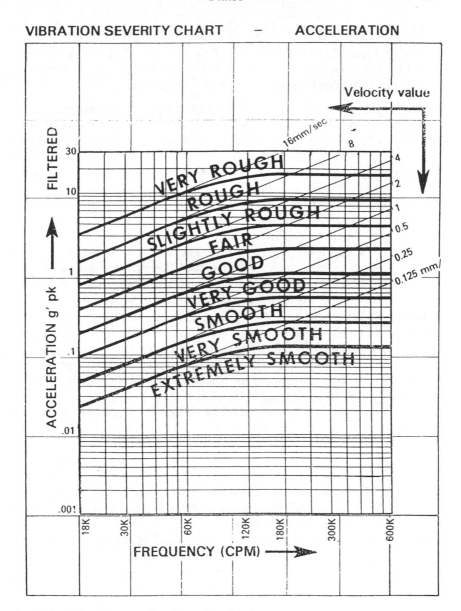

Fig. 11.4 Vibration severity chart. (Entek IRD)

It can therefore be stated that:

When measuring overall vibration levels, i.e. when the frequencies of that vibration are unknown, velocity is the only measurement parameter that can be applied directly to a general chart of severity.

Under conditions of dynamic stress, displacement alone may be a better indicator of severity. Earlier we discussed the effects of repeated bending relating to the failure of a piece of sheet metal, but the sheet metal analogy did not demonstrate very well one property of most rigid machinery components, that is brittleness (the tendency to break or snap when stressed beyond a given limit). For example, consider a slowly rotating machine such as a mine hoist drum, rotating at 50 revolutions per minute (RPM) with a vibration of 2500 microns peak-to-peak displacement caused by rotor imbalance.

In terms of vibration velocity, 2500 microns at 50 CPM is only 6.5 millimetres per second peak, which would be considered from Figure 11.3, as slightly rough for general machinery, and little cause for concern. However, keep in mind that the components of this machine are being deflected 2500 microns or 2.5 mm. Under these conditions, failure may occur due to stress (displacement) rather than fatigue (velocity).

Acceleration is closely related to force, and even though the displacement and velocity may be small, relatively large forces can occur at high frequencies. For example, consider a measured vibration of 25 microns at 6000 CPM. This corresponds to a velocity reading of 7.8 mm per second peak, which may be considered as rough for general machinery as shown in Figure 11.3. This also corresponds to an acceleration reading of 0.5 g. Next consider a vibration of 0.25 microns at a frequency of 600 000 CPM. This vibration also corresponds to a velocity reading of 7.8 mm per second peak but an acceleration reading of 50 g. At 6000 CPM, failure may occur due to fatigue (velocity); however, at the higher frequency of 600 000 CPM, failure will almost certainly result from the applied force (acceleration).

Such an excessive force may result in the breakdown of the lubrication film in the machine bearings with disastrous results. This is not to say that all failures of journal type bearings are the result of very high frequency vibration. More common causes of such failures include contaminated lubricant, bearing overload, misalignment, imbalance etc. The purpose here is to illustrate that the forces are significant at high frequencies, and that this may not be detected if only the velocity is measured.

11.5　VIBRATION ANALYSIS

There are literally hundreds of specific mechanical and operational problems that can cause a machine to exhibit excessive vibration. Obviously, when a vibration problem exists, a detailed analysis of the

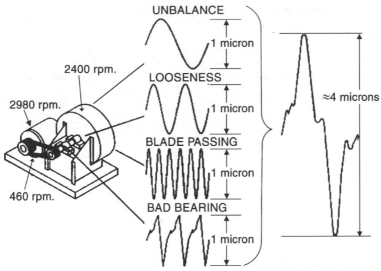

Fig. 11.5 Most machinery vibration is complex, consisting of many frequencies.

vibration should be performed to identify or pinpoint the specific cause. This is where knowing the frequency of vibration is most important. Vibration frequency is an analysis or diagnostic tool.

The forces that cause vibration are usually generated through the rotating motion of the machines parts. Because these forces change in direction or amplitude according to the rotational speed (RPM) of the machine components, it follows that most vibration problems will have frequencies that are directly related to the rotational speeds.

To illustrate the importance of vibration frequency, assume that a machine, such as the one shown in Figure 11.5, consisting of a fan operating at 2400 RPM and belt (460 RPM) driven by a motor operating at 2980 RPM, is vibrating excessively at a measured frequency of 2400 CPM (1 × fan RPM), this clearly indicates that the fan is the source of the vibration and not the motor or belts. Knowing this simple fact has eliminated literally hundreds of other possible causes of vibration.

Table 11.1 is a chart that lists the most common vibration frequencies as they relate to rotating speed (RPM), along with the common causes for each frequency. To illustrate how this chart is used, if the frequency of excessive vibration was found to be 2400 CPM or 1 × RPM of the belt driven fan described in the above example, the possible causes listed on the chart would be:

- unbalance;
- eccentric pulley;
- misalignment – this could be misalignment of the fan bearings or misalignment of the motor and fan pulleys;

Table 11.1   Vibration frequencies and the likely causes

Frequency in terms of rpm	Most likely causes	Other possible causes and remarks
1 × rpm	Imbalance	1. Eccentric journals, gears or pulleys 2. Misalignment or bent shaft – if high axial vibration 3. Bad belts if rpm of belt 4. Resonance 5. Reciprocating forces 6. Electrical problems
2 × rpm	Mechanical looseness	1. Misalignment if high axial vibration 2. Reciprocating forces 3. Resonance 4. Bad belts if 2 × rpm of belt
3 × rpm	Misalignment	Usually a combination of misalignment and excessive axial clearances (looseness)
Less than 1 × rpm	Oil whirl (less than $\frac{1}{2}$ rpm)	1. Bad drive belts 2. Background vibration 3. Sub-harmonic resonance 4. Beat vibration
Synchronous (a–c line frequency)	Electrical problems	Common electrical problems include broken rotor bars, eccentric rotor, imbalanced phases in poly-phase systems, unequal air gap
2 × synchronous frequency	Torque pulses	Rare unless resonance is excited
Many times rpm (harmonically related frequency)	Bad gears Aerodynamic forces Hydraulic forces Mechanical looseness	Gear teeth × rpm of bad gear Number of fan blades × rpm Number of impeller vanes × rpm May occur at 2, 3, 4 and sometimes higher harmonics if severely loose
High frequency (not harmonically related)	Reciprocating forces Bad antifriction bearings	1. Bearing vibration may be unsteady – amplitude and frequency. 2. Cavitation, circulation and flow turbulence cause random, high frequency vibration 3. Improper lubrication of journal bearings (friction excited vibration) 4. Rubbing

- bent shaft;
- looseness;
- distortion – soft feet or piping strain;
- bad belts – if belt RPM;
- resonance;
- reciprocating forces;
- electrical problems.

Using this simple chart, along with the fact that the frequency of excessive vibration is 2400 CPM (1 × fan RPM) has reduced the number of possible causes from literally hundreds to only ten possible causes.

A little common sense can reduce this number of possible causes even further. First, since the vibration frequency is not a fundamental or multiple of the rotating speed (460 RPM) of the drive belts, belt problems can be eliminated as a possible cause. Secondly, since this is not a reciprocating machine such as reciprocating compressor or engine, the possibility of reciprocating forces can be eliminated from the remaining list. Finally, since the frequency is not related to the drive motor in any way, the possibility of electrical problems can be eliminated. Now, the number of possible causes of excessive vibration has been reduced to only seven by simply knowing that the vibration frequency is 1 × RPM of the fan.

The first step in the analysis process is therefore to relate the rotating speed of the machine components to the known frequency of vibration. Additional tests and measurements can be taken to further reduce the number of possible causes of a vibration problem. Vibration analysis is actually a process of elimination.

Of course, not all machinery problems will generate vibration at a frequency equal to the rotating speed (1 × RPM) of the machine. For example, referring to the chart in Table 11.2 it can be seen that some problems such as looseness, misalignment, resonance and reciprocating forces can often generate vibration at frequencies of 2×, 3× and sometimes higher multiples of RPM. Problems with gears usually result in vibration at frequencies related to the 'gear mesh' frequency or the product of the number of teeth on the gear multiplied by the gear RPM. Aerodynamic and hydraulic problems with fans and pumps will normally show vibration frequencies which are the product of the machine RPM times the number of blades or impeller vanes. In addition, not all problems will result in vibration frequencies that are exact multiples of the rotating speed of the machine. The non-harmonicaly related vibration frequencies generated by flaws or defects in rolling-element bearings are a good example.

Table 11.2 Vibration identification

Cause	Amplitude	Frequency	Phase	Remarks
Imbalance	Proportional to imbalance Largest in radial direction	1 × rpm	Single reference mark	Most common cause of vibration
Misalignment of couplings or bearings, and bent shaft	Large in axial direction, 50% or more of radial vibration	1 × rpm usual, 2 and 3 × rpm sometimes	Single double or triple	Best found by appearance of large axial vibration. Use dial indicators or other method for positive diagnosis. If sleeve bearing machine and no coupling misalignment balance the rotor
Bad bearings–anti-friction type	Unsteady–use velocity measurement if possible	Very high, several × rpm	Erratic	Bearing responsible most likely the one nearest point of largest high-frequency vibration
Eccentric journals	Usually not large	1 × rpm	Single mark	If on gears largest vibration in line with gear centers. If on motor or generator vibration disappears when power is turned off. If on pump or blower attempts to balance
Bad gears or gear noise	Low–use velocity measure if possible	Very high–gear teeth × rpm	Erratic	...
Mechanical looseness		2 × rpm	Two reference marks Slightly erratic	Usually accompanied by unbalance and/or misalignment
Bad drive belts	Erratic or pulsing	1, 2, 3 and 4 × rpm of belts	One or two depending on frequency. Usually unsteady	Strobe light best tool to freeze faulty belt
Electrical	Disappears when power is turned off	1 × rpm or 1 or 2 × synchronous frequency	Single or rotating double mark	If vibration amplitude drops off instantly when power is turned off, cause is electrical
Aerodynamic or hydraulic forces	...	1 × rpm or number of blades on fan or impeller × rpm	...	Rare as a cause of trouble except in cases of resonance
Reciprocating forces	...	1, 2 and higher orders × rpm	...	Inherent in reciprocating machines; can only be reduced by design changes or isolation

11.6 COMPLEX VIBRATION

Figure 11.1b illustrates the response of a spring-mass system to a single exciting force (unbalance). The result is a vibration having only one frequency. This is called a simple vibration. In reality, machines will often have several causes of vibration, with each cause having its own unique frequency. Whenever more than one frequency of vibration is present, the result is called a complex vibration, Figure 11.5, shows a fan with several different vibration frequencies resulting from different defects. Of course, each vibration frequency present can readily be identified along with its cause using standard analysis equipment and techniques.

Several terms are used to describe various frequencies of vibration, and it is important for the vibration technician to understand these terms in order to communicate effectively with others in the field of machinery vibration. Some of the more common terms describing vibration frequencies are listed below along with their definitions.

Predominant frequency   Predominant frequency is the frequency of vibration having the highest amplitude or magnitude.

Synchronous frequency   Synchronous frequency is the vibration frequency that occurs at 1 × RPM.

Subsynchronous frequency   Subsynchronous frequency is vibration occurring at a frequency below 1 × RPM. A vibration that occurs at 1/2 × RPM would be called a subsynchronous frequency.

Fundamental frequency   Fundamental frequency is the lowest or first frequency normally associated with a particular problem or cause. For example, the product of the number of teeth on a gear times the RPM of the gear would be the fundamental gear-mesh frequency. On the other hand, coupling misalignment can generate vibration at frequencies of 1 × , 2 × and sometimes 3 × RPM. In this case, 1 × RPM would be called the fundamental frequency.

Harmonic frequency   A harmonic frequency is a frequency that is an exact, whole number multiple of a fundamental frequency. For example, a vibration that occurs at a frequency of two times the fundamental gear mesh frequency would be called the first harmonic of gear mesh frequency. A vibration at 2 × RPM due to, say, misalignment, would be referred to as the first harmonic of the running speed frequency (1 × RPM).

Order frequency   An order frequency is the same as a harmonic frequency.

Subharmonic frequency   A subharmonic frequency is an exact submultiple (1/2, 1/3, 1/4, etc.) of a fundamental frequency. For example, a vibration with a frequency of exactly 1/2 the fundamental gear-mesh frequency would be called a subharmonic of the gear mesh frequency. Vibration at frequencies of exactly 1/2, 1/3 or 1/4 of the rotating speed (1 × RPM) frequency would also be called subharmonic frequencies; and these can also be called subsynchronous frequencies. However, not all subsynchronous frequencies are subharmonics. For example, a vibration with a frequency of 43% of the running speed (1 × RPM) frequency is a subsynchronous frequency but it is not a subharmonic.

11.7   VIBRATION SEVERITY

Since vibration amplitude (displacement, velocity or acceleration) is a measure of the severity of the defects in a machine, the next question is, 'how much vibration is too much?' To answer this question, it is important to keep in mind that the objective is to detect defects in their early stages for correction via scheduled maintenance. The goal is not to find out how much vibration a machine will stand before failure, but to get fair warning of incipient defects so that they may be diagnosed and eliminated before a failure occurs. Absolute vibration tolerances or limits for any given machine are not possible. That is, it is impossible to select a vibration limit which, if exceeded, will result in immediate machinery failure. The development of mechanical failure is far too complex for such limits to exist.

However, it would be impossible to effectively utilize vibration as an indicator of machinery condition, unless some guidelines are available. Thus, the years of experience accumulated by those familiar with machinery and machinery vibration, have been utilized to provide the empirical, though realistic guidelines required. We have seen that vibration velocity provides a direct measure of machinery condition for the intermediate vibration frequencies of 600 to 120 000 CPM. Thus the velocity values in Figures 11.3 and 11.4 provide such a guide for overall velocity readings.

The guidance offered in Figures 11.3 and 11.4 apply to machinery such as motors, fans, blowers, pumps and general rotating machinery where vibration does not directly influence the quality of a finished product. Readings should be taken directly on the bearings or structure of the machine in question. Of course, the vibration tolerances suggested in the charts will not be applicable to all machines. For example, some machines such as reciprocating compressors, hammer mills or rock and coal crushers will inherently have high levels of vibration. Therefore, the values selected using these guidelines should only be used, so long as

◆ ENTEK IRD

TENTATIVE GUIDE TO VIBRATION TOLERANCES FOR MACHINE TOOLS

TYPE OF MACHINE	Displacement of vibration as read with pickup on spindle bearing housing in the direction of cut.
• Grinders	**Tolerance Range**
Thread Grinder	0.25 to 1.5 microns
Profile or Contour Grinder	0.76 to 2.0 microns
Cylindrical Grinder	0.76 to 2.5 microns
Surface Grinder (vertical reading)	0.76 to 5.0 microns
Gardner or Besly Type	1.3 to 5.0 microns
Centreless	1.0 to 2.5 microns
• Boring Machine	1.5 to 2.5 microns
• Lathes	5.0 to 25.0 microns

These values come from the experience of IRD personnel who have been trouble shooting machine tools for over 30 years with the IRD equipment. They merely indicate the range in which satisfactory parts have been produced and will vary depending upon size and finish tolerances.

Fig. 11.6   Vibration tolerances for machine tools.

experience, maintenance records and historical data prove them to be valid for a particular application.

For machines such as grinders and other precision machine tools, where vibration can affect the quality of a finished product, the guide to vibration tolerances for machine tools in Figure 11.6 may be used. Applying vibration tolerances to machine tools is rather easy because they can be based on the machine's ability to produce a certain size of finish tolerance. The values shown in Figure 11.6 are the result of years of experience with vibration analysis of machine tools and indicate the vibration levels which have produced satisfactory parts. Of course, these values may vary depending on the specific size and finish tolerances required. A comparison of the normal pattern of vibration on the machine and the quality of finish, plus the size control required, will reveal what level of vibration is acceptable. The first time the quality of finish or size control deteriorates, an unacceptable vibration level would be indicated. The initial values selected from Figure 11.6 can then be modified to new, more realistic levels.

In 1974, the International Standards Organisation (ISO) pulished ISO Standard 2372 (BS4675) 'Mechanical Vibration of Machines with Operation speeds between 10 and 200 revolutions per second. The objective of this standard is to establish realistic guidelines for acceptable vibration levels... with respect to reliability, safety and human perception.'

This standard differs from the general severity standards previously referred to, in that it establishes six broad classifications for various types of machinery. Within the standard descriptions of various machines facilitate classification by type, power rating and mounting method. Thus, to use ISO 2372, it is first necessary to classify the machine, then compare the vibration level obtained with the levels given for the class of machine within the standard. The severity of the machine condition being indicated by the letter A, B, C, or D. The standard defines these letters as:

A–good, B–acceptable, C–still acceptable, D–not acceptable.

Making a decision to shut down a machine is often a very difficult one indeed, especially when it involves downtime of critical machinery. Therefore, when establishing shutdown levels of machinery vibration, experience and factors such as rate of deterioration, safety, labour costs, downtime costs and the importance of a machine's operation to company profit must also be considered.

Another important aspect of ISO 2372 is that, it introduced velocity RMS (root mean square) as the standard unit of measurement. Velocity RMS is now considered, in most cases, to be a more effective indication of the damaging energy in a vibration.

11.8   VELOCITY RMS

In the preceding discussions, it was explained that vibration displacement is measured in mils or microns peak-to-peak, velocity in in/sec or mm/sec peak and acceleration in G's peak. Most modern vibration instruments can now be set to read either velocity peak or velocity RMS.

Root-mean-square (RMS) is basically a measure of average vibration amplitude as a function of time. The vibration waveform in Figure 11.7a is a sinusoidal (simple) waveform, typical of that generated by unbalance. This is a velocity waveform and has a true peak value of 1.0 mm/ sec-peak. Its RMS value can best be visualized by drawing a rectangle around the upper portion of the waveform as shown. The RMS value is that portion of the rectangle that is 'occupied' or by the vibration signal or the shaded area shown in Figure 11.7a. For a sinusoidal vibration waveform, the RMS value is 0.707 times the peak value. For the sine wave in Figure 11.7a, 70.7% of the rectangle is 'filled' or occupied by the

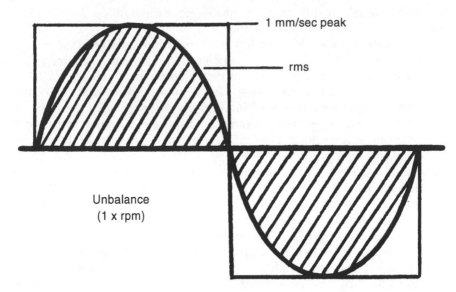

Fig. 11.7a The root-mean-square (RMS) value of the vibration is the area of the rectangle 'occupied' by the vibration signal.

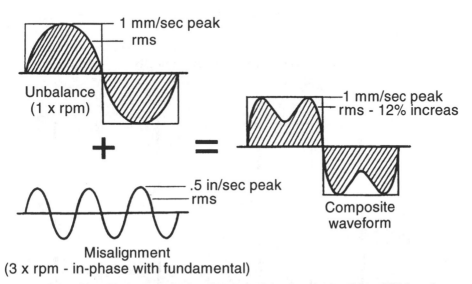

Fig. 11.7b Although the true peak velocities of the fundamental (1 × RPM) and composite vibrations are the same, the RMS velocity increased approximately 12% due to the addition of the 3 × RPM vibration.

vibration signal. In this case the RMS velocity would be 1.0 mm/sec-peak × 0.707 = 0.707 mm/sec-RMS.

The reason for the ISO decision to standardize on RMS response is explained by referring to the complex vibration shown in Figure 11.7b. In this example there are two frequency components; one at 1 × RPM with a true peak velocity value of 1.0 mm/sec-peak and a second frequency component at 3 × RPM with a true peak velocity value of 0.5 mm/sec-peak. These two vibration components combine to produce the 'composite' waveform illustrated. Note that the true peak velocity value of the composite waveform is 1.0 mm/sec-peak or the same amplitude as the 1 × RPM component alone. By comparison, the RMS amplitude of the composite waveform or the area under the rectangle, shows a 12% increase. In other words, if true peak measurements are used, it is possible that, depending on the frequency components present, the true peak amplitude of the composite waveform could actually be less than the peak amplitude of the predominant component.

Although there are definite reasons to use RMS responding instruments instead of true peak measurements, the use of RMS detection actually creates a problem when it comes to detecting short-duration spikes or pulses. The vibration produced by gear and rolling-element bearing problems will usually be in the form of short duration pulses. Figure 11.8 illustrates the type of vibration generated by a bearing defect such as a flaw on a raceway. The only energy and resultant vibration generated by the flaw is when a rolling element passes over the defect, resulting in a short duration spike. Although the true peak amplitudes generated can be most significant, RMS values of the spike pulses are typically very small when compared to other sources of vibration such as unbalance and misalignment. This can readily be seen in Figure 11.8 where a rectangle has been drawn around one cycle period of the spike

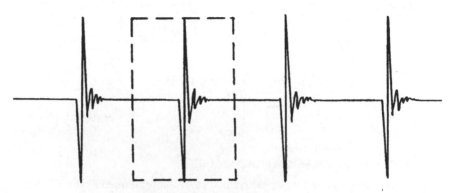

Fig. 11.8 Spike-pulse vibrations caused by problems such as bearing and gear defects may have very low RMS values but high peak values.

pulse frequency. It should be obvious that the area of the rectangle 'occupied' by the spike pulse (e.g. its RMS value) is very small compared to the true peak value (Milne, 1994). Experience has shown in many cases that rolling element bearings, even in advanced stages of failure, may show little if any increase in displacement, velocity or acceleration measurements.

From the preceding discussion, it should be apparent that overall measurements of displacement, velocity and even acceleration are not well suited for detecting rolling-element bearing deterioration and other problems that cause spike-pulse signals. The problem is basically the limitations of RMS responding circuitry combined with other inherent sources of vibration such as unbalance and misalignment that tend to overshadow or dominate the bearing vibration. However, the solution is simple. Design a circuit that utilizes true peak-to-peak detecting response instead of RMS detection.

11.9 HIGH-FREQUENCY DETECTION SYSTEMS

Most modern instruments designed to monitor vibration now have a special measurement parameter (in addition to the standard engineering units of displacement, velocity and acceleration) which is specifically designed to detect the incipient failure of rolling element bearings and gears. Most of these systems operate on the same basic principles but each instrument vendor has a slightly different detection system and unit of amplitude. They are designed to sense the very small amplitude impacts generated within the audiosonic/ultrasonic frequency range by the microscopic surface flaws on bearing elements and gears. The basic features of the SPIKE ENERGY (abbreviated 'g-SE™') approach developed by Entek IRD are as follows.

1. Since the frequencies of bearing vibration are very high, and vibration acceleration tends to emphasize higher frequencies, a vibration acceleration signal is utilized from an accelerometer transducer.
2. Incorporate a 'band-pass' frequency filter that will electronically filter out frequencies below 5 KHz (300 000 CPM) and above 50 KHz (3 000 000 CPM). The lower cut-off frequency of 5 KHz (300 000 CPM) filters out or ignores most other inherent sources of vibration such as unbalance, misalignment, aerodynamic and hydraulic pulsations, electrical frequencies etc. that tend to dominate or 'hide' the vibration from bearing defects.
3. Since the spike-pulse signals generated by bearing defects have very low RMS values, a true peak-to-peak detecting circuit is incorporated instead of an RMS detecting circuit.

Entek IRD instruments therefore, give a reading that is the product of impact amplitude, pulse rate and high-frequency random vibration energy. These three parameters are electronically combined into a single Figure of merit called 'gSE™', or acceleration units of spike energy™.

When establishing a programme to monitor the condition of rolling element bearings, various methods can be used. The primary method is 'trending'. In this method, the machine bearings are measured period- ically and their g-SE levels recorded. No change in level over a period of time indicates a good bearing, while a significant upward trend indicates a deteriorating bearing.

A 'comparison' method may also be used. That is, the g-SE levels of similar machines are measured, and any levels that significantly depart from the norm are singled out for further analysis and more frequent monitoring. This method rapidly leads to the establishment of criteria levels that distinguish good and bad bearings. It should be noted from the General Severity Chart shown in Figure 11.9 that the g-SE™ levels depend primarily on machine rotational speed (RPM). Levels will typically double for each doubling of RPM. From a vibration severity standpoint however, low-speed bearings can usually tolerate more damage than high-speed bearings. Low-speed bearings also tend to deteriorate more slowly than high speed.

When making bearing checks, it must be kept in mind that there are sources other than bearings that produce spike energy. Some of the more common sources that are likely to be encountered are:

(a) gears,
(b) cavitation,
(c) rubbing or
(d) striking of metal parts such as seals and coupling guards.

These sources, if close to the bearing being measured, should be checked to avoid possible misinterpretation of the data. It should also be evident that a programme to detect incipient gear defects can be established in the same manner as that described above for rolling element bearings. Figure 11.9 shows a typical rolling element bearing g-SE severity chart. This chart can be used as an aid in establishing g-SE severity criteria. It is necessary due to the individual characteristics of each machine system, related to machine type, speed, load, bearing type etc. This chart should be used as a guide. By plotting the g-SE levels of the machines to be included within a bearing check programme on the chart, severity criteria can be readily developed and tailored to suit the individual machines involved.

Although spike energy signal amplitudes are a primary indicator of the severity of a bearing defect, a number of factors can also affect the actual measurement levels. These are:

(a) distance of the probe to the bearing defect;
(b) number of interfaces between probe and defect;

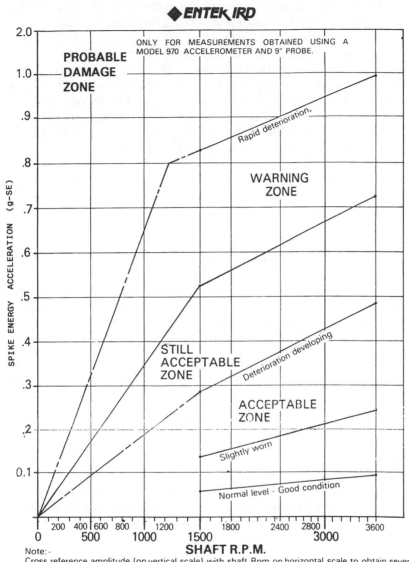

Fig. 11.9 Chart for the evaluation of rolling bearings.

(c) rolling element velocity, that is the RPM and bearing size;
(d) load on bearing;
(e) location of defect on bearing, inner race, outer race, rolling element, or cage;
(f) type defect, that is discrete, distributed, or intermittent; the latter being caused by ball defect spiralling in the raceway;
(g) measurement point relative to the bearing load zone.

Note that the load zone in the final factor above, is that angular region of the bearing that is carrying the machine load. It may be in one direction if the load is caused by a fixed direction force, such as resonance, or rotating if the load is caused by imbalance.

Because these factors affect the amplitudes, caution must be exercised in correlating spike energy signal levels with physical defects. As an example, if a defect occurs on an outer race and measurement is made in the load zone, the level measured can be higher than a similar inner race defect measured out of the load zone. The reason is that the outer race transmission path has fewer interfaces to reduce the defect energy signal levels in comparison with the inner race. Also the defect forces within the zone will be higher than those outside the zone. In spite of these variables there has been considerable consistency in the capability of spike energy to detect bearing defects.

11.10   VIBRATION AND PREDICTIVE MAINTENANCE

From the foregoing discussion, it can be concluded that it is normal for machines to vibrate and make noise. Even machines in the best of condition will have some vibration and noise associated with their operation due to minor defects. It is therefore important to remember that:

1. every machine will have a level of vibration and noise which is regarded as inherent normal;
2. when machinery noise and vibration increase or become excessive, some mechanical defect or operating problem is usually the reason;
3. each mechanical defect generates vibration and noise in its own unique way.

It is the last point which makes it possible to positively identify a problem by measuring and recording the machinery noise and vibration characteristics. If a problem can be detected and analysed early, before extensive damage occurs then:

1. shutdown of the system for repairs can be scheduled for a convenient time;
2. a work schedule together with requirements for manpower, tools and replacement parts can be prepared before the system is shutdown;
3. Extensive damage to the machine resulting from forced failure is minimized.

As a consequence of the above, repair times can be kept short, resulting in less machinery downtime. If excessive vibration is reduced, machine operating life is extended, and time and money are not wasted dismantling machines which are already operating smoothly. A four-stage

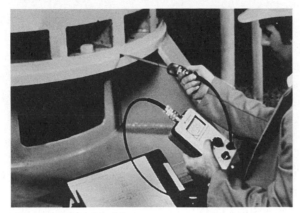

Fig. 11.10   IRD portable hand-held meter.

programme of detection, analysis, correction and verification is required
to monitor the condition of machinery using vibration monitoring and
analysis.

The detection stage of a typical programme would first require
periodic manual vibration measurements on each critical machine. This
is done using either a portable hand-held meter as shown in Figure 11.10
with readings recorded manually on hard copy data sheets similar to
that shown in Figure 11.11. Alternatively, a programmable data collec-
tor/analyser such as the one shown in Figure 11.12a may be used to
download the readings to a computer with dedicated vibration monitoring and analysis software. This type of monitoring is carried out to detect
any increase in vibration over a period of time and to warn of developing defects. Periodic monitoring has the advantage of providing a history
of the machine's condition at a relatively low cost and with flexibility,
one man can take many readings on many machines in a short space of
time. However, some high-performance machines, such as steam and
gas turbines, or high-speed centrifugal pumps and compressors are not
well suited to periodic monitoring. This is because they develop problems very quickly, with little or no preliminary warning. For these
systems, on line, continuous, automatic monitoring is required using
hard-wired vibration sensors located at critical points on the machine.
When the vibration exceeds pre-set levels, a warning may sound or the
machine shutdown automatically.

Vibration monitoring is, however, only the detection part of the total
programme to control machinery condition. Once a significant change in
the vibration level has been detected, the next step is to determine the
nature of the problem. This is the purpose of analysis to pinpoint a
specific machinery problem by identifying its unique vibration characteristics. To undertake this analysis, an instrument such as the analyser
shown in Figure 11.12a is required. The instrument has all the facilities

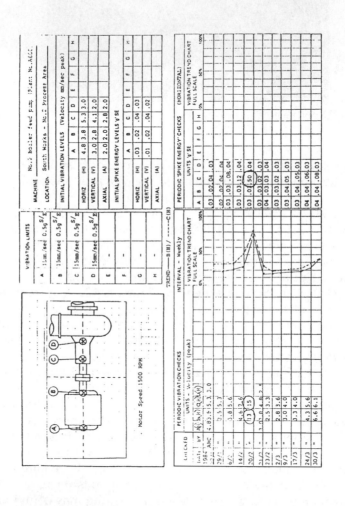

Fig. 11.11 PMP data sheet.

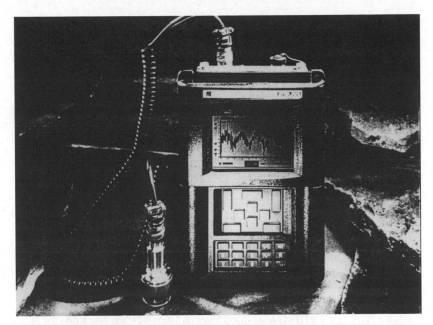

Fig. 11.12a Entek IRD 'Dataline' data collector/analyser: this type of instrument can also be used for on-site analysis and balancing.

Fig. 11.12b Data collector/analyser instruments are used with a dedicated software package for periodic monitoring and analysis of vibration.

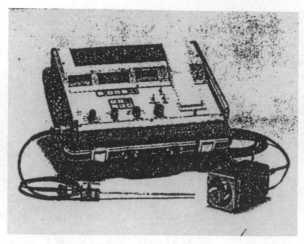

Fig. 11.12c   Microprocessor analyser-balancer.

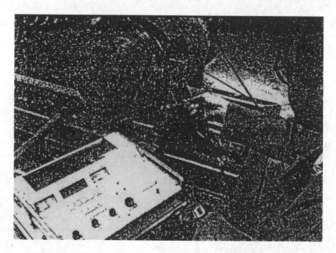

Fig. 11.12d   Analyser-balancer in use for in-place balancing of an electric motor drive pulley.

necessary to measure the detailed characteristics of vibration. By comparing this information with what is known about the machine, its speed or range of speeds, what it does, and the various causes of vibration, the defect can be readily pinpointed and corrections prescribed. With machinery problems detected and diagnosed in their early stages, correction can be scheduled for a convenient time. In the case of imbalance, correction can sometimes be made in-situ using the same data collector/ analyser as shown in Figure 11.12a.

A predictive maintenance programme requires two important skills. The first, which maintenance departments already have, is the knowledge of how machines work, difficulties common to these machines, and how to repair them. The second skill is the ability to recognize and pinpoint operational and mechanical defects accurately and early in their development. Vibration measurement is the key to this skill. Accordingly there are eight basic steps required to put a vibration-based predictive maintenance programme in place. These are:

1. purchase the monitoring instrumentation;
2. train personnel to carry out the programme;
3. list the critical machines to be included in the programme;
4. select periodic vibration checks points;
5. determine each machine's condition and normal level of vibration;
6. establish acceptable levels of machinery vibration;
7. select the interval for periodic checks;
8. start a data record system.

The above steps are simple, and should be implemented sequentially to ensure a successful programme. Care should be taken to include initially, only those machines which are critical to production or its immediate support, or have a poor record of serviceability and/or high maintenance costs.

11.11   MONITORING MACHINE VIBRATION

The machines selected for the programme should then be regularly monitored to identify the normal condition of the machine and to show any increase in vibration when trouble occurs. Most machines have their own unique kind of defects. For instance, fans and blowers are subjected to abrasive wear from the air and materials in the air. Thus machines of this type need to be watched for increasing vibration due to imbalance. On the other hand, a pump is more likely to develop misalignment due to hydraulic conditions, load and other factors during operation. Thus, the points selected for making periodic vibration checks should be selected with close attention to the type of defects a machine is likely to develop.

Normal practice is to select the bearing and pickup direction revealing the highest level of vibration. For machines equipped with rolling bearings, periodic checks should be taken at each bearing using the spike energy technique, to ensure positive detection of early failure. Each checkpoint on a machine should be clearly marked so that all subsequent readings are made at precisely the same location. Remember the purpose of periodic vibration checks is to detect trouble in its early stages. The interval between checks must be short enough to provide a reasonable assurance that the growth in vibration severity is not overlooked. On the other hand, it would not be economical to schedule periodic checks too frequently. Hence there must be some reasonable compromise.

It is important to note that vibration readings should be taken immediately after machinery repairs, overhaul or significant modifications have been made, to re-establish normal vibration levels. While experience and factors such as safety, labour cost, downtime cost and the importance of a machine's operation to company profits must be considered, it is also necessary to establish a safe vibration limit for each machine in the programme, which will provide adequate warning when trouble is present. Since the mechanical condition of a machine is not known initially, it will be necessary to measure the machine's vibration and compare the observed readings with pre-established acceptable levels of machinery vibration. These initial readings can be taken using a portable meter to identify the unfiltered or overall velocity measurements in the horizontal, vertical and axial directions at each bearing of the machine.

If low levels of vibration are found it is usually safe to assume the machine is in good condition, if the overall vibration readings are considered high or excessive, however, this indicates that some mechanical problem exists and a complete vibration analysis should be performed. Such an analysis consists of:

(a) machinery vibration signatures taken in the horizontal, vertical and axial directions at each bearing point on the machine; imbalance will show up predominantly in the horizontal or vertical directions, whereas misalignment will reveal relatively high amplitudes in the axial direction.

(b) a comparison of the acquired data with the characteristics for various causes of vibration known to afflict the machine, and which cause failure, thereby positively identifying the problem.

With the problem identified, correction can then be scheduled for a convenient time. All vibration readings should be recorded on forms similar to those shown in Figure 11.11 or on trend charts or spreadsheets using dedicated software packages. As the data recording system is used to relay important information to decision-makers it must be effective without being excessive. Paperwork should be kept to a minimum. The use of modern computer technology to generate automated management reports is a real time-saver.

Occasionally, the solution to a machinery noise or vibration problem may require that special techniques or instrumentation be used. For example, the data collector/spectrum analyser, shown in Figure 11.12b provides a display of the total amplitude versus frequency (FFT) or the time waveform in 'real time'. This ability to observe vibration or noise almost instantaneously is useful for evaluating transient noise and vibration while the machine undergoes operational changes. This provides information that would be difficult to obtain in any other way.

Some modern computer vibration monitoring software, may now incorporate an 'expert' automated diagnostic program.

There are two basic forms of expert system, 'interactive' and automated'. The interactive system produces the desired analysis by a question and answer routine. This can be very time-consuming and is only as good as the knowledge base. The fully automated system, on the other hand, will detect any machines in alarm, retrieve the machine mechanical details from the machine database, compare the new spectrum with the baseline spectrum and then identify those frequencies that have changed. It will then compare these frequencies with the knowledge base and then write the analysis report. All with no human intervention (Milne, 1994).

11.12 CONCLUSIONS

A programme of preventive maintenance using vibration monitoring can result in definite benefits. These include prolonged machinery life, minimum unscheduled downtime, reduced maintenance costs, elimination of unnecessary overhauls, no requirement for standby equipment, quieter operation, increased machinery safety, quality improvement and customer satisfaction.

When selecting the measurement parameter for analysis, there are two factors to consider. First, why is the reading being taken? Secondly, what is the frequency of the vibration? With these two criteria in mind, the following recommendations are offered.

1. For general machinery analysis, use velocity whenever possible. Amplitude readings in velocity are directly comparable to severity without the need to cross refer amplitude with frequency. Where vibration frequencies are very low (approximately below 600 CPM), where stress is important, and where velocity readings may be quite low, displacement should be used. For very high frequencies (approximately above 120 000 CPM), where applied forces are important and where velocity and displacement readings are quite low, use acceleration.
2. For periodic vibration checks or continuous monitoring use velocity where possible, especially on machines where a number of potential problems may arise, such as imbalance, misalignment etc. A velocity reading will detect a significant development of a defect from any one of these sources. If displacement readings alone are taken, the relatively small increase in displacement from a high frequency source, such as a bearing problem, may be hidden by the larger imbalance or

misalignment vibration. However, displacement may be used if the machine problems are relatively low in frequency.
3. For periodic checks on machines with gearing or rolling element bearings use spike energy measurement wherever possible. This provides the best means of detecting the rate of change, and state of change, associated with very high-frequency low amplitude signals.

11.13 ACKNOWLEDGEMENTS

The author wishes to acknowledge with grateful thanks, the use of Entek IRD International information and previously published material, in the preparation of this chapter.

11.14 REFERENCES

Entek IRD International (1996) *Vibration Technology Course One–Training Manual.*
Fox, R.L. (1977) Preventive maintenance of rotating machinery using vibration detection, *Iron and Steel Engineer*, **54**(4), April.
ISO Standard–ISO 2372 (1974) The measurement of machinery vibration, *International Standards Organisation (ISO)*, Geneva, 1974 (BS 4675 Part 1 1974).
Note: This standard has now been superseded by ISO Standard, 10816-1:1995 (BS 7854 Part 1:1996). All references made in this chapter to the previous standard apply equally to the new standard.
Milne, R. (1994) *Amethyst: An Expert System for the Diagnosis of Rotating Machinery*, Intelligent Applications Ltd.

CHAPTER 12

Common vibration monitoring techniques

Joseph Mathew

Centre for Condition Monitoring, Monash University,
Department of Mechanical Engineering, Wellington Road,
Clayton, Victoria 3168, Australia

12.1 INTRODUCTION

It is generally accepted that maintenance can be performed via one of three different approaches. Breakdown maintenance is practised when machines are operated until they fail, without any maintenance being conducted upon them until the failure occurs. Scheduled maintenance, on the other hand, is concerned with interrupting production on machines at regular intervals for maintenance. This is done in order to reduce the number of unplanned stoppages that can arise as a result of a breakdown maintenance strategy. However, the selection of an optimum maintenance interval is a difficult task in many situations, even when statistical probability theory is used. Too frequent maintenance work, resulting from a short interval, can be a waste of resources and also of valuable production time.

In addition, it can increase the risk of damage to machinery, a possibility that may arise due to the chance of human error during reassembly. On the other hand, too long an interval would result in more failures than desired due to premature breakdown. Consequently, a more acceptable maintenance policy would be to determine the maintenance interval by monitoring the actual condition of machinery. This third approach certainly results in increased plant availability, which in turn produces a greater reward from the capital invested. Other benefits include reduced maintenance costs and improved safety, particularly on machinery where failure may constitute a health or physical hazard.

Handbook of Condition Monitoring
Edited by A. Davies
Published in 1998 by Chapman & Hall, London. ISBN 0 412 61320 4.

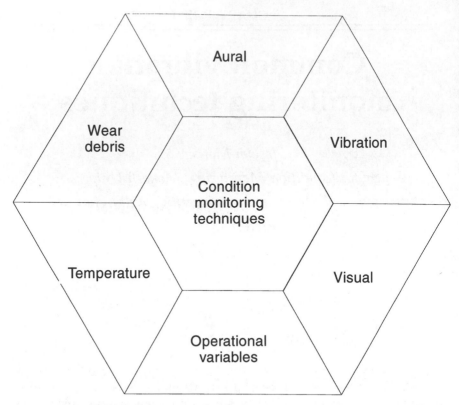

Fig. 12.1   Condition monitoring techniques.

The clear advantages offered by the application of the 'on-condition' approach has in recent times led to the development of a vast number of techniques for monitoring machinery condition (Collacott, 1977; Neale, 1979). These can be broadly classified under the categories described in Figure 12.1. Aural and visual monitoring techniques are considered to be basic forms of surveying machine condition. It is commonly accepted that skilled personnel, who have an intimate knowledge of machines, are capable of indentifying a potential failure by simply listening to the sounds of distress emitted by a nearby machine.

There are those who have enhanced these skills by being able to 'listen' to machine vibrations. The method consists of placing a spanner or rod against the machine and the ear or earmuffs. An extension to the technique involves the use of a stethoscope. These instruments ranging from simple rod cum tube and earpiece devices to those that pick up the rod or tube vibrations using a microphone. The signal obtained from the microphone can then be either amplified or filtered before being fed into a headset (Marson and Campbell, 1981).

Visual monitoring, as outlined in a previous chapter of this book, can sometimes provide a direct indication of the machine's condition without the need for further analysis. As mentioned earlier, the available techniques can range from using a simple magnifying glass or low-power microscope to a light assisted device such as Borescope or Stroboscope (Neale, 1979). Other forms of visual monitoring include the use of dye penetrants to provide a clearer definition of any cracks occurring on the machine's surface, and the use of heat sensitive or thermographic paints. Such paints do need to be selected to match the temperatures to be measured, and this would involve some a priori knowledge of the likely surface temperatures.

The monitoring of the operational variables in machinery is sometimes also known as performance or duty cycle monitoring (Collacott, 1977). The objective of this technique is to assess each machine's performance with regard to its intended duty. Any major deviations from expected or design values indicate the existence of a problem which usually relates to machine malfunction. However, in many machines, the monitoring of their operating variables tends to be less sensitive than other more direct methods, in detecting incipient or early failure. Quite often the failure in machine elements has to be well advanced before a measurable change is produced in its operational variables.

The use of temperature monitoring, which has also been outlined previously, consists of measuring of the operational temperature and the temperature of component surfaces (Neale, 1979). Monitoring operational temperature can be considered as a subset of the operational variables case discussed above. The monitoring of component temperatures has been found to relate to wear occurring in machine elements, particularly in journal bearings, where lubrication is either inadequate or absent (Mathew and Alfredson, 1989). Techniques for monitoring temperature of machine components, as we have seen, can include the use of optical pyrometers, thermocouples, thermography and resistance thermometers.

Wear debris are generated at relative moving surfaces of load-bearing machine elements. Hence, it is possible to assess the condition of these surfaces if the wear debris are collected and analysed. The techniques currently in use are wide-ranging, and these are described in a later section of this book. However it is important to acknowledge the existence of these techniques, and to note that they can be used successfully in conjunction with, say, vibration monitoring, to provide corroborative evidence of machinery failures (Kuhnell and Stecki, 1985; Mathew and Stecki, 1987).

As we have seen, equipment vibrations arise from the cyclic excitation forces which are present within a machine. These forces are sometimes built into the design of the equipment, or can be due to real changes in the dynamic properties of the individual machine elements due to wear or failure. These excitation forces may be transmitted to adjacent components or the adjoining machine structure, thus causing parts of the equipment remote from the source to vibrate accordingly to varying degrees.

The basic measurement of machinery vibration has been discussed in a previous chapter and as outlined can be made using a wide array of transducers (Collacott, 1979). Accordingly, these are not discussed here. However, it is important to mention that the piezoelectric accelerometer is probably the most popular measurement transducer in use today. This is due to its wide frequency and dynamic range. It should also be noted that the acceleration signals obtained from these transducers are sometimes integrated to produce velocity or even displacement values for different applications. These signals can then be processed in a number of alternative ways to highlight different aspects of the data, which may then be used in the detection and diagnosis of machine condition. The various techniques can be broadly classified under the categories shown in Figure 12.2 and these will now be described in the sections that follow.

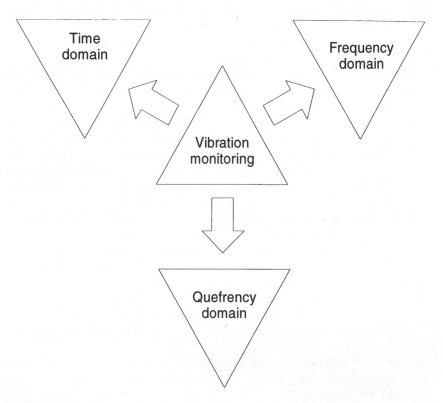

Fig. 12.2 Vibration monitoring techniques.

12.2 THE TIME DOMAIN

Time domain vibration signals, if understood properly, can yield enormous amounts of information. Further analysis is usually carried out so that some characteristics not readily observed visually are highlighted. Several techniques have either been proposed, or have been used in machine condition monitoring and these are shown in Figure 12.3.

Waveform analysis consists of recording the time history of the event on a storage oscilloscope or a real-time analyser. Apart from an obvious fundamental appreciation of the signal, such as if the signal was sinusoidal or random, it is particularly useful in the study of non-steady conditions and short transient impulses. Discrete damage occurring in gears and bearings, such as broken teeth on the former and cracks in the inner and outer races of the latter, can be identified relatively easily (Alfredson and Mathew, 1985a). An example is shown in Figure 12.4, where the waveform of the casing vibration acceleration of a single-stage gearbox with a broken tooth on the pinion is presented.

The pinion in this example was directly coupled to a 5.6 kiloWatt (kW), 2865 revolutions per minute (RPM), alternating current (AC) electric motor. Under nominal load, a shaft speed of approximately 3000

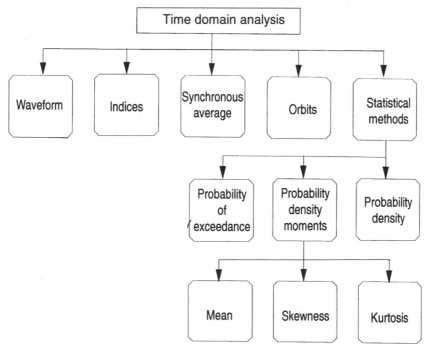

Fig. 12.3 Time domain analysis tree diagram of analysis techniques.

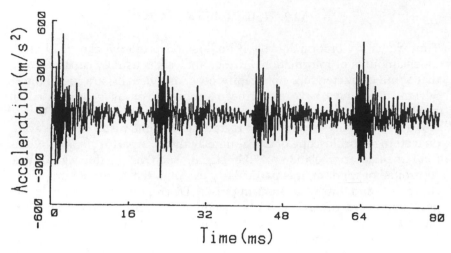

Fig. 12.4　Time domain waveform of a damaged gear.

revolutions per minute was obtained. Hence a single discrete fault on the pinion, such as the broken tooth, would produce pulses in the time domain with a period of occurrence of 20 milliseconds (ms) or so. This feature is clearly evident in the figure. Waveform analysis can also be useful in identifying beats and vibrations that are non-synchronous with shaft speeds, and in machine coast down analysis waveforms can indicate the occurrence of resonances (Collacott, 1979).

Indices have also been used in vibration analysis. The peak level, RMS level and their ratio Crest factor are often used to quantify the time signal. The peak level is not a statistical quantity and hence may not be reliable in detecting damage in continuously operating systems. The RMS value, however, is more satisfactory for steady-state applications. Even so, the values of either of these parameters tend to be governed by the amplitude of large components occurring in the time signal. Hence, unreliable data would be obtained when monitoring machinery where the time signal contains information pertaining to more than one element, say a multistage gearbox where the time signal would contain information from the high- and low-speed gears as well as the bearings.

The crest factor, defined as the ratio of the peak to RMS levels, has been proposed as a trending parameter as it includes both parameters (Braun, 1980). However, investigations by the author have shown that this parameter usually increases marginally with incipient failure, and subsequently decreases due to the gradually increasing RMS value typical of progressive failure. Quite often, the trend recorded by this parameter has been found to be similar to another time domain parameter, the Kurtosis factor. Evidence of this similarity will be produced later when the Kurtosis factor is discussed.

Synchronous averaging is the time signal averaged over a large number of cycles, and synchronous with the running speed of the machine. This technique not only removes the background noise, but also periodic events not exactly synchronous with the machine being monitored. It is especially useful in gear-vibration diagnosis where multiple shafts are present. All components not synchronous with the shaft of interest can be deleted. A typical measurement setup would consist of a transducer, usually an accelerometer, a tachometer which would produce the reference pulse and a signal averager. Signals of shafts not synchronous with the reference shaft can also be averaged, if the repetitive rate of the reference pulses are altered with the aid of a pulse frequency multiplier (Collacott, 1979).

Patterns known as Lissagous figures are obtained by displaying time waveforms obtained from two transducers whose outputs are phase shifted by 90°, on an oscilloscope where the time base is substituted with the signal from one of the probes. When shaft relative displacement probes are used, the pattern so obtained is the shaft orbit, and this can be used to indicate journal bearing wear, shaft misalignment, shaft imbalance, lubrication instabilities in hydrodynamic bearings and shaft rub (Collacott, 1979). The techniques of proximity analysis are well established, particularly with applications to turbomachinery. Eddy current transducers are commonly used in addition to the appropriate signal conditioners. However, care needs to be exercised when mounting these eddy current devices, so that the effects of mechanical and electrical runout are minimized (Anon).

If this is not done, then spurious signals will be obtained during measurements. An example of orbit analysis, used in monitoring journal bearing wear, is presented in Figure 12.5. Here two eddy current proximity probes were placed at 90° radially along the shaft. Mechanical glitch was minimized by accurate machining of the target area of the probes. Electrical glitch was minimized by pushing an aluminium sleeve onto the target area. In this particular test, the bearing inner diameter was deliberately enlarged by boring out bearing material. The results clearly showed that the orbit diameter increased particularly in the vertical direction, indicating that this rotor-bearing system was stiffer in the horizontal direction. Hence, the added bearing clearance due to the simulated wear condition resulted in larger vertical relative displacements.

Statistical analysis can also be carried out on time domain data. The probability density, for example, is the probability of finding instantaneous values within a certain amplitude interval, divided by the size of the interval. All signals will have a characteristic probability density curve shape. These curves if derived from machinery vibration signals can subsequently be used in monitoring machine condition. Examples of the application of this parameter to rolling element bearing monitoring are readily available (Alfredson and Mathew, 1985a; Braun, 1980; Dyer

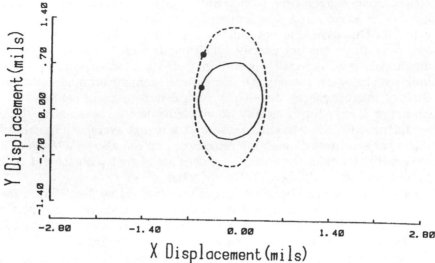

Fig. 12.5 Orbit monitoring.

and Stewart, 1978). A typical example of damage detection in a high-speed rolling element bearing is shown in Figure 12.6.

Damage was induced by producing a small radial groove on the outer race of the bearing. This action resulted in the vibration acceleration waveform looking not unlike Figure 12.4. The waveform when the bearing was in good condition was characterized by a random signal, with a probability density curve that was similar to the normal distribution or bell-shaped curve. Note that the expression of the vertical axis in the logarithmic scale tends to produce a distorted shape when compared with probability curves expressed in linear scales. This is deliberately carried out to highlight changes at low probabilities if and when they occur.

The horizontal axis in Figure 12.6 is the vibration acceleration signal normalized to the standard deviation. The probability density curve for the damaged bearing when compared with the curve for the good bearing is seen to be significantly different in shape. The high levels of probability density at the medium and the large spread at low probabilities are characteristics of highly impulsive time domain waveforms. The probability density technique is useful in the diagnosis of machine condition, as it is based on comparisons of shape variations rather than amplitude variations.

Probability density curves can also be trended like any other trending parameter if the data can be presented in the form of waterfall or cascade diagrams. For example, Figure 12.7 shows the effect of shaft rubbing on the housing acceleration of a journal bearing. Baseline curves, occurring

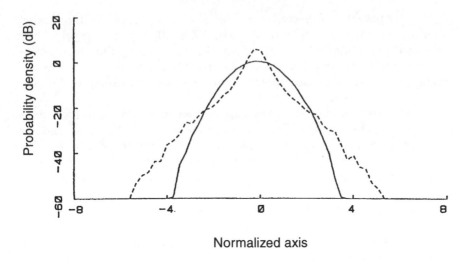

Fig. 12.6 Probability distribution of bearing vibration acceleration amplitude.

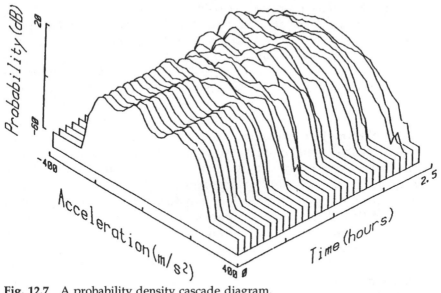

Fig. 12.7 A probability density cascade diagram.

earlier in the cascade diagram, are clearly shown to undergo distinct alterations in both shape and amplitude, indicating a malfunction of some sort. Presentation of data in this form can readily provide a large portion of the required information for condition monitoring in quite a short space of time.

The probability of exceedance has also been used in monitoring bearing condition (Dyer and Stewart, 1978). It is the integral of the probability density curve and gives the probability that the instantaneous vibration amplitude exceeds any particular amplitude. Again, these curves can be monitored to provide an indication of bearing failure.

The shape of the probability density curve can be described by a series of single-number indices. These are the moments of the curve and are analogous to mechanical moments about the centroid of a plane. The first and second moments are well-known as the mean and the mean square. Odd moments relate information about the position of the peak density relative to the median value. Even moments indicate the spread in distribution. Usually moments greater than two are normalized by removing the mean and dividing by the standard deviation raised to the order of the moment. The third moment is Skewness and the fourth moment is Kurtosis. For practical signals the odd moments are usually close to zero whereas the higher even moments are sensitive to impulsiveness in the signal.

Kurtosis has been selected as a compromise measure between the insensitive lower moments and the over-sensitive higher moments. This parameter has been proposed as being sensitive to failure in rolling element bearings (Dyer and Stewart, 1978). However, an independent evaluation of this technique (Mathew and Alfredson, 1984) has shown that high Kurtosis values would only be obtained if the original time waveform was of an impulsive nature. This characteristic does arise in some forms of failure in bearings such as cracked races. Spalls occurring on the edges of the rolling elements of barrel or spherical roller bearings can also cause relatively large pulses in the time domain waveform. When applied to the previous example of a bearing where the outer race was damaged, the trend of the Kurtosis factor is as shown in Figure 12.8.

The Crest factor trends are also shown in this figure. Both of these parameters produced values of approximately '3' which indicated that the waveform was generally random in nature when the bearing was in good condition. Trends changed dramatically especially for the Kurtosis factor when damage was introduced. A peak value of '13' or so was attained by the Kurtosis parameter, signifying that the shape of the probability density curve had changed appreciably. Progression of the damage did not show continued increases in the value of Kurtosis. Instead the converse was true, indicating that the impulsive content in the waveform gradually decreased. It is acknowledged, though, that in this particular example, trend increases were recorded towards the termination of the test.

Further tests were also conducted where the bearings were subjected to overload and loss of lubrication. In only a few cases was the Kurtosis

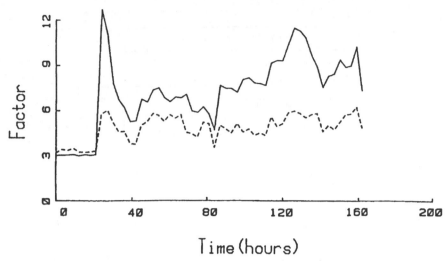

Time (hours)

Fig. 12.8 Comparison of Kurtosis and Crest factor trends for outer race damage at 21 hours of operation.

technique successful in detecting damage. Even in these successful instances, the amplitude of this parameter decreased significantly after an initial rise, suggesting that the waveform of the time domain signal became increasingly more random as failure progressed. These results clearly show that the Kurtosis parameter, although useful in some forms of failure, could not be relied on as a trending parameter for the purposes of prognosticating bearing condition.

The trends recorded by the Crest factor were surprisingly similar to that recorded by the Kurtosis. Of the several tests conducted, the largest discrepancy occurred in the example presented previously. In some cases, the trends recorded by both parameters were almost identical. The implications of this are that simple Crest factor meters, which cost significantly less than Kurtosis meters, can be used in place of the latter for the monitoring of rolling element bearing condition.

12.3   THE FREQUENCY DOMAIN

Digital fast Fourier analysis of the line waveform has become the most popular method of deriving the frequency domain signal. The signature spectrum so obtained can provide valuable information with regards to machine condition (Collacott, 1979). Enveloping or demodulating the time waveform prior to performing the Fast Fourier Transform (FFT) is also gaining popularity and is included in the following discussion on

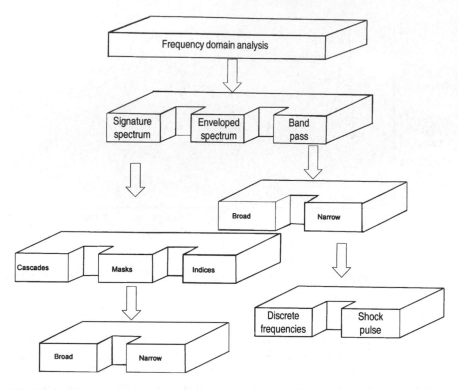

Fig. 12.9 The frequency domain analysis techniques.

frequency domain techniques. Finally, spectral information can also be derived using analogue filter sets tuned to passing the information only in bands of interest. Each of these techniques can be related as shown in Figure 12.9.

The vibrational characteristics of any machine are to some extent unique, due to the various transfer characteristics of the machine. The method of assembly, mounting and installation of the machine all play a part in its vibration response. Consequently, when a machine has been commissioned, a signature spectrum should be obtained under normal running conditions. This signature will provide a basis for later comparison in order to locate those frequencies in which significant increases in vibration level have occurred. Figure 12.10 is an example of how effective signature spectral comparisons can be in both detecting and diagnosing failure.

Failure was induced on this occasion by removing lubrication on several occasions in a gearbox. The final spectra clearly recorded increases throughout the frequency range. The arrows indicate the positions of those components related to the gear mesh component and two

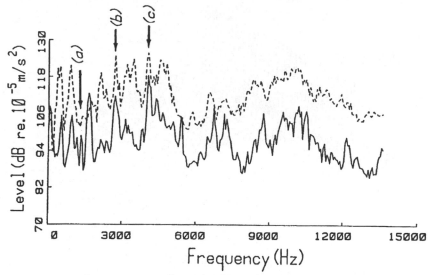

Fig. 12.10 Gearbox casing acceleration signature spectra.

harmonics. Indeed this technique of spectral comparisons is widely used. Often an operator records on-site data on a tape recorder/data logger and analyses them on site, or in the laboratory/office using digital real-time fast Fourier analysis. Manual or computerized comparisons can then be made. More recently, the process has been automated somewhat by using portable microprocessor instrumentation which has both memory and intelligence. These devices simplify the data-acquisition process and the amount of housekeeping required, as reliable interfaces with appropriate computer systems are available.

Signature spectra can also be compared by plotting successive spectra with respect to time in the form of cascade plots. In this fashion, a developing fault can sometimes be easily recognized (Mathew and Alfredson, 1984). An example of a spectral cascade is presented in Figure 12.11, where an eddy current transducer was used to monitor the shaft relative displacement of the input shaft of a single-stage gearbox. The gears were subjected to an overload condition and the signals of the shaft displacement transducer were monitored regularly. The spectral cascade obtained showed that significant amplitude increases occurred towards the end of the test, clearly indicating that a change in gearbox condition had occurred. This type of presentation can also be used to illustrate spectra derived from acceleration transducers, provided the spectra do not contain too much information. In many cases, the harmonic content provided by acceleration transducers can be significant. This, coupled with their large frequency response, normally produces cascade plots that are cluttered and therefore of limited use.

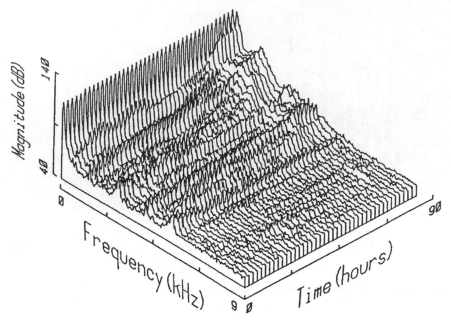

Fig. 12.11 Spectral cascades.

The amount of information present in a cascade plot can be reduced if each spectral change that occurred is expressed in a single number. Various indices have been proposed (Alfredson and Mathew, 1985b). By and large, these spectral indices have been found to be more sensitive than time domain indices such as peak and RMS levels in the monitoring of rolling element bearings (Mathew and Alfredson, 1984), journal bearings (Mathew and Alfredson, 1989) and more recently gears (Mathew and Alfredson, 1987). In some cases, the differences in trends recorded by the frequency domain parameters can be quite startling. Figure 12.12 depicts a comparison of trends of time and frequency domain parameters when applied to monitoring journal bearing condition.

Lubrication was removed approximately 0.5 hours after the start of the test. More damage was then induced to the bearing by increases in the load to the bearing on two occasions as shown in the figure. The frequency parameter Matched Filter RMS recorded increases in trends throughout the test duration. The instances when reductions were recorded occurred during applications of the load. Although amplitudes of the frequency domain parameter changed by as much as 15 decibels (dB), the performance recorded by the time domain parameters was disappointing. The reason for this difference in the trends of these

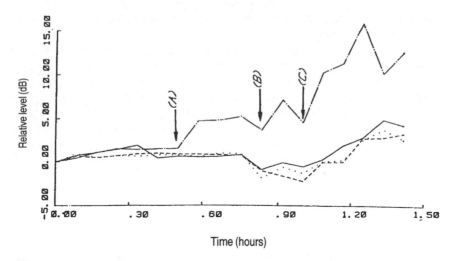

Time (hours)

Fig. 12.12 Trend parameters: (a) terminate lubricant supply; (b) impose light load; (c) increase load.

parameters and the implications thereof, particularly for monitoring journal-bearing condition, have been discussed previously (Mathew and Alfredson, 1989).

An alternative method of trying to evaluate changes occurring in the signature spectrum is to form a spectral mask. This is derived from the baseline signature, plus an allowable tolerance limit (Randall, 1981a). Regular comparisons of new spectra with this mask will indicate if a problem is occurring. Sometimes, the narrow bandwidth spectrum is converted to a constant percentage bandwidth spectrum to compensate for speed variations. Broad masks are used if large speed variations are encountered, while narrow masks are used if only minor fluctuations in speed are anticipated. It is to be noted that a considerable amount of experience is required in the drawing up of accurate maintenance limit masks, as vibrational spectral amplitudes can also change under the varying duty cycles a machine may experience.

The enveloped spectrum has been shown to be particularly useful in monitoring machine elements that fail by producing relatively short duration impulses (Prasad, Ghosh and Biswas, 1985), a feature typical of incipient damage in rolling element bearings. Often, incipient damage in rolling element bearings cannot be detected using signature spectral comparisons, as the energy contributions by these impulses are usually swamped by the vibration components of more dominant elements, or the extraneous influences of nearby machinery. The technique involves, first, a high-pass filtering operation to remove dominating low-fre-

quency components in the spectrum. The resulting signal is then rectified partially or fully.

A normal frequency spectrum is then derived using either a real-time
analyser or a computer. If the latter is used, then a low-pass filter is also
incorporated to prevent digital aliasing. The spectrum so obtained is
sometimes also called a demodulated spectrum. Bearing damage in
complex machinery has been detected by this technique (Harting, 1978).
However, caution must be exercised when monitoring rolling element
bearings with a more progressed failure mode, as these high-frequency
impulses are usually not present in data collected from such bearings.
Hence, the technique may not be successful in detecting gross damage
in bearings.

Pass band analysis is yet another technique of reducing the quantity
of data made available in a spectrum to manageable proportions. The
technique involves monitoring only a band of frequencies either broad
or narrow, in which defect frequencies of components are anticipated
(Neale, 1979). The monitoring of a narrow-pass band is sometimes
known as discrete frequency monitoring. Many of the existing vibration
monitoring software packages that incorporate trending also allow
discrete frequency monitoring. Hence, for example, the trend of the gear
mesh frequencies or bearing defect frequencies could be monitored.
Again, this approach if used on its own can present problems. For
example, with reference to the gear mesh component and its harmonics
shown in Figure 12.10, larger amplitude increases were recorded by
many other frequency components in the spectrum. Hence, trending a
parameter that describes all components in some way would provide a
better measure of failure occurring.

The shock pulse method (Beercheck, 1976) may be considered to be a
specialized application of characteristic frequency monitoring. Failure of
high-speed rolling element bearings results in energy being emitted at
ultrasonic frequencies. This technique uses an accelerometer tuned
mechanically and electrically to a frequency of 32 kiloHertz (kHz) to
detect these distress signals. It is a relatively simple measurement and is
widely used in industry to monitor high-speed bearings.

12.4   THE QUEFRENCY DOMAIN

Quefrency is the abscissa for the cepstrum which is defined as the
spectrum of the logarithm of the power spectrum. It is used to highlight
periodicities that occur in the spectrum in the same manner as the
spectrum is used to highlight periodic components occurring in the time
domain signal. The derivation of the cepstrum is not a trivial matter and
care must be exercised when doing so (Randall, 1981b). Cepstrum
analysis has predominantly been used in the analysis of gearbox vibra-

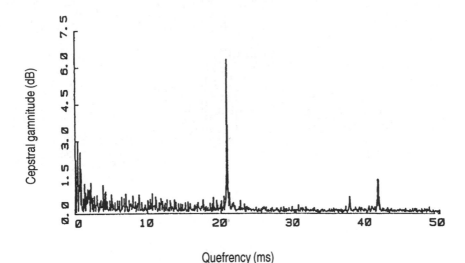

Fig. 12.13 A gearbox casing acceleration cepstrum.

tions (Randall, 1980). Strong components can often be detected in the cepstrum due to the presence of modulation components or sidebands in the spectrum.

An example is presented in Figure 12.13, which shows the cepstrum of a single stage 1:1 gearbox casing acceleration signal. The large component at 20.75 ms corresponds to the shaft fundamental modulation component, and that at 41.5 ms is a 'rahmonic' or the equivalent to a harmonic in spectral terms. The 'gamnitudes', which are similar to the magnitude of spectral components, are, as shown in the figure, the vertical measurement of the signal. These components usually increase with increasing wear in gears, although there have been instances when cepstral quefrencies have recorded decreases in gamnitude. An illustration of this condition is presented in Figure 12.14.

The bandwidth of the cepstral components were broadened so that gamnitude variations could be readily discerned. The diagram was obtained using casing acceleration signals from the signal-stage gearbox which was subjected to an overload condition. The initial increase in gamnitude of the fundamental component and its rahmonic was due to pitting in the pitch line of the spur gears. Continued operation caused further damage such as scoring of the addendum and dedendum of the teeth. The test was finally discontinued due to excessive vibration and acoustic noise.

Signature spectral comparisons showed that the base spectral noise level increased significantly with advanced damage and tended to overwhelm the shaft modulation components. Hence, the emphasis of the modulation components in the spectrum was reduced, resulting in

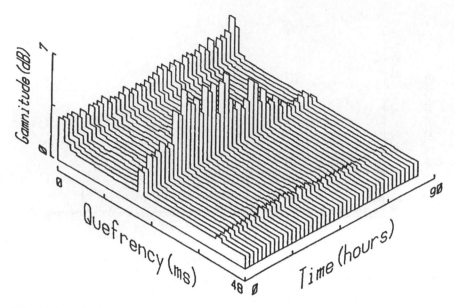

Fig. 12.14 A cepstral cascade diagram.

the lowering of the gamnitudes of associated quefrency components. Other forms of cepstral analysis include comparisons of signature cepstra, and single number indices derived from successive cepstra in much the same way as signature spectral techniques and spectral indices are used. However, one must be prepared to accept reducing values which may be indicative of advanced failure modes.

12.5 CONCLUSIONS

It must now be evident to the reader that condition monitoring encompasses measurement and analysis techniques that belong to a wide variety of scientific disciplines. This chapter has been predominantly concerned with vibration analysis, and has attempted to present an overview of the many techniques either being researched or currently in use in industry. In the final analysis it must be emphasized that the majority of these techniques work best at detecting a single symptom of the machine malfunction. For example, spectrometric oil analysis can only provide information on the concentrations of wear metals in an oil sample. Particle counting on the other hand only provides information about the size and distribution of wear debris. Vibration monitoring is well suited towards the monitoring of rotating machinery but can present difficulties when attempting to monitor reciprocating machinery.

Even within each discipline, subsets of techniques are available. For example, the major portion of this chapter has been dedicated to various vibration analysis techniques suitable for monitoring machine condition. No one technique can provide all the answers. The Kurtosis technique may be suitable for monitoring rolling element bearings but is unsuitable for monitoring journal bearings. Similarly, cepstrum analysis is a useful tool in diagnosing gear failures, but so far, it has not been employed in detecting rolling element bearing failures and with good reason.

Consequently, it would appear that the detection, diagnosis and prognosis of wear or failure occurring in machinery can be greatly enhanced if a 'syndrome' approach is adopted, in which a combination of symptoms identified by appropriate techniques is used to identify failure. This approach has not been widely adopted, and this has been partly due to the increased costs associated with utilizing a wide array of techniques. Yet another reason for the low use of such an approach is the strong preference for individual techniques, as exercised by researchers and engineers currently involved in condition monitoring.

12.6 REFERENCES

Alfredson, R.J. and Mathew, J. (1985a) Time domain methods for monitoring the condition of rolling element bearings, *Transactions of the Institution of Engineers*, Australia, Vol. ME10, July, 102–7.

Alfredson, R.J. and Mathew, J. (1985b) Frequency domain methods for monitoring rolling element bearings, *Transactions of the Institution of Engineers*, Australia, Vol. ME10, July, 108–12.

Anon, (undated) Glitch–Definition and methods for correction including shaft burnishing to remove electrical runout, Bently Nevada Applications Note.

Beercheck, R.C. (1976) Listening for the sounds of bearing trouble, *Machine Design*, **48**, 82–6.

Braun, S.G. (1980) The signature analysis of sonic bearing vibrations, *IEEE Transactions on Sonics and Ultrasonics*, Vol. SU-27, **6**, November, 317–28.

Collacott, R.A. (1977) *Mechanical Fault Diagnosis and Condition Monitoring*, Chapman & Hall, London, UK.

Collacott, R.A. (1979) *Vibration Monitoring and Diagnosis*, John Wiley & Sons, New York, USA.

Dyer, D. and Stewart, R.M. (1978) Detection of rolling element bearing damage by statistical vibration analysis, *ASME Transactions–Journal of Mechanical Design*, **100**(2), 229–35.

Harting, D.R. (1978) Demodulated resonance analysis–a powerful incipient detection technique, *ISA Transactions*, **17**(1), 35–40.

Kuhnell, B.T. and Stecki, J.S. (1985) Correlation of vibration, wear debris and oil analysis in rolling element bearing condition monitoring, *Maintenance Management International*, **5**, 105–15.

Marson, R. and Campbell, S. (1981) Versatile electronic stethoscopes, *Electronics Today International*, **10**, August, 30–3.

Mathew, J. and Alfredson, R.J. (1984) The condition monitoring of rolling element bearings using vibration analysis, *ASME Transactions–Journal of Vibration, Acoustics, Stress and Reliability in Design*, **106**, 447–53.

Mathew, J. and Alfredson, R.J. (1987) The monitoring of gearbox vibration operating under steady conditions, *Proceedings of the 11th Biennial ASME Conference on Mechanical Vibration and Noise, Mechanical Signature Analysis*, Boston, 27–30 Sept., 47–54.

Mathew, J. and Alfredson, R.J. (1989) The condition monitoring of journal bearings using vibration and temperature analysis, *Journal of Condition Monitoring*, **2**(3), 193–212.

Mathew, J. and Stecki, J.S. (1987) Comparison of vibration and ferrographic techniques in application to high speed gears operating under steady and variable conditions, *The Transactions of the ASLE*, August, 105–15.

Neale, M. (1979) *A Guide to the Condition Monitoring of Machinery*, HMSO, London, UK.

Prasad, H. Ghosh, M. and Biswas, S. (1985) Diagnostic monitoring of rolling element bearings by the frequency resonance techniques, *Transactions of the ASLE*, **28**(4), 439–48.

Randall, R.B. (1980) Advances in the application of cepstrum analysis to gearbox diagnosis, *2nd International Conference on Vibrations in Rotating Machinery*, I. Mech. E, Cambridge, 169–74.

Randall, R.B. (1981a) Computer assisted incipient fault detection on rotating and reciprocating machines, *Noise and Vibration Control Worldwide*, **12**, Sept., 230–4.

Randall, R.B. (1981b) Cepstrum analysis, *Bruel and Kjaer Technical Review*, No. 3.

APPENDIX   EQUATIONS FOR COMMON VIBRATION MONITORING TECHNIQUES

1. Peak level (time domain), L_{pk}

$$L_{pk} = 20\log(A_{max}/10^{-5}),\ dB \qquad (12.1)$$

where:

A_{max} = maximum value in the time domain;

10^{-5} = reference acceleration in ms^{-2}.

2. RMS level (time domain), L_{rms}

$$L_{rms} = 20\log\left(\left(\sqrt{\left((1/M)\left(\sum_{i=1}^{M}(A_i^2)\right)\right)}\right)\middle/10^{-5}\right),\ dB \qquad (12.2)$$

where:

A_i = amplitude of the ith digitized point in the time domain;
M = number of points in the time domain.

3. Crest factor, C_f

$$C_f = A_{max}/A_{rms} \tag{12.3}$$

where:

$$A_{rms} = \left(\sqrt{(1/M)} \left(\sum_{i=1}^{M} (A_i^2) \right) \right)$$

4. Matched filter RMS (frequency domain), Mf_{rms}

$$Mf_{rms} = 10 \log \left((1/N) \left(\sum_{i=1}^{N} (F_i/F_{iref})^2 \right) \right), \text{ dB} \tag{12.4}$$

where:

F_i = amplitude of the ith point in the frequency spectrum;
F_{iref} = amplitude of the ith point in the baseline spectrum;
N = number of points in the frequency spectrum.

5. Kurtosis factor (time domain), K_f

$$K_f = \int_{-\infty}^{\infty} ((x - \bar{x})^4 p(x) dx)/\sigma^4 \tag{12.5}$$

where:

x = mean value;
$p(x)$ = probability density function of x;
σ = standard deviation of x.

6. Kurtosis (digital approximation), K_{fd}

$$K_{fd} = \left((1/M) \left(\sum_{i=1}^{M} (A_i - m)^4 \right) \right) / \left((1/M) \left(\sum_{i=1}^{M} (A_i - m)^2 \right) \right)^2 \tag{12.6}$$

where:

m = mean value of the digital time series.

CHAPTER 13

Fundamentals of vibroacoustical condition monitoring

C. Cempel

Technical University of Poznan, ul Piotrowo 3,
PL-60-965 Poznan, Poland
e.mail: cempel@put.poznan.pl

13.1   INTRODUCTION

There are several modes of wear in machines and mostly they are connected with vibroacoustical phenomena. Either as causes, effects, or both (Collacott, 1977; Braun, 1985 and Cempel, 1991). For machines, and/or other mechanical Operating Systems (OS), wear means internal energy dissipation, which results in damage and usually a deterioration of condition. As this dissipation is immediate to system operation, we can apply the theory of Energy Transforming Systems (ETS), which is referred to below, together with the concept of symptom reliability. These concepts, which are specially designed for the condition monitoring (CM) of machines, or energy processors (EP), and their sets, are introduced and linked with the theory of condition evolution of EP. This theory is also briefly outlined and applied to the assessment of machine residual life and condition forecasting.

Throughout the chapter the concept of two time scales is extensively used, as initially introduced to diagnostics in Cempel (1991). It defines:

- **Dynamic time**—as short time, order of seconds, where all dynamic and vibroacoustical phenomena are observed and analysed.
- **Lifetime of system or its part**—as long time, weeks, months, years, etc where the time of system operation is counted, and wear phenomena/system evolution occur.

Handbook of Condition Monitoring
Edited by A. Davies
Published in 1998 by Chapman & Hall, London. ISBN 0 412 61320 4.

Although we are dealing with machines in operation, very often the words 'system', 'system operation' or 'operating system' will be used instead, depending on the context.

13.2   WEAR AND VIBROACOUSTIC PHENOMENA IN MACHINES

Operation of machine and equipment means energy transformation. A small part of that energy dissipates inside a system due to various types of wear processes, leading to the degradation of the condition of a system, and to its breakdown. Below, we will analyse this phenomenon from the energy point of view, and for its direct application to condition monitoring (Fitch, 1992). Wear in mechanical systems is intimately linked with their operation and occurs in many ways. It is always closely related to dynamic phenomena such as vibration and acoustics, which includes noise and ultrasound, and also to the energy flow through the system, including its dissipation inside the machine.

Taking this into account, the main type of wear one should first consider is the fatigue phenomenon. If this occurs in a machine's structural part, it causes a loss of integrity. If it occurs on the surface, it can cause pitting and/or spalling of moving parts, i.e. bearings, gears etc. Alternatively 'fretting' can occur, which is the fatigue corrosion of movable structural joints, especially those operating in a corrosive atmosphere. The energy dissipated inside the system due to the various types of fatigue wear can be calculated using the formulae given below. Some of these are of a qualitative nature, due to a broad meaning of the coefficients.

- **Volumetric fatigue**   Morrow's hypothesis gives the following way to calculate the dissipated energy (Kliman, 1985):

$$E_{d1} = 3(\sigma_a^{1+b}/k)^{1/b}n, \, n = \int f(\theta)d\theta \simeq f.\theta, \tag{13.1}$$

where k, and b are material constants, σ_a is the alternative stress amplitude, n is the number of load cycles, f is the instantaneous or average excitation frequency, and θ is the lifetime of the system or the component under consideration.

- **Surface fatigue**   This energy value can be calculated according to the Palmgren relationship as energy dissipated due to pitting and spalling (Luczak and Mazur, 1981):

$$E_{d2} = c_1 p^3 n, \tag{13.2}$$

where additionally c_1 is a material coefficient, and p the unit pressure in the mechanism.

- **Corrosive fatigue or fretting** The wearing energy here is proportional to the volume V_0 of the removed material (Luczak and Mazur, 1981; Poltzer and Meissner, 1983).

$$E_{d3} = c_2 V_0 = c_z(((k_0\sqrt{p} - k_1 p)/f) - k_2 \hat{A})n, \tag{13.3}$$

where additionally c_z is a proportionality constant, k_0, k_1, k_2, are material constants, and $\hat{A}$ is the vibration amplitude.
- **Adhesive/abrasive** This is the second important type of wear which proceeds in every rotating or sliding structural joint. The intensity of this type of wear depends on lubrication quality, the unit pressure and on the relative vibration amplitude. The energy dissipated by frictional wear is proportional to the volume of the abrasively removed material. This is governed by Archard's law (Poltzer and Meissner, 1983; Engel, 1976).

$$E_{d4} = c_3 V_0 = c_3 k_3 p \hat{V} R_e^{-1} \theta, \tag{13.4}$$

where c_3 is a proportionality constant, k_3 a material constant, R_e, the yield stress of the material, and $\hat{V}$ is the amplitude of sliding velocity in a mechanism.
- **Erosion** This is caused by such phenomena as cavitation, corrosion, the impact of fluid stream, ions or other particles. With the exclusion of erosion by radiation, mechanical vibration intensifies erosion and is sometimes the causes of it, for example in the case of cavitation. The dissipated energy for cavitational wear may be assessed by:

$$E_{d5} = c_5 V_0 = c_5 B \hat{V}^{b(\theta)} \theta. \tag{13.5}$$

Where in addition c_5 is the proportionality constant, B a material constant, and $b(\theta)$ the cavitational exponent. The same type of energy formula can be invented for other types of erosion, but with a different interpretation of $\hat{V}$, as the velocity amplitude of the flow of a liquid, or a jet of particles etc.
- **Creep** This is the important type of wear in mechanical systems. It is particularly prevalent at high temperature and for complex mechanical loads. This distortional phenomenon is highly dependent on temperature, mean working stress and vibration. Particularly at high frequencies f. It may be acknowledged here that the dissipated energy is proportional to the creep strain ε_{cy}, highly dependent on the stress σ, the temperature T and the frequency f of ultrasound (Juvinal, 1967; Severdenko, 1976).

$$E_{d6} = c_6 \varepsilon_{cy} = c_4((\sigma/E) + e(T, f)\sigma^d \theta). \tag{13.6}$$

Where, in addition c_6 is a proportionality constant, $e(T, f)$ the creep function, E Young's modulus, and d a material constant.

Having specified the main types of wear and associated dissipated energy (Fitch, 1992), we can conclude that these energies are event dependent (n) like fatigue processes, and lifetime (θ) dependent for other forms of wear; hence one can write in general:

$$E_d = \begin{cases} An - \text{event type wear} \\ B\theta - \text{lifetime wear} \end{cases} \tag{13.7}$$

We have to acknowledge further that each load parameter has its own mean and dynamic component, with subscript a.

$\sigma = \sigma_m + \sigma_a -$ stress

$p = p_m + p_a -$ unit pressure   (13.8)

$V_s = V_m + V_a -$ velocity of fluid stream or rigid body.

These dynamic components are proportional to the total power V of the associated dynamic phenomena observed externally. This, in the simplest case, may be only the power associated with the vibration. Alternatively, it may be with all of the vibroacoustic phenomena vibration, noise, pulsation etc. and also with all the dynamic residual processes taking place in the mechanical system. As the result of the above considerations, one can write for the dissipated energies E_{d1} until E_{d6} the following general function:

$$E_{di} = E_{ti}(\theta \text{ or } n, V, k_i, R_e, E, \ldots, V_m, p_m, \sigma_m, T, \ldots). \tag{13.9}$$

The variables specified in the last relation can be understood in terms of system behaviour, as dependent on the design quality Q, the manufacturing quality M, the usage or operational load intensity O, the place of the system installation P, and the renewal quality R. Hence the last relationship may be rewritten as:

$$E_{di} = E_{di}(\theta \text{ or } n, V, Q, M, O, P, R). \tag{13.10}$$

Now as the number of cycles n and the lifetime θ is related by the frequency f, (equation 13.1), we will indicate below the life θ only. For each particular part of the system, as well as for each system, one unique mode of complex wear takes place. Hence the total energy dissipated during the system operation, being a function of life θ and the power of residual processes $V(\theta)$, can be written as:

$$E_d(\theta) = \Sigma a_i E_{di}(\theta, V, Q, M, O, P, R)$$
$$= E_d(\theta, V, Q, M, O, P, R), \tag{13.11}$$

where a_i are the weighting coefficients, which are dependent on the complexity of the wear mode in a given case of mechanical system operation. For the total system in operation, the intensity of wear or damage can be defined simply as, the dissipated power or time rate of energy dissipation N_d. Hence, denoting the design and usage parameters just described by the logistic vector: $L = (Q, M, O, P, R)$, we can write:

$$E_d(\theta, V, L) = \int_0^\theta N_d(\tau_0, V, L)d\tau_0$$

$$= \Sigma N_{di}(\theta_i, V, L)\Delta\theta_i \qquad (13.12)$$

$$= \Sigma\Delta E_{di}(\theta_i, V, L),$$

$N_d = dE_d(\theta, V, L)/d\theta.$

The last relationship is of particular help in cases of event-type damage or wear, giving the discrete increase $\Delta E_{di}(\theta, V, L)$, which is similar to the case of cyclic phenomena, equation (13.7), i.e. fatigue, earthquake etc. At this point, two remarks have to be made concerning the equations of the different forms of wear:

- Firstly, the term energy is not understood here in the strict thermodynamic sense (Glansdorff and Prigogine, 1971).
- Secondly, the energy formulae presented for the different types of wear have a statistical meaning, as they do not reflect exactly the history of the load, which is important in some cases (Juvinal, 1967).

Operating mechanical equipment, as well as their parts, are open systems with respect to energy flow and dissipation. Hence we can treat them as dissipative systems in terms of General System Theory (Waelchli, 1992; Glansdorff and Prigogine, 1971). Moreover, every mechanical system breaks down as the result of wear, i.e. energy dissipation. From this simple observation, it can be concluded that the energy dissipation capacity of each mechanical system is finite. This postulate is typical of an existing system. Hence, designating again these influence parameters as components of the logistic vector $L^T = (Q, M, P, L, R)$, one can write qualitatively the limits of energy dissipation as:

$$0 \leqslant E_d(\theta, V, L) \leqslant E_{db}(L). \qquad (13.13)$$

Thus for a new system, or its parts, we assume the dissipative energy is equal to zero, and its limit or breakdown value is $E_{db}(.)$. Here E_{db} depends at least on the system quality, the level of its design, manufacture, load sequence/intensity, and maintenance operations, if any are indicated. Having in mind the variety of mechanical equipment, their component parts, and parametric dependence on dissipated energy from L. It may be seen that we need a special dimensionless measure of damage, system evolution or wear development. We can define this dimensionless measure of damage development or advancement as (Cempel and Natke, 1992; Cempel, 1993a):

$$D \equiv E_d(\theta, L)/E_{db}(L) = D(\theta, L) \qquad (13.14)$$

New system $= 0 \leqslant D \leqslant 1 =$ at the breakdown

This can be defined as a local and/or global damage measure (Natke and Cempel, 1995). It is of great interest as to what the particular forms of the above measure can take for the different types of wear. Taking into account events and continuous forms of wear separately, equation (13.7), we can find (Cempel, 1991; Cempel and Natke, 1992):

$$
D = \begin{cases}
\theta/\theta_b & \text{--for simple continuous wear and any} \\
& \text{wear in an average sense.} \\
\Sigma_i(\theta_i/\theta_{bi}) & \text{--for multilevel continuous wear} \\
& \text{(including creep).} \\
n/n_b & \text{--for simple cyclic wear.} \\
\Sigma_i(n_i/n_{bi}) & \text{--for multilevel cyclic wear.} \\
\Sigma_i(\theta_i/\theta_{bi}) + \Sigma_j(n_j/n_{bj}) & \text{--for intermittent, continuous} \\
& \text{and cyclic wear.} \\
(*) & \text{--any combination of the above in real} \\
& \text{operation of the system.}
\end{cases}
\tag{13.15}
$$

Hence we can see the damage development measure is equivalent, for example, to the Odquist–Katchanov law of creep for the continuous type of wear, to the Palmgren–Miner fatigue law for cyclic wear and to Archard's law of friction. Thus, the damage measures above are introduced. They are simple to interpret and have a good experimental base. The measure of damage advancement in terms of system behaviour is simply called 'technical condition', or just 'condition' when applied to mechanical or civil engineering systems. However, such defined values are hard to measure directly, if at all. Although, they can sometimes be measured quite easily on samples in a laboratory condition. Hence, for technical applications we need to use symptoms S of the system condition, which are measurable physical quantities covariable with the damage measure. This means we should know the symptom operator $S \equiv \phi(D) \equiv \phi(\theta)$.

13.3 MACHINES AS ENERGY PROCESSORS

Let us consider a model of the energy transforming system or energy processor, which can describe a variety of operating systems, machines, structures etc, transforming some energy to fulfil its design mission. Making no limiting assumptions concerning the energy exchange with the system environment, and referring to (Cempel and Natke, 1993; Cempel, 1993a; Cempel, 1990a; Cempel and Natke, 1992), we can present the ETS module and its energy flow as in Figure 13.1. As may be seen in the figure, there is one power input equivalent to all the inputs, energy and material supply, control, environment interaction etc. and there are

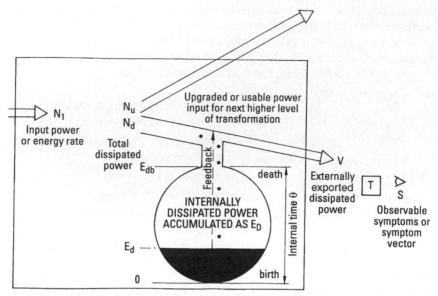

Fig. 13.1 Energy transforming system (ETS), or processor (EP), with its basic structure.

two outputs of the EP. The first output gives the desired product or mission of the system, and the second represents the externally dissipated power – the price of energy transformation during the system mission. There are four basic postulates, or constitutional assumptions, which must fulfil this model:

1. First, the stationary input power, energy time rate N_i must be balanced by the upgraded power N_u and the dissipated power N_d. The dissipated power divides into two streams, the internally accumulated N_{da} and externally exported V, Hence we can write:

$$N_i = N_u + N_d, \; N_d = N_{da} + V, \tag{13.16}$$

2. The accumulated internal part of dissipated energy E_{da}, is the function of the system life time θ as the integral from N_{da} and is limited by the dissipative capacity of the system itself E_{db}, so it gives:

$$E_{da}(\theta, V) = \int_0^\theta N_{da}(\theta)\, d\theta$$

$$= \int_0^\theta [N_d(\theta) - V(\theta)]\, d\theta \leqslant E_{db} \tag{13.17}$$

3. There is very important destructive feedback in the system, i.e. the accumulated dissipated energy E_{da}, this is evidence of damage

evolution in the system, and controls the dissipative power flow exported outside and vice versa. The greater the dissipation intensity V is, the greater is the accumulation $E_{da}(\theta)$. Therefore the differential increase of $dE_{da}(\theta)$ must be proportional to the corresponding increase of $V(\theta)$, i.e. $dV(\theta)$. Hence it follows that:

$$dV(\theta) = \beta dE_{da}(\theta, V), \quad \beta = \text{const} > 0. \tag{13.18}$$

4. Finally, the internal structure of the system must be immune, or a constant, over the system's lifetime θ.

$$\partial N_d / \partial V = \alpha = \text{const}, \quad V = \alpha^{-1} N_d + \text{const1}, \quad \alpha > 1. \tag{13.19}$$

Using the above assumptions and the destructive feedback relation, equation (13.18) (Cempel, 1993a), one can obtain the differential equation for externally dissipated power. This is the only quantity accessible for partial observation, as may be seen from Figure 13.1.

$$dV/V = \beta(\alpha - 1)d\theta/(1 - \beta(\alpha - 1)\theta). \tag{13.20}$$

Introducing the breakdown time of the system, when the denominator in the above equation tends to zero gives:

$$\theta_b = [\beta(\alpha - 1)]^{-1},$$

and also acknowledging that the dimensionless lifetime was proved to be a measure of the damage (Cempel and Natke, 1993):

$$D \equiv E_{da}(\theta)/E_{db} = \theta/\theta_b, \tag{13.21}$$

we obtain the differential equation and the respective solution:

$$dV/V = dD/(1 - D), \quad V = V_0(1 - D)^{-1}. \tag{13.22}$$

Correspondingly, the total dissipated power in a system, machine, will be:

$$N_d(\theta) = N_{d0}(1 - \theta/\theta_b)^{-1} = N_{d0}(1 - D)^{-1}.$$

It is worthwhile remembering here that, according to our solution, equation 13.22, the externally dissipated power V tends to infinity when the system approaches breakdown. The same is true with respect to the total dissipated power N_d. However, in reality, there will be no infinite power, because the balance of power, equation (13.16), must be maintained. Hence we can only say that, at the breakdown moment, all the power delivered to the system N_i is dissipated in the destructive process. This is exactly what one can observe in the case of breakdown of any technical system.

Now the question is: how to apply the above model of condition evaluation of an operating system in practice? Initially, the trouble is that no one can observe or measure directly any of the model governing quantities D, N_d, or V which describe the ETS life. As seen from Figure

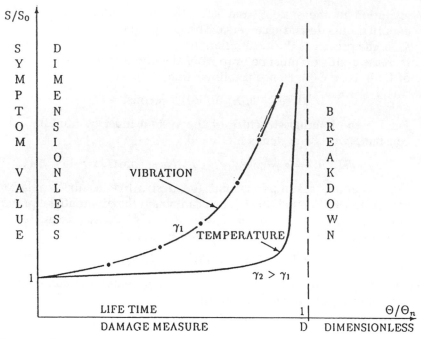

Fig. 13.2 Symptom life curve of an energy processor (EP) as a model of a system in operation (OS) for two values of the shape coefficient.

13.1, we can only observe a T-filtered part of externally exported power V. This is, for example, the vibration amplitude for the whole dynamic process, or the temperature for the heat generated etc. Hence, let us assume we are observing the life symptom $S(\theta)$ of our ETS evolution, which is the function of the externally dissipated power $V(\theta)$ as below equation (13.23):

$$S/S_0 \equiv \Phi(V/V_0) = \Phi((1 - \theta/\theta_b)^{-1}) = \Phi((1 - D)^{-1})$$

$$= \Phi((\Delta D)^{-1}) \qquad (13.23)$$

Here $S_0 = \Phi(1)$ is the initial value of the symptom, $D = \theta/\theta_b$ is the dimensionless life of the ETS, and more importantly $\Delta D = 1 - D$ is the residual system life or its damage capacity, $\Phi(*)$ is the unknown or postulated life-symptom operator, not necessarily of vibroacoustical origin. It can be inferred from the above, that through the observation of the ETS or EP life symptom values S, Figure 13.2, we can assess the residual life of a given system, or unit, using equations (13.21) and (13.23) because:

$$\Delta D(S) = \Phi^{-1}((1 - D)^{-1} = 1 - D = 1 - \theta/\theta_b$$

$$= (\theta_b - \theta)/\theta_b = 1 - (n/N_b).$$

Of course it is true if we know or have identified the symptom operator $\Phi(*)$. The forms of this symptom operator can be different dependent on the wear modes' participation factors for the given ETS. So usually it should be identified by some proper diagnostic experiment. In the simplest case we can try to approximate this unknown by some exponent function as below:

$$S/S_0 \equiv \Phi((\Delta D)^{-1}) \simeq (V/V_0)^{1/\gamma} = (\Delta D)^{-1/\gamma}, \text{ so } \Delta D = (S/S_0)^{-\gamma}, \gamma > 0.$$

$$(13.24)$$

This gives us a Pareto symptom life curve for the deterministic case, and respective Pareto symptom distribution for the probabilistic case of symptom condition monitoring. We will return to these important models and problems later in the chapter.

For now, let us illustrate what we have found in equation (13.24), namely the life behaviour of symptoms, or the condition of evolving systems. Figure 13.2 shows the change of symptom value against life, the so-called symptom life curve for the case of a Pareto-like symptom operator. Two values of the shape coefficients γ_1 and γ_2 of the symptom operator are shown. It can be seen that the smaller the shape coefficient value, the greater is the life sensitivity of the symptom (Cempel, 1990b).

For Pareto-like symptoms of condition, the determination or recalculation of system damage capacity, when some symptom of value S has been measured, is very easy. This is shown in Figure 13.3 for the case of a Paretian symptom life curve, and is true for the other symptoms as well, with a respective change of the shape and scale of the graph. It

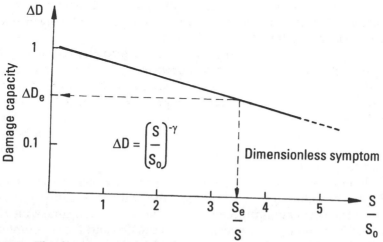

Fig. 13.3 The damage capacity for the energy processor with Paretian symptom life curve.

is worth reminding ourselves here that the damage capacity for the Paretian life curve will be a straight line when using a log-log scale for the graph, with γ as the shape factor.

The second question which is emerging now is how can the knowledge of the life symptom behaviour of an energy processor or an ETS model of a system be used in the diagnostics of the OS? For a simple system in operation, it may be a straightforward equivalence of one fault one life symptom. For a more complicated operating system, we can apply this concept in accordance with reliability theory, to look for the weakest element of the system (O'Connor, 1985). Hence we should first look for such a symptom which will describe only the technical condition of the weakest element in the OS.

13.4   MACHINERY AS A HIERARCHY OF ENERGY PROCESSORS

The theory presented here has a great inference power, and can be generalized far beyond the domain of the machines or operating systems considered here (Winiwarter and Cempel, 1992, 1995). It has been generalized lately into the theory of energy transforming systems (Cempel, 1993a, 1993c; Cempel and Natke, 1993, 1994), where new fractal-like abilities and behaviour have been revealed. The theory gives a physically based method for the reliability and diagnostic modelling of operating systems, and will now be outlined below.

Operating systems are mostly of complex design, function, energy flow etc. as postulated in the General System Theory (GST), this complexity is a fundamentalized reality of most systems (Bertalanffy, 1973). From this point of view an OS consists of many subsystems or assemblies, subassemblies, parts etc. with their specific energy flows and dissipations. Hence the question arises, should every subsystem be treated as an energy transforming system?

Considering this more deeply, we will come to the conclusion that every subsystem of an OS with a real or virtual boundary can be treated as an ETS. Moreover, sometimes subsystems must be treated as ETS, because their failure probability is high, substantially lowering the mission completion for the OS. Consequently, from the point of view of the energy flow, OS must be treated as an ensemble of ETS units, organized hierarchically according to their design goal, specifications and functional properties.

We should now be aware that each ETS module has its own internal life θ_i, and breakdown time θ_{bi}. Due to that, all the dynamic processes of energy transformation proceed in a multidimensional time scale. At a meta-time (t) given by the environment of the OS, and the set of internal times θ_i. Figure 13.4 is an attempt to explain this complex situation of the interdependency of subsystem lives and energy flows in the OS. Looking at this figure we can note several important conclusions.

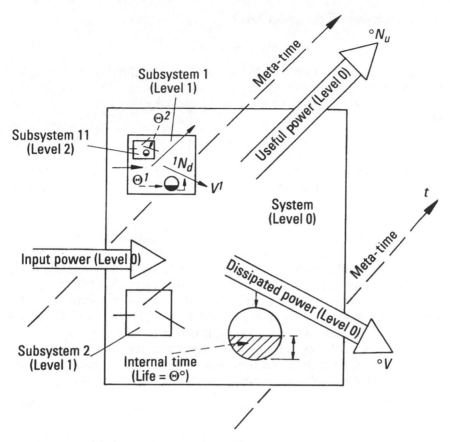

Fig. 13.4 Operating system as the hierarchical structure of ETS modules and the resulting interdependence of internal times immersed in the metatime flow.

First we can see the self-similarity of energy-transforming processes at each level of the hierarchy of the OS. In other words, we can observe the self-similar, or fractal-like energy transformation processes on every level of decomposition of OS energy flow (Winiwarter and Cempel, 1995). Each level of energy transformation defines its own internal time scale, and the whole OS is immersed in the meta-time (t) coming from the environment or meta-system. Secondly, on each level of the hierarchical ETS model we can establish an energy balance equation, which will take into account all the energy streams down from the given level of hierarchy. For the situation depicted in Figure 13.4 we can initially write the energy balance at the OS level as:

$$^{0}N_i = {}^{0}N_u + {}^{0}N_d. \tag{13.25}$$

The upgraded energy component, is a weighted sum of the energies being upgraded on the lower level:

$$^0N_u = \Sigma_i^k c_j \, _j^1 N_u \qquad (13.26)$$

The dissipated energy on the OS level is also a weighted sum of energies dissipated on the lower level $a_j \, _j^1 N_d$, and the energy dissipated on the level 0 due to the transformation of upgraded energy from the lower level 1.

$$^0N_d = \Sigma_1^k a_j \, _j^1 N_d + \Sigma_1^k b_j \, _j^1 N_u, \qquad (13.27)$$

where of course, $a_j < 1$, $b_j + c_j = 1$, $j = 1 \ldots$, and the power streams $_j^1 N_d$, $_j^1 N_u$, and others can be decomposed down to the required level of organization.

From the point of view of the reliability of an operating system and its diagnostics, it can be seen that now we have a good tool for creating suitable models of the OS. It can also be seen that each part of the exported external power stream from each subsystem etc. contains all the information from the lower level of system hierarchy. What remains unclear is the observability of these energy flows, either directly or by some symptoms of the type $S(_1^j V)$, together with the problem of the method of decoding the desired condition related information.

It is now evident how to build our symptom vector S with components $S^T = \{S_1, \ldots, S_k\}$ for the condition monitoring of the operating system. Essentially, there exists two ways of performing this:

1. According to ascending values of breakdown times θ_{bi} of ETS modules.
2. According to ascending values of the probabilities of mission completion P_k with respect to possible k-th ETS failure (Blanchard and Fabricky, 1990).

So denoting $S_j = \Phi_j(^j V)$ as a condition symptom for the j-th ETS we can create the symptom vector for condition monitoring of our OS in the form:

$$S^T = \{S_1, S_2, \ldots, S_k\} \qquad (13.28)$$

only if the values of the subsystem breakdown time in the OS is:

$$^1\theta_b < {}^2\theta_b, \ldots, {}^k\theta_b \qquad (13.29)$$

Alternatively we can use the succession of mission completion probabilities with no failure of k-th ETS subsystem respectively:

$$P_1(^1\theta_b > \theta) < P_2(^2\theta_b > \theta) < \cdots < P_k(^k\theta_b < \theta) \qquad (13.30)$$

where θ is the system life needed for mission completion. Analysing the above equations, one can say that this is the physically based method for

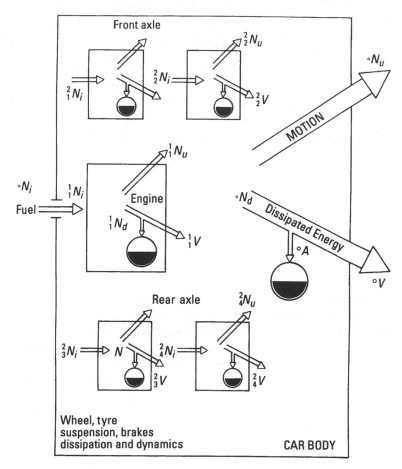

Fig. 13.5 Energy flow and dissipation in a car as an OS, for the creation of its reliability and diagnostic model.

constructing the reliability and diagnostic model of the OS, through the analysis of dissipative energy flow, which has a direct link and feedback with the system dynamics and the vibration as well. This is because the greater the intensity of the energy flow, the lower is the breakdown time of the system component.

In order to show the potential and the modelling power of the ETS-theory, when applied to OS, let us consider the car as an OS. The main energy input here is fuel, so omitting the road influence, uphill/downhill, we can equate the primary input power ^0N_i as coming from the fuel $^1_1N_i$, as in Figure 13.5. The upgraded power of the engine $^1_1N_u$ gives the input to each of the four subsystems, consisting of wheels, suspension and brake. So we can write the balances of power for the zero-level as:

$$^0N_i = {}^0N_u + {}^0N_d = {}^0N_u + ({}^0V + {}^0A), \tag{13.31}$$

and for the first level of the engine as:

$$^1_1N_i = {}_1N_u + {}^1_1N_d, \tag{13.32}$$

and for the second level of wheels etc. as:

$$^1_1N_u = \Sigma_1^n {}^2_j N_i + {}^0N_d - {}^0V. \tag{13.33}$$

The power dissipated by the whole car, Figure 13.5 is:

$$^0N_d = \Sigma_1^4 {}^2_j N_d + {}^1_1 N_d + {}^0A, \tag{13.34}$$

and externally dissipated power can be expressed by:

$$^0V = {}^0N_d - {}^0A = \Sigma_1^4 {}^2_j N_d + {}^1_1 N_d = \Sigma_1^4 ({}^2_j V + {}^2_j A) + {}^1_1 V + {}^1_1 A. \tag{13.35}$$

Now it is evident from the above, that all the information required for the condition assessment of subsystems, and the total system (A, V) are accessible, either from the flow of the upgraded or dissipated power. Accordingly, and dependent on how convenient it is to measure, we would observe either the OS output or the subsystem outputs.

We next deal with the problem of the construction of a symptom vector for OS condition assessment. Let us consider the probability of mission completion with respect to no failure in the given subsystem (ETS) of the car model. As the engine is very prone to failure, while the four-wheel subsystems are equivalent to each other in this respect, and the car body is almost insensitive to failure, we can write:

$$^1_1P < {}^2_1P \sim {}^2_2P \sim {}^2_3P \sim {}^2_4P < {}^0_0P. \tag{13.36}$$

Hence, for the condition assessment of the car, at this decomposition level of reliability, we will take five symptoms only as below:

$$S_1 = \Phi_1({}^1_1V); \ S_2 = \Phi_2({}^2_1V); \ S_3 = \Phi_3({}^2_2V);$$

$$S_4 = \Phi_4({}^2_3V); \ S_5 = \Phi_5({}^2_4V). \tag{13.37}$$

The question of what form of the symptom operator $\Phi(*)$, one should choose for a given subsystem is a matter for diagnostic technology, and can be found elsewhere (Cempel, 1991). Thus, it seems that the method of reliability and diagnostic modelling of any OS, by using the ETS approach, and supported additionally by operational symptom reliability of the OS, has been illustrated in detail. It seems to be a proper combination of approaches, but of course, it needs validation in every concrete case.

13.5 SYMPTOM RELIABILITY–CONTINUOUS OPERATING SYSTEMS

As mentioned previously, if a set of systems, or units, of the same type are in use, these units generally differ in origin, by having different histories via the variables Q, M, O, P, R. Moreover, the life span of many systems is greater than the rational time of experimental observation. Hence, we are in the position of having to infer the condition of a particular system with only a few statistical characteristics. The appropriate symptom is designated by $S(\theta, L)$, and we will observe it over $N \gg 1$ units of the same type. We must remember also that L is the vector of generally unknown components, $L = \{Q, M, O, P, R, \ldots\}$, which have to be treated as random variables due to the lack of information. So, proper statistics are needed for the description of this random variable $S(*)$, and to assess the condition of the given unit.

Hence, we now need to describe a special diagnostic experiment for that purpose. As we know, in order to produce some goods or meet other demands, we normally use a number of identical units. When we run this set of systems, we have several possibilities for its use and observation. Firstly, we can use each unit only up to its breakdown, with no immediate replacement or repair. Secondly, we can use it almost up to breakdown and then replace by a new one, or by a repaired unit just before the anticipated breakdown, i.e. when the unit's 'good' condition ends. The first model of operation is common in all branches of engineering, and the second one, with both diagnostics and renewal, is in common use in critical manufacturing and processing industries, plus the utility services via condition monitoring.

Let us first consider the use of a group of identical systems, or units, without renewal, such as we encounter frequently in non-critical areas of machine operation in engineering. Here there is no condition monitoring of the systems in use, and we count the system life time θ in order to assess the reliability of a given unit, and the availability of a machinery set. As should be obvious, this is lifetime based reliability or the system will fulfil its mission with demanded lifetime $\theta < \theta_b$, where θ_b is time to system breakdown. Lifetime-based reliability can be defined as below (O'Connor, 1985; Aven, 1992):

$$R_\theta(\theta) \equiv P(\theta_b > \theta) = P(\theta_b - \theta > 0), \quad 0 \leqslant \theta < \infty. \tag{13.38}$$

By transforming this directly to the symptom domain with the help of the symptom $\Phi(*)$ operator, equation (13.23), assumed to be monotonously increasing, we have:

$$R_\theta(\theta) = P(\theta_b > \theta) = P[\Phi^{-1}(S_b) > \Phi^{-1}(S)] = P(S_b > S) \equiv R_\theta(S)$$

with the $\theta = \Phi^{-1}(S)$, and $S_b = \Phi(\theta_b)$ as the symptom value of the breakdown. This is not exactly known, and accordingly treated as a random variable. For simplicity the same probability may hold both for θ and S, and this is true when the symptom is a linear function of life. Now knowing the condition evolution of operating systems as energy processors, equation (13.23), we can also transform this definition of reliability as below:

$$R_\theta(\theta) \equiv P(\theta_b > \theta)$$

$$= P(\theta_b - \theta > 0) = P((\Delta\theta_b/\theta_b) > 0)$$

$$= P(\Delta D > 0). \tag{13.39}$$

Hence we can say that lifetime-based reliability is equivalent to the probability that the system residual life ΔD, here being a random quantity, is greater than zero.

When using machine condition monitoring in critical industries we have a different situation with respect to the role of the reliability of systems in continuous operation. Here in order to minimize the number of unexplained breakdowns, and their associated loss, we are additionally implementing the symptom limit value S_1, calculated by means of statistical decision theory (Cempel, 1990b). We can thus describe the use of condition monitoring for critical operating systems in the following way.

A set of N systems, or units, is in continuous operation and has symptom condition monitoring. Thus, symptom S is measured periodically and the symptom limit value S_1 is imposed in such a way that it must be $S < S_1$ for all units in operation. When the measured symptom of a unit has exceeded the limit $S \geqslant S_1$, the unit is excluded from operation and substituted by a new or repaired one. In addition, some units can randomly approach the symptom breakdown value S_b, which also excludes them from operating in a natural way, and thus incur immediate substitution as above. As S_1 is the limit of the good condition, we may state that all units fulfilling the censoring rule $S < S_1$ are in the 'good' condition. Additionally, there is some probability that very few units will fail in the near future. Having clarified this situation, we can define the symptom reliability for critical systems in operation (Cempel, Natke and Yao, 1995).

Symptom reliability $R(S)$, is the probability that a unit classified as being in 'good' condition, will be in operation with the measured or demanded symptom value S. Treating it initially as a conditional probability with S_1 we can write:

$$R(S) \equiv P(S_b > S | S < S_1) \equiv P_G(S_b > S). \tag{13.40}$$

In order to explain better the meaning of this defined quantity, let us assume that we have measured this set of units at a calendar time t_0, obtaining a set of readings S, and that all the limitations as in definition equation (13.40) have been fulfilled. A short time later at $t_0 + \delta t$ it will be $S + \delta S$, with arbitrary small increments $\delta t, \delta S$, and a few units may approach the symptom breakdown value $S + \delta S = S_b$. Hence, following our definition in equation 13.40, we can also now write it as a conditional probability when it was previously $S < S_b$.

$$P_G(S + \delta S \leqslant S_b | S < S_b) = P_{GG}(S + \delta S < S_b | S < S_b)$$
$$+ P_{GF}(S + \delta S = S_b | S < S_b)$$
$$= P_{GG}(S + \delta S) + P_{GF}(S + \delta S) \qquad (13.41)$$

Here the partial $P_{GG}(*)$ is the probability that the unit will not fail with the symptom increment δS, and $P_{GF}(*)$, is the probability of failure when the symptom increases. As the symptom increment may be arbitrarily small, we can understand the last partition of the probability of 'good' condition, as the extension of the definition of symptom reliability. For further clarification, it will be useful to assess the probability of failure $P_{GF}(*)$ at $S + \delta S$, when the unit is in 'good' condition with symptom S. Treating it again as a conditional probability we can write and calculate the following:

$$P_{GF}(S+\delta S)=P_G(S+\delta S=S_b|S<S_b) = ((P_G(S < S_b \leqslant S + \delta S))/(P_G(S < S_b)))$$
$$= (P_G(S) - P_G(S + \delta S))/(P_G(S))$$
$$= (R(S) - R(S + \delta S))/(R(S))$$
$$= \lambda(S)\delta S - \cdots \simeq \lambda(S)\delta S. \qquad (13.42)$$

Here we have defined a new quantity, the symptom failure intensity of a set of units in continuous operation, measured in a relative number of failures per unit increment of the symptom S, for example millimetres/ second.

$$\lambda(S) \equiv -(\mathrm{d}\ln R(S))/(\mathrm{d}S).$$

This is analogous to the life-based failure intensity calculated by (Aven, 1992). As may be seen, by treating $P_{GF}(S)$ as a conditional probability, we can assess its value in the vicinity of S by using Taylor's expansion of the known symptom reliability $R(S)$. However, the question is still open concerning the relation of the defined symptom reliability to the life-based reliability. In order to answer this, let us again consider the definition in equation (13.40), in terms of damage capacity ΔD. In this case for simplicity a linear relationship between ΔD and S is assumed.

$$R(S) \equiv P(S_b > S | S < S < S_1) = P(\Delta D > 0 | \Delta D > \Delta D_1) \neq R_\theta(S). \qquad (13.43)$$

It can be seen that both definitions will be equivalent only if $\Delta D_1 = 0$, that means if the symptom limit value is the natural breakdown value, $S_1 = S_b$. Hence the life-based and the symptom-based reliability are not equivalent to each other, and we can treat $R(S)$ as the generalization of

$R_\theta(\theta)$. Finally we can present fully the definition of our symptom reliability and the respective probability densities as below:

$$R(S) \equiv P_G(S) = P_{GG}(S) + P_{GF}(S),$$

with:

$$P_G(S) = P_{GG}(S) + P_{GF}(S), \text{ and } R(S) = \int_S^\infty P_G(S_u)dS_u. \tag{13.44}$$

Going more deeply into the normalization of the above distribution and the respective relative frequencies we find:

$$\int_{-\infty}^\infty P_G(S)dS = 1 = G + F.$$

Normalization is thus provided with respect to the total set described by $P_G(*)$, which means with respect to all $N = N_G$ units in our diagnostic experiment, this being the sum of N_{GG} and N_{GF}. At the same time we have obtained above the other measure of reliability, namely the symptom availability G. This partial probability can now be understood as the average ratio of units that will not fail after measurement at t_0, say before the next inspection:

$$G = \int_{-\infty}^\infty P_{GG}(S_u)dS_\mu \simeq (N_{GG})/(N_{GG} + N_{GF}).$$

and similarly, the average portion of running units which will fail at $S + \delta S$, before the next inspection is:

$$F = \int_{-\infty}^\infty P_{GF}(S_u)dS_u \simeq (N_{GF})/(N_{GG} + N_{GF}),$$

where, of course, $N = N_{GG} + N_{GF}$ is the respective number of all the units in continuous operation. In summary, we propose to use in condition monitoring the symptom reliability calculated directly from the symptom database of running N machines, or units. This means using the empirical histogram:

$$P_G(S) \simeq (N_G(S + \Delta S \geqslant S_e \geqslant S))/(N\Delta S)$$

as symptom Probability Density Function (PDF), and calculating the symptom reliability from the integral:

$$R(S) \equiv \int_S^\infty P_G(S_u)dS_u. \tag{13.45}$$

We already know the operating unit may deteriorate in accordance with our model of damage evolution, the ETS model, and we also know what kind of data we can gather, by observing the distribution of symptoms over a group of systems in operation. Let us now return to further possibilities for calculating the survivor function (reliability) expressed by the above equation.

13.6 CONDITION ASSESSMENT, SYSTEM RESIDUAL LIFE AND SYMPTOM RELIABILITY

It is worth remembering here that if there is a set of systems of the same type, but of different life history, then during their operation one can observe a family of symptom life curves, see equation 13.24:

$$S(\theta, D, M, O, P, R) = (S_0(Q, M, O, P, R))/[1 - \theta/\theta_b(Q, M, O, P, R)]^{1/\gamma}$$

$$S(\theta, L) = S_0(L)/[1 - (\theta/\theta_b(L))]^{1/\gamma} = S_0(L)/[1 - D(L)]^{1/\gamma} \qquad (13.46)$$

Looking at this, symptom life curve is obviously dependent on several parameters, or even processes. For example, the load history O. Hence, we can conclude that the condition assessment of mechanical systems should be supported by statistics, applied to a suitably chosen symptom during routine diagnostic monitoring. However, another important question remains unanswered, how do we find the life symptom operator for the case of the usage of a set of systems or machines in operation, knowing only the PDF $P_G(S)$, from the empirical histogram? It is possible to answer this question by observing the life behaviour of a group of ETSs of the same type, assuming that each unit of the set is at a different life stage of its damage evolution process. This means that the distribution of a particular unit's life will begin with birth, ($\theta = 0$, $D = 0$), and continue to death ($\theta = 0$, $D = 1$). We will consider this reliability/diagnostic experiment later.

From the observation of reliability $R(S)$, of the set of operating systems, we know the behaviour of the ETS set in the symptom domain S. Hence, let us transform this knowledge to the lifetime domain θ, or the damage domain D. From probability theory (Papoulis, 1965), we know how to transform these domains assuming that $S(D)$ increases monotonically:

$$P_G(S) = P_G(D)dD/dS = P_G(D)(dS/dD)^{-1}, \qquad (13.47)$$

The first term gives us the density of the distribution of ETS life evolution in our sample, and the second describes its life symptom evolution. This we have already found by use of ETS theory and by means of equations (13.23) and (13.24).

In general this domain relationship produces several possibilities for inference. Initially, we know the pdf $P_G(S)$ from the histogram of the condition monitoring database, and assuming the densities $P_G(\theta)$, or $P_G(D)$, we can calculate the average symptom life curve $S(\theta)$, or $S(D)$ for the given machine system type at a given operation site. Secondly, knowing the damage advancement distribution $P_G(D)$ in the running machine set, for a given machine type, or known $S(\theta)$, we can calculate the symptom distribution to be observed in the future condition monitoring system. This is done, in order to choose the best possible symptom operator $\Phi(*)$ for condition assessment and forecasting. It is worthwhile remembering here that with the knowledge of symptom life curve $S(\theta)$, and assumed way of observing $P_G(\theta)$, we can calculate the symptom reliability directly. By performing these operations, for the assumed damage allocation density $P(D)$ in a sample of ETS, we can propose in general for symptom related density and with a_k as the normalization factor (Cempel, 1993a):

$$P(D) = a_k(S(D)/S_0)^k, \, k < \gamma(a_k = 1, \text{ if } k = 0) \tag{13.48}$$

We have already found the life symptom evolution of our ETS module, equation (13.24), or in general, equation (13.23). So we can also calculate the needed derivative dS/dD, which for a Pareto-type symptom life curve is:

$$dS/dD = 1/\gamma(S_0/S)^{-\gamma+1}, \, S \geqslant S_0, \, \gamma > 0, \tag{13.49}$$

Now by inserting equation (13.48) and equation (13.49) into the reliability integral equation (13.33) we obtain:

$$R(S) = \int_S^\infty P_G(S)dS = \int_{Se}^\infty \gamma a_k(S_0/S_e)(S_0/S_e)^{\gamma-k-1}dS_e$$

$$= ((\gamma a_k)/(\gamma - k))(S_0/S)^{\gamma-k}, \, \gamma > k. \tag{13.50}$$

In the above case, we have found the Pareto type of symptom reliability or survivor function, as can be seen directly in equation (13.50). It is also true, in the general case of reliability, as can be shown, that the sum of Paretian distributions maintains its property. For these cases of life distribution in our sample we thus obtain Paretian reliability, if of course the observed life symptom is of a Pareto type. Hence we can assume in the simple case of Paretian symptom reliability, with $k = 0$ in equation (13.50):

$$R(S) \equiv P(S_b > S) = (S_0/S)^\gamma, \, \gamma > 0, \, S \geqslant S_0 \tag{13.51}$$

For Paretian reliability, we can prove a more valuable relationship. By putting the Paretian life curve:

$$S/S_0 = (1 - D)^{-1/\gamma} = (\Delta D)^{-1/\gamma}$$

into equation (13.51) we obtain:

$$R(S) = (S_0/S)^\gamma = 1 - D \equiv \Delta D(S). \tag{13.52}$$

This means that our symptom reliability is **equivalent to the system residual life**, or its damage capacity $\Delta D(S)$. This seems to be a very important statement, especially from the practical point of view. It means that, knowing the symptom reliability $R(S)$ for the given population of ETS or operating systems, and by measuring the symptom value for the given unit, we can assess the residual life of that unit, because we have:

$$\Delta D = 1 - D = 1 - (\theta/\theta_b) = (\Delta\theta/\theta_b), \qquad \text{the residual life for continuous wear (life).} \qquad (13.53)$$

$$\Delta D = 1 - D = 1 - (n/N_b) = (\Delta n_b/N_b), \qquad \text{the residual life for cyclic wear (life), or for uniformly spaced measurements along the calendar time (for average wear).}$$

with n and N_b as the current limit of cycles, or measurements, for the life θ respectively.

This possibility, that in general the symptom reliability can be treated as equivalent to the residual system life, sometimes with a small error, is of great advantage to ETS theory. Moreover, we can prove the validity of relationships in equation (13.52) for other types of symptom reliability functions, e.g. Weibull, Fréchet, Uniform and Exponential. It is even possible to recalculate the respective symptom oprators $\Phi(V/V_0)$ and the life-symptom curves $\Phi[(1 - D)^{-1}]$ of the ETS theory for each reliability model mentioned above, see equation (13.23). These possibilities are shown in Table 13.1 (Cempel, 1993a).

It may be noted from Table 13.1 that all the proposed symptom operators $\Phi(*)$ give monotonous growth of the life symptom curves, sometimes with asymptotic behaviour at $D = 1$, Weibull, Fréchet, and Pareto. The life sensitivity of symptoms $S(D)$ depends reciprocally on the value of the shape exponent γ, which is a maximum for the Fréchet and Weibull reliability type and its respective life curves. It must be noted once more that for uniform observation, $P_G(D) = $ constant, the known symptom life curve is $S(D)$, and determine also the same type of the symptom reliability $R(S)$.

The respective reliabilities, shown in Table 13.1, belong to the so-called longtailed distributions, and are generated by the densities with right-hand skewness. It concerns, in particular, the Fréchet distribution and its asymptotic expansion Pareto, typically used in the statistics of evolutionary systems. However, our main goal is to assess the machine system present, and future condition, by observing some life symptoms. In order to do this in the symptom domain, we require some symptom limit values S_1, or standard values for comparison. In the domain of damage measure, we will assess the residual or remaining life, as shown already by the relationship in equation (13.24). But in order to determine the symptom limit value S_1, for the breakdown, or alarm value S_a, we have to apply statistical decision theory (Birger, 1978), namely the Neyman–Pearson rule of risk assessment.

Table 13.1 Models of the life symptom operators, life symptom curves and the respective symptom reliabilities as generated by the theory of ETS

Symptom operator $V > V_0,\ \gamma > 0$ $\Phi\left(\dfrac{V}{V_0}\right) \simeq$	Symptom life curve $S_0 > 0,\ \dfrac{S(D)}{S_0} =$	Symptom reliability $R(S) =$ or residual life	Remarks Symptom model
$\left(\ln\dfrac{V}{V_0}\right)^{1/\gamma}$	$[-\ln(1-D)]^{1/\gamma}$	$\exp-\left(\dfrac{S}{S_0}\right)^{\gamma}$	Weibull, $S \geqslant 0$
$\left[-\ln\left(1-\dfrac{V_0}{V}\right)\right]^{-1/\gamma}$	$[-\ln D]^{-1/\gamma}$	$1-\exp-\left(\dfrac{S}{S_0}\right)^{-\gamma}$	Fréchet, $S \geqslant 0$
$\left(\dfrac{V}{V_0}\right)^{1/\gamma}$	$(1-D)^{-1/\gamma}$	$\left(\dfrac{S}{S_0}\right)^{-\gamma}$	Pareto, $S \geqslant S_0$ (asymptotics of Fréchet)
$1+\left(1-\dfrac{V_0}{V}\right)\cdot\dfrac{1}{\gamma}$	$1+\dfrac{1}{\gamma}D$	$1+\left(1-\dfrac{S}{S_0}\right)\cdot\gamma$	Uniform, $S \geqslant S_0$ $S \sim S_0$ (approximation of Pareto and exp(*))
$\exp\left[\dfrac{1}{\gamma}\left(1-\dfrac{V_0}{V}\right)\right]$	$\exp\left(\dfrac{1}{\gamma}\cdot D\right)$	$1-\gamma\ln\dfrac{S}{S_0}$	Exponential, $S \geqslant S_0$

If we know the availability of the machine population, or the machine group, $0 < G < 1$, and if we specify the allowable risk, $0 < A \ll 1$, of false condition assessment, we can calculate the limit values for breakdown S^1 and alarm S_a as follows:

$$G \cdot R(S_1) = A$$
$$G \cdot R(S_a) = cA \tag{13.54}$$

Note that for operating systems without renewal or like structures, we should take $G = 1$.

As may be seen, the symptom reliability $R(S)$ is used here again, and we should specify only the allowable probability of uneeded repair A, allowable false decision, and the value of the coefficient c. Usually, dependent on the age and the maintenance quality of machines, $A = (0.02\ldots0.1)$, which corresponds to $(2\%\ldots10\%)$, and the alarm coefficient $c = (2\ldots4)$ (Birger, 1978). This means we are ready to repair needlessly $(2\ldots10)\%$ of our machinery stock in order to avoid breakdown. This rule of the symptom limit value calculation, gives good results in the vibration condition monitoring of machines (Cempel,

1990b).

Concerning the residual life assessment, we will start from the relationship in equation (13.52). Having measured the life symptom value S_e of some machine we can write:

$$R(S_e) = 1 - D_e \equiv \Delta D_e. \tag{13.55}$$

On the other hand, and from the above considerations, we also can set the safety limit of the remaining life, because it turns out from equation 13.54:

$$R(S_1) = A/G \equiv \Delta D_1 - \text{limit remaining life.} \tag{13.56}$$

Now we are prepared to write the basic relation for the condition assessment of the system:

Good condition:

$$\text{if } S_e < S_1 \text{ or } \Delta D_e > \Delta D_1.$$

Faulty condition: $\hspace{8cm}$ (13.57)

$$\text{if } S_e \geqslant S_1 \text{ or } \Delta D_e \leqslant \Delta D_1.$$

This is the way of condition assessment, and in the case of the periodic monitoring of an operating system, with the life step $d\theta$ the breakdown time can be calculated as below:

$$\theta_b = N_b d\theta = (n_e/1 - R_e)d\theta, \tag{13.58}$$

and the respective residual number of periodic observations:

$$\Delta N_b = (n_e R_e)/(1 - R_e), \text{ or in another way,}$$

$$\Delta N_b \equiv N_b - n_e = (n_e \Delta D_e)/(1 - \Delta D_e). \tag{13.59}$$

Where $R_e = R(S_e)$ and n_e is the successive number of given observations with the value S_e, when counting has started with new objects, i.e. for $\theta = 0, n = 0$. This is the assessment of the remaining life, but we can also make the prognosis of the next symptom value S_p for the specific chosen model of symptom operator, and the forecasting step $\Delta D_p \leqslant \Delta D_e$. Based on equation (13.23) and Table 13.1, we can write:

$$S_p/S_0 \equiv (S(D_e + \Delta D_p)/S_0) = \Phi[(1 - D_e - \Delta D_p)^{-1}]$$

$$= \Phi[(1 - ((n_e + p)/N_b))^{-1}], \tag{31.60}$$

where $\Delta D_p = (p.d\theta)/(N_b.d\theta)$ is the dimensionless forecasting step transformable to the number of observations p, if the condition monitoring is periodic. For the symptom models of Weibull or Fréchet type, we can predict the next symptom value using the simple formula:

$$S_{n_e + p} = S_{n_e}(1 + (pN_b))/(n_e \gamma \Delta N_b), p = 1, 2, \ldots \ll n_e \leqslant N_b \tag{13.61}$$

where p is the condition forecasting horizon in terms of the number of measurements.

13.7 EXAMPLE OF MACHINE LIFE PREDICTION BY ETS THEORY

As examples of the application of the ETS concept, two cases from the mechanical engineering area will be shown here. By utilizing this theory, several computer programs have been prepared to transform the symptom database, and to look for a most appropriate symptom reliability distribution. For example, Weibull, Fréchet etc. (Cempel, 1993b) and a polynomial approximation (Cempel and Ziolkowski, 1995). The program calculates some measure of goodness of fit, like the coefficient of determination R^2 and χ^2 measure, the exponent γ of given distribution, the symptom limit value S_1, the residual life ΔD_e, and the residual observation number before breakdown ΔN_b; see equation 13.59 and neighbours.

These series of programs have been prepared using Matlab, and as a first example, according to ETS theory, the data shown in Figure 13.6 for Diesel engines (Tomaszewski, 1985) were analysed. Here we can see that the Weibull symptom distribution is most appropriate, that is also partly true for the polynomial approximation. The upper-left figure shows the observed behaviour of the engine number sil54 measured at point d1, with the symptom, RMS-vibration acceleration of the casting. The

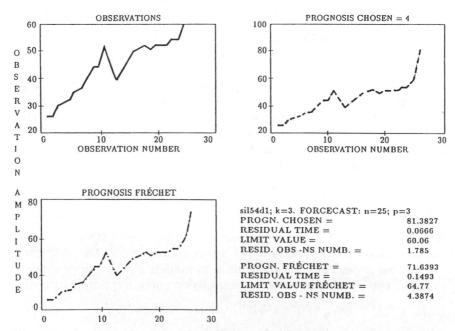

Fig. 13.6 An example of Diesel engine diagnosis by the computer program cem8.m based on ETS theory.

upper-right figure gives us the forecast of the engine behaviour at the forecasting horizon $p = 3$, and the lower-right the same for the Fréchet distribution.

There is also shown, the abbreviated message of the program concerning the symptom limit values, forecasted symptom values, and the residual time and observation number for the both polynomial and Fréchet forecasts. It can be stated that the polynomial forecast is much more accurate here, as the engine failed shortly afterwards. There are more examples of the accurate forecasting of the life of diesel engines in Tomaszewski (1985) but we will not display them here.

The next example of an application of ETS theory, which uses the above mentioned diagnostic program, is where the results of a simulation concerning the symptom operator and its shape factors are taken into account. This is because in modelling the system condition evolution, either by simulation or in practical cases of condition monitoring, one of the most important tasks is to choose the functional form of a condition degradation model. We must look for the possible form of the operator $\Phi(*)$, describing the condition evolution of the energy processor model, and to holistic modelling, wherein the polynomial model of degradation is proposed. Which one to choose, and how to assess the shape factors γ, α is the question.

As usual we can proceed by choosing heuristically the simplest possible model, for example, linear $S/S_0 = 1 + 1/\gamma D$, $D = \theta/\theta_b$, and by trial and error method or utilizing some additional data, corrosion speed in given environment for example, to assess the shape factor γ or α. Secondly, having some symptom data from the real case of condition monitoring, we can use a computer program to make the job easy and more objective. Let us consider this problem in connection with two particular machine diagnostic tasks. First, the monitoring of ball-bearing condition in small electric motors, which are used in the textile industry, and secondly, checking railroad diesel engines used for the main drive of trains.

In the first case, a large number of units were under condition monitoring with the root mean square acceleration amplitude as the symptom, sample sps 12m–298 units. We can assume in a first approach that we have the case of uniform distribution of damaged advancement in the sample, i.e. $P(D) =$ constant. So one can obtain in this way the average symptom life curve. This was done by the already mentioned computer program DEM8.M, and the results are shown in the first part of Table 13.2.

One can see from there, that the Weibull and Fréchet models can be used as they have almost the same determination coefficient and χ^2 measure of goodness of fit. But due to the historical connotation of Weibull's name to ball bearings, the last one was chosen as the model for condition forecasting and residual life assessment (O'Connor, 1985).

Table 13.2 The choice of symptom models and their shapes coefficients γ for two cases in machine condition monitoring

	Symptom distribut.	Shape factor γ	Scale factor β	Determin. coeffic. $R^2\%$	χ^2	Symptom model choice
Small	Weibull	0.715	11.483	98.0	4.13	1
electric	Frechet	2.034	14.414	96.8	2.89	2
motors	Pareto	1.648	7.823	90.5	24.338	4
(sps 12 m)	Uniform	0.03	1	59.2	71.151	5
(N = 298	Exponent.	0.318	1	89.3	14.661	3
units)						

Symptom: $\dfrac{S}{S_0} = [-\ln(-1-D)]^{1/\gamma}$,

$$D = \frac{\theta}{\theta_b}, \; \gamma = 0.715$$

Weibull model,   (chosen as best).

Reliability: $R(S) = \exp - \left(\dfrac{S}{S_0}\right)^{\gamma}$

Diesel	Weibull	4.117	2.001	96.7	11.476	2
Engines	Fréchet	4.221	1.561	80.0	27.044	4
(sild1)	Pareto	2.882	1.27	64.2	50.642	5
(N = 56	Uniform	0.67	1	97.3	9.697	1
readings)	Exponent.	1.066	1	93.5	15.777	3

Symptom: $\dfrac{S}{S_0} = 1 + \dfrac{1}{\gamma}D, \; D = \dfrac{\theta}{\theta_b}, \; \gamma = 0.67$

Uniform model,   (chosen as best).

Reliability: $R(S) = 1 + \gamma\left(1 - \dfrac{S}{S_0}\right)$

The second case concerns 56 symptom readings from four diesel engines which are currently in use. Here the average acceleration amplitude was used as the symptom of condition, being measured at the top of the first cylinder head (Tomaszewski, 1985). The readings were measured every 10 000 kilometres, sample sildl–56 readings.

This data was also processed by the 'DEM8.M' program, and the results are shown in the last part of Table 13.2 (Natke and Cempel, 1995). One can see here that the Weibull model of the symptom operator is also at the top of the list, although the uniform model is better, and this is not only due to its simplicity. Hence, looking for the forecast of the next symptom value, or for the shape of the degradation function, equation (13.61), in a simulation, we will use the linear model of wear and symptom evolution.

13.8 CONCLUSIONS

Using the system approach, it was shown above by analytical consideration supported by some experimental results, that the following conclusions can be presented as shown below:

1. Machine operation can be modelled as an energy processor with finite dissipation capacity.
2. The evolution of condition of EP, i.e. its internal damage growth, gives the direct connection to vibroacoustical symptoms as used in machine condition monitoring.
3. Introduction of symptom-based reliability, connected with EP model evolution, allows the determination of residual life and future condition, on the basis of condition monitoring measurements.
4. This concept is easy to implement, via a computer program, for on-line assessment of residual life and the future state of a machine, provided we observe some appropriate symptom of fault and/or overall condition.
5. The observed symptom must be well-chosen by a proper experiment, characterized by good sensitivity to the condition evolution of the diagnosed system, and have a small sensitivity to external disturbances.

13.9 REFERENCES

Aven, T. (1992) *Reliability and Risk Analysis*, Elsevier Applied Science, London, UK.

Bertalanffy, L. von. (1973) *General System Theory*, G. Braziler (ed.), New York, USA.

Birger, I. Ya (1978) *Technical Diagnostics* (in Russian), Nauka, Moscow, Russia.

Blanchard, B.S. and Fabricky, W.J. (1990) *Systems Engineering and Analysis*, Prentice-Hall, New York, USA.

Braun, S. (ed.) (1985) *Mechanical Signature Analysis – Theory and Applications*, Academic Press, New York, USA.

Cempel, C. (1990a) The generalized tribovibroacoustical (TVA) machine model in plant diagnostic environment, *Bulletin of the Polish Academy of Sciences* (Technical Science), **38**(1–12), 39–49.

Cempel, C. (1990b) Limit value in practice of vibration diagnosis, *Mechanical System and Signal Processing*, **4**(6), 438–93.

Cempel, C. (1991) *Vibroacoustic Condition Monitoring*, Ellis Horwood, Chichester, UK.

Cempel, C. (1992) Damage initiation and evolution in operating systems, *Bulletin of Polish Academy of Sciences* (Technical Science), **40**(3), 204–14.

Cempel, C. (1993a) Theory of energy transforming systems and their applications in diagnostics of operating systems, *Applied Mathematics and Computer Science*, **3**(3), 533–48.

Cempel, C. (1993b) Machine degradation model–its generalization and diagnostic application, *Proceedings of Symposium–IUTAM*, Wupertal, August.

Cempel, C. (1993c) Dynamics, life, diagnostics; holistic approach to modelling of operating systems, *Proceedings of Symposium–Modelirung und Wirklichkeit*, Hanover, 7–8 October.

Cempel, C. and Natke, H.G. (1992) Damage evolution and diagnosis in operating systems, *Proceedings of Symposium – Safety Assessment of Time Variant and Nonlinear Structures using System Identification Approaches*, Lambrecht, Vieweg, Braun-schweig.

Cempel, C. and Natke, H.G. (1993) System life cycle – system life: the model based technical diagnostics, a view on holistic modelling, *System Research*, 53–63.

Cempel, C. and Natke, H.G. (1994) *An Introduction to the Holistic Dynamics of Operating Systems*, *Proceedings of Conference – SAFEPROCESS*, Helsinki, Finland, June.

Cempel, C., Natke, H.G. and Yao, J.P.T. (1995) Damage capacity and symptom based reliability as advanced signal processing procedures, *Proceedings of Conference – Structural Damage Assessment using Advanced Signal Processing Procedures*, Pescara, Italy, May.

Cempel, C. and Ziolkowski, A. (1995) The new method of symptom based processing for machinery condition monitoring and forecasting, *Mechanical Systems and Signal Processing*, 9(2), 129–37.

Collacot, R.A. (1977) *Mechanical Fault Diagnosis and Condition Monitoring*, Chapman & Hall, London, UK.

Engel, P.A. (1976) *Impact Wear of Materials*, Elsevier, New York, USA.

Fitch, E.C. (1992) *Proactive Maintenance for Mechanical Systems*, Elsevier Advanced Technology, London, UK.

Glansdorff, P. and Prigogine, I. (1971) *Thermodynamic Theory of Structure, Stability and Fluctuations*, Wiley Interscience, London, UK.

Juvinal, R.C. (1967) *Stress, Strain and Strength*, McGraw-Hill, New York, USA.

Kliman, V. (1985) Fatigue life estimation under random loading using energy criterion, *International Journal of Fatigue*, 1, 39–44.

Luczak, A. and Mazur, T. (1981) *Physical Wear of Machinery Elements* (in Polish), WNT Warsaw, Poland.

Natke, H.G. and Cempel, C. (1979) *Model Aided Diagnosis of Mechanical Systems, Fundamentals, Detection, Localization, Assessment*, Springer-Verlag, Berlin, Heidelberg, New York.

O'Connor, P.D.T. (1985) *Practical Reliability Engineering*, John Wiley, Chichester, UK.

Papoulis, A. (1965) *Probability, Random Variables, Stochastic Processes*, McGraw-Hill, New York, USA.

Poltzer, G. and Meissner, F. (1983) *The Foundation of Friction and Wear* (in German), VEB Leipzig, Germany.

Severdenko, V.P. (1976) *Ultrasound and Plasticity* (in Russian), Minsk, Russia.

Tomaszewski, F. (1985) *Vibration Diagnostics of Diesel Railroad Engines* (in Polish), PhD thesis, Technical University of Poznan, Poland.

Waelchli, F. (1992) Eleven Theses of General System Theory (GST), *System Research*, 9(4), 3–8.

Winiwarter, P. and Cempel, C. (1992) Life symptoms; open energy transformation systems with limited internal structuring capacity, and evolution, *Systems Research*, **9**(4), 9–34.

Winiwarter, P. and Cempel, C. (1995) Birth and death energy processors, *Proceedings of Symposium – System Science Society*, Amsterdam, The Netherlands, June.

CHAPTER 14

Commercial applications of vibration monitoring

P. Shrieve and J. Hill

ATL Consulting Services Ltd, Condition Monitoring Division, 36–38 The Avenue, Southampton, SO17 1XN, UK

14.1 INTRODUCTION

To remain competitive in today's marketplace, many companies are seeking methods and techniques to achieve maximum operating efficiency. In ensuring such efficiency, their operations and maintenance philosophies must be cost-effective, and thereby permit the company to achieve maximum profitability. The maintenance of large industrial assets has previously been considered as a fixed overhead by the majority of companies, and as a consequence, carried each year. However, attitudes have changed drastically in this area over the past five years, and maintenance is now being considered as a self-sufficient business centre of primary importance with respect to company profitability.

Accordingly, although maintenance is now seen as necessary to ensure the continued, safe and profitable operation of industrial machinery and plant, industrial economics demands that higher plant availability and reliability are achieved in the most cost-effective manner. Maintenance programmes must therefore, achieve minimum plant component failures at a minimum cost, and this has led to the growth of cost-effective maintenance 'on condition' supported by a condition monitoring programme.

Condition monitoring is no longer a new technique and over the past ten years many companies have implemented condition-based monitoring systems as a means of reducing maintenance activity and expenditure. Maintenance constitutes a significant percentage of the operating costs of industrial machinery, and unless addressed specifically, main-

Handbook of Condition Monitoring
Edited by A. Davies
Published in 1998 by Chapman & Hall, London. ISBN 0 412 61320 4.

tenance costs will continue to rise, driven by the desire to over-maintain and to inspect plant according to manufacturers' recommendations. It is well known for example:

That significant time and money is spent in inspecting, replacing and monitoring machinery that is running perfectly well.

By knowing at any time the condition of a machine, a judgement may be made on whether a planned maintenance request can be deferred. If such a deferment can be made, then a saving in personnel and maintenance costs will result. Thus condition monitoring can be seen not to replace planned maintenance, but instead:

Allows the time factor in a planned maintenance system to become variable rather than remain fixed.

Many factors influence the maintenance requirements of a plant, such as the manufacturing process, the 'built-in' redundancy of machinery, and the type and reliability of the equipment. It is correct therefore to assume that maintenance requirements vary dramatically from plant to plant, with plant economics being the governing factor. This being the case, it is obvious that the condition monitoring requirements will also vary from plant to plant for the very same reasons. Accordingly, it is therefore of vital importance that condition monitoring is fully integrated into plant maintenance procedures.

However, during the implementation of condition monitoring, few companies have evaluated the achievable cost/benefit in order to determine the correct mix of maintenance techniques, i.e. breakdown, planned, condition-based etc. within the overall maintenance programme. It is important to note that condition monitoring carries significant costs and is therefore:

Only commercially viable if the potential cost savings exceed the cost of the monitoring.

As an example, it has been estimated that 90% of offshore installations in the North Sea utilize condition monitoring (Shrieve, 1994). Techniques range from visual inspections to establish seal conditions, to detailed vibration and performance monitoring to determine the operational condition of large rotating machinery. Given the location and operating conditions of the equipment, the viability of condition monitoring is assured. However, not all applications are as clear-cut as this, and a cost/benefit analysis should always be carried out for every potential application.

Condition monitoring techniques, as outlined in this book, can be wide and varied, but all have an equal importance within an overall condition monitoring/condition-based maintenance strategy. In the sections which follow, the focus is on the use of vibration monitoring techniques, and examples are given of commercial applications. The equal importance of all techniques should always be remembered, however, and, as shown in the case studies which follow, the choice of technique and the correct interpretation of the data very much depend on the application.

14.2 GAS COMPRESSOR TRAIN MONITORING

The problem outlined in this application was discovered by ATL engineers during the offshore commissioning phase of two gas compressor trains in 1994 (Anon, 1994). Each compressor train consists of a 5 megaWatt (mW) gas turbine driving a four-stage centrifugal compressor, via a single helical gearbox. Although no full factory string tests were carried out, the gas turbines and compressors underwent performance test runs in the factory.

The compressor train vibration protection system comprised orthogonally mounted proximity probes mounted at each journal bearing, with the exception of the gearbox non-drive bearings which did not have these probes. In addition, two accelerometers were mounted externally on the gearbox casing. Upon startup, higher than anticipated overall vibration levels were experienced on the gearbox high-speed shaft. The amplitudes were 1.4 mils (36 micrometres) peak-to-peak, with the 1 × component of 0.35 mils (8.9 micrometres).

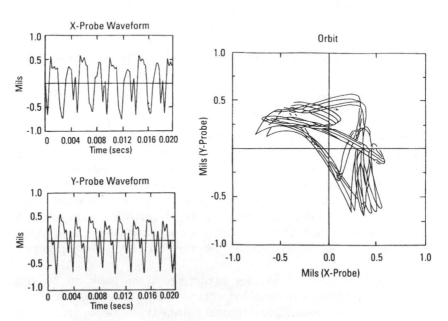

Fig. 14.1 Gearbox output shaft waveforms/orbit.

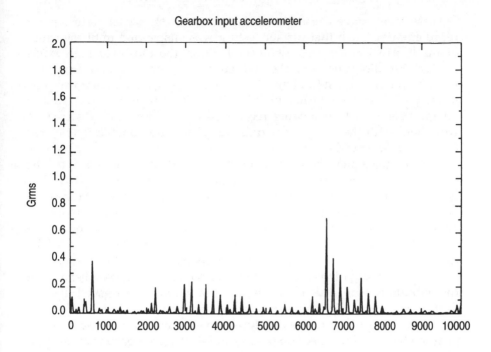

Fig. 14.2 Gearbox input accelerometer.

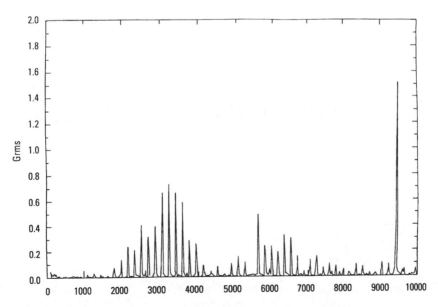

Fig. 14.3 Gearbox output accelerometer.

These levels gave cause for concern to the compressor train vendor representative, such that further testing was suspended until the cause of the high vibration was established. From the indicated high levels throughout the rundown, the evidence of a scratch and from the waveform data shown in Figure 14.1, the ATL engineer on site was able to establish that the indicated high levels of vibration were due to runout on the probe track. Following receipt of this information, discussions were held with the compressor train suppliers and commissioning was recommenced that day.

On the subsequent run, with the speed and load increased on the compressor, it was apparent that the gearbox accelerometers, particularly on the output shaft, were reading higher than anticipated. The spectral analysis, shown in Figures 14.2 and 14.3, indicated that the frequency content was complex and showed many frequency components. The separation of these frequency components was observed to be that of the input shaft speed, and changed accordingly as the speed was altered.

Further investigation carried out by the ATL engineer, established that the vibration was due to a disturbance being experienced by the high-speed shaft. This was emanating from the low-speed shaft as shown in Figure 14.4. It was concluded that damage was present on the gear teeth of the low-speed shaft, and the damaged area was causing an impact type disturbance on the high-speed shaft. It was not possible to identify the type of damage or the long-term effects from the data captured. Accordingly ATL recommended that the gear teeth on the low-speed shaft should be checked for signs of damage at the earliest opportunity.

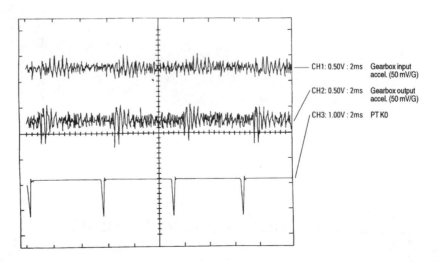

Fig. 14.4   Gearbox accelerometer waveforms.

The follow-up inspection to the internals of the gearbox indicated that corrosion was present, both on the gearbox output shaft probe track, and on a section of the low-speed shaft gear teeth, the latter had caused some pitting. Accordingly, the gearbox was maintained in service, as a standby, until the new gearbox internals were made available and replacement effected.

14.3 POWER GENERATION TRAIN MONITORING (a)

In this application, baseline vibration surveys were carried out on the power generation trains of an offshore installation during the onshore commissioning phase (Anon, 1994). Each power generation train consisted of a 25 megaWatt (mW) aero-derivative gas turbine driving an alternator, through a single helical, parallel shaft gearbox.

The input to the gearbox was via a flexible coupling, while the output from the gearbox used a quill shaft directly coupled to the alternator. The permanently installed vibration protection system fitted to the machine train consisted of accelerometers, on the gas turbine, and eddy

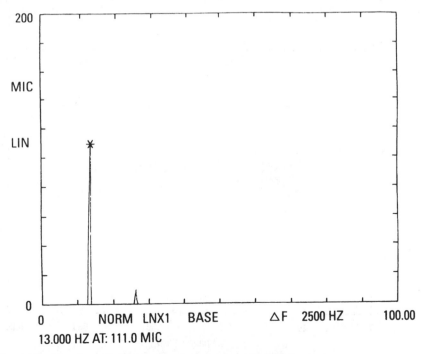

Fig. 14.5 Gearbox pinion free-end spectra at 750 RPM.

current probes on all journal bearings throughout the power turbine, gearbox and alternator.

During the initial startup the unit tripped due to high vibration on the gearbox pinion free end bearing at approximately 750 Revolutions Per Minute (RPM). The ATL engineer was able to replay the recorded data to establish that, at the trip point, the overall amplitude of vibration was in excess of 120 micrometres peak-to-peak. The predominant frequency component was at the 1 × rotational speed of the low-speed shaft and this is shown in Figure 14.5.

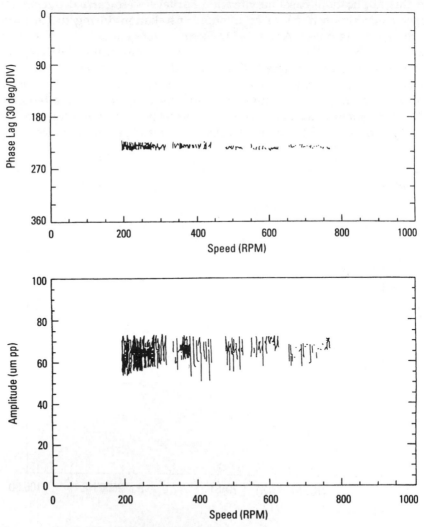

Fig. 14.6 Bode plot of alternator drive-end bearing.

As mentioned in an earlier chapter, the common causes of vibration at
1 × rotational speed are unbalance or misalignment. Analysis of the data
for all of the low-speed measurement locations indicated that amplitudes
were high throughout, and all at the 1 × rotational frequency component. The rundown data showed that at the 1 × frequency, filtered
amplitude and phase angle remained almost constant throughout the
coastdown.

This data is illustrated with the Bode plot for the alternator drive end
bearing in Figure 14.6. The problem was diagnosed as a possible bowed
shaft combined with some misalignment between the gearbox and
alternator. It was recommended that the coupling assembly should be
inspected and the alignment checked.

After investigation, it was found that the fit between a spigot and
recess on the faces of the gearbox/alternator coupling hubs was too
tight. As a consequence, the low-speed shaft was being forced into a
misaligned condition when the coupling faces were bolted together. By
rotating one of the shafts through 180° and then recoupling, the misalignment was minimized to allow commissioning to continue.

On restart, the vibration amplitudes were high but remained acceptable for commissioning of associated systems. The commissioning was

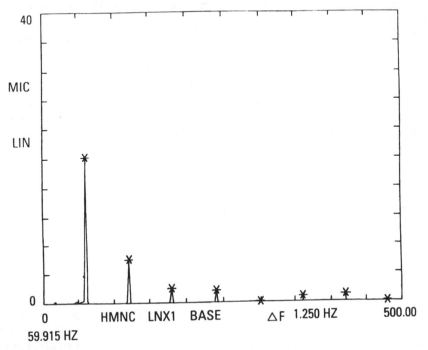

Fig. 14.7 Gearbox pinion free end at RPM after shaft rotation.

completed with the maximum overall vibration amplitude of 40 micrometres peak-to-peak observed at the gearbox pinion free end bearing, 27
micrometres of this was due to the $1 \times$ rotational frequency component
as shown in Figure 14.7.

14.4   POWER GENERATION TRAIN MONITORING (b)

This case considers a main power generation train on an offshore
installation after commissioning and several months of operation (Anon,
1994). The generator package consisted of an industrial gas turbine
driving an alternator through an epicyclic reduction gearbox. A permanent vibration protection system was installed in the form of orthogonally mounted eddy current probes on the gas generator, power
turbine and alternator.

No baseline vibration survey was carried out immediately after commissioning, as electrical loads were low. The installed on-line condition
monitoring system indicated acceptable vibration amplitudes at this
time. However, as platform commissioning continued, and electrical
load was increased, alarms began to be generated by the machine
vibration protection system.

These alarms occurred on startup and were considered by the platform personnel as instrument warnings, not mechanically related warnings. However, it was apparent from the on-line condition monitoring
system that the vibration alarms were not spurious and further investigation was required.

In order to evaluate the condition of the main power generation train
a more detailed vibration survey was carried out. The data for the gas
generator bearings, when the speed was increased from idle governor to
synchronous speed, demonstrated that the predominant frequency was
at the $1 \times$ rotational speed component as shown in Figure 14.8.

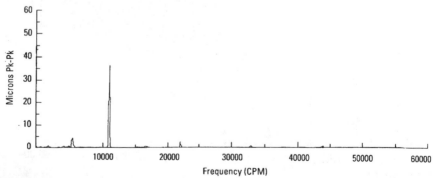

Fig. 14.8   Gas generator rear-end bearing spectra between idle and synchronous
speed.

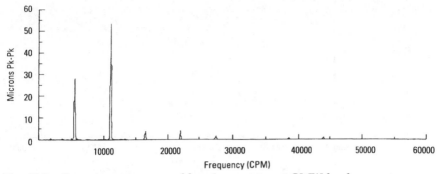

Fig. 14.9   Gas generator rear-end bearing spectra at OMW load.

Just prior to synchronous speed being reached, the overall vibration amplitude at the gas generator rear end bearing was seen to increase dramatically, from approximately 15 micrometres peak-to-peak, at idle, to 55 micrometres peak-to-peak at synchronous speed. This increase was due to the appearance of a frequency component at gas generator 0.5 × rotational speed as illustrated in Figures 14.9 and 14.10.

As the electrical load was increased the amplitude of the 0.5 × frequency component was observed to reduce, such that at 1.1 megaWatts its amplitude was less than 5 micrometres peak-to-peak. This is shown in Figure 14.11. The ATL engineer involved in the commissioning diagnosed these characteristics as a looseness at the gas generator rear bearing. It was considered that the most probable area for such looseness was at the interface between the housing and the bearing assembly.

Accordingly, it was recommended that the rear bearing of the gas generator be inspected for looseness. The simplest solution was to

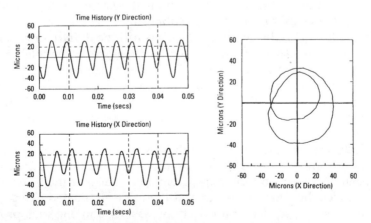

Fig. 14.10   Gas generator rear-end bearing orbit at OMW load.

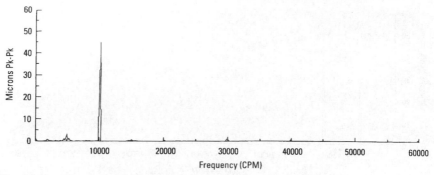

Fig. 14.11   Gas generator rear-end bearing at 1.1 MW load.

remove the gas turbine and conduct the inspection in the manufacturer's works. The resultant inspection of the gas turbine, at the manufacturer's plant, showed that excessive clearance of the bottom half shell of the rear bearing was present.

14.5   AERODERIVATIVE GAS TURBINE MONITORING

As outlined above, many aeroderivative industrial gas turbines are in use worldwide for power generation and gas compression. Such engines have been developed by manufacturers to a high level of component reliability. This has been driven by advanced materials technology, precision engineering and the stringent safety standards of the airline industry. In airline service, these engines spend 90% of their life breathing ultra-clean air at altitude and burn a relatively high grade kerosene. In addition, the maintenance environment enables a high level of 'hygiene', particularly in the area of lubricants.

Contrast this with the aeroderivative gas turbine in industrial service. The environment can be extreme such as desert, arctic or offshore and just about anything else in between. Fuel quality can be at best variable, and in the case of diesel, particularly in remote locations, contamination is a major problem. While maintenance standards on most industrial complexes are extremely high, lubrication 'hygiene' and quality will always be a problem.

The industrial application of aeroengines has therefore led to a whole series of reliability problems, such as those outlined above, and which have been addressed by manufacturers and users. In many cases the problems have been designed out and therefore eliminated. Some problems do, however, persist for a variety of reasons, and one such problem is bearing wear and failure.

In aeroderivative gas turbines rolling contact bearings are utilized throughout for both rotor main bearings and thrust bearings. Some engine types use roller bearings, while others use balls. The LM2500 gas generator for instance has bilobed ball bearings, and the RB211 has squeeze film roller bearings, to quote some special examples. In the latter a thin film of oil is introduced between the bearing outer race and the bearing housing. This is claimed to reduce vibration and increase bearing life.

However, as shown in the earlier case studies, the majority of aeroderivative gas turbine installations, in keeping with most large capital plant, will be fitted with some form of vibration supervisory system. The primary function of these systems is to provide alarm indications and ultimately trip the machine in the event of catastrophic damage. For aeroderivative gas turbines, such supervisory systems are casing mounted vibration transducers, and in the most rudimentary protection systems, velocity sensors are attached to the casing of the gas turbine.

In such systems, overall vibration velocity is monitored, either Root Mean Square (RMS) or peak-to-peak, and compared against preset alarm and trip values selected by the user. There are still in existence some installations where the vibration velocity signal is integrated, to provide a vibration displacement as the alarm/trip indicator. However, in the latest supervisory vibration monitoring systems casing mounted accelerometers are used, the signal being integrated to give vibration velocity for alarm/trip indication. An option with this latter system, is the incorporation of tracking filters to assess the vibration levels at shaft rotational speeds.

The protection system utilizing velocity transducers, particularly those which integrate the velocity signal to displacement are, to say the least, very ineffective at detecting the early onset of faults. The only fault which will be 'visible' to this type of system will be gross imbalance, probably consequential to some other fault, and a severe degree of damage to the blading and rotor assembly is likely to occur. These systems are by no means ideal for machinery protection.

The accelerometer-based protection systems, and particularly those with tracking filters, will provide a slightly higher level of protection. This is owing to the accelerometer providing a superior frequency response, and the ability to alarm and trip on vibration energy within a specified frequency band governed by shaft speed. As stated in earlier chapters of this book, simple vibration level measurement, either in wide or narrow frequency bands, will not react significantly until a state of severe damage exists, usually consequential to the primary fault.

It can be concluded, therefore, that current vibration-based gas turbine protection systems react only to gross imbalance, which is usually consequential to other more subtle faults such as blade loss or bearing damage. The vibration characteristics of these failures produce very low-level and discrete modifications to the machinery vibration signal. It is impossible therefore for supervisory systems employing casing-mounted vibration transducers to detect and react to early bearing damage in aeroderivative gas turbines. This is in contrast to heavy industrial gas turbines with tilting pad or journal type bearings. For these, proximity probes in each bearing will provide very good overall protection for both rotor and bearing related faults.

Aeroderivative gas turbine operations have in many instances experienced bearing failures which have resulted in dramatic consequential rotor and stator damage. The installed vibration-based protection system in most cases did not even register any increased vibration until overall shaft and blading damage caused gross imbalance. In these situations, the ability to diagnose the primary fault becomes masked by the secondary effects. Accordingly, and to overcome this problem, ATL has evolved via several gas turbine monitoring projects, a reliable system for bearing monitoring and this will now be outlined below (Smith, 1990).

14.6   THE ATL APPROACH

The primary objective in all project cases has been to provide a monitoring system to detect early damage to rolling contact bearings in aeroderivative gas turbines. The need has been driven by repeated failures which, owing to the fact that they were not detected at an early stage, resulted in severe consequential damage to the gas generator rotor and blading. Several gas turbine manufacturers have addressed the problems experienced, and produced modified bearing designs. However, end-user unease has resulted in a need for simple, reliable, condition monitoring systems for aeroderivative gas turbine bearings.

The approach to this requirement has been based on the following:

- that the most reliable technique for detecting damage in rolling contact bearings is vibration monitoring;
- that the vibration signals from casing mounted accelerometers do not enable bearing faults to be 'seen' at the early damage stage.

Vibration monitoring has proved to be a reliable indicator of rolling element bearing damage in a range of rotating machinery where the accelerometer can be attached close to the bearing. In a gas turbine, however, a casing-mounted accelerometer can be in excess of 0.4 metres from the bearing location. At such distances even with the most advanced signal processing, the extraction from the vibration signal of information on bearing condition is unreliable. The ability to locate an accelerometer close to the bearing will obviously overcome this problem.

In the projects in which ATL has been involved, accelerometers have been located on bearing support structures within the 'cold' portions of the aeroderivative gas turbines. For a recent project it was decided that each engine of a two-engine facility would be fitted with accelerometers to monitor the condition of LP compressor bearing. This particular bearing had a history of failure. Temperature measurements were chosen in addition to vibration monitoring, as there was evidence to suggest that bearing temperature increases were associated with previous failures. This it was felt would provide valuable corroboration with the vibration measurements.

The accelerometers and thermocouples were fitted during a routine engine overhaul by the licensed overhaul company. In each case two Vibro-meter CA160 accelerometers were fitted to the bearing support structure together with six thermocouples attached to the outer raceway face of the bearing. The thermocouples were spaced at approximately 60° intervals around the outer raceway face. Each accelerometer was fixed to the bearing support structure within 4 centimetres of the outer race of the bearing. Fixing was effected by three bolts per accelerometer, and all were lock-wired for safety.

The signal cables for the thermocouples were bundled and pulled through one of the support struts to a junction box fitted to the gas turbine casing. The armoured accelerometer leads were pulled through another support strut to a junction box located on the skid. The accelerometer and thermocouple signals were fed to a standard supervisory alarm/trip package supplied by Bently-Nevada. All the raw alternating current (AC) vibration signals and the 4–20 milliamp temperature signals were acquired from the supervisory system for processing in the diagnostic routine. A functional block diagram of the scheme is shown in Figure 14.12.

The supervisory system was utilized as an additional safety measure for the gas generator, and the alarm and trip levels were set accordingly. This system is rack-mounted alongside the existing gas generator supervisory system which utilizes accelerometers on each machine's casing. The diagnostic system consists of an industrial rack mounted 286 personal computer (PC) with keyboard and remote monitor, and the expansion slots of the PC contained the following cards for data acquisition and signal processing.

- a two-channel low-pass filter;
- a two-channel envelope circuit;
- a two-channel Fast Fourier Transform (FFT) card;
- a 16 channel Analog to Digital (A/D) multiplexer card.

The raw AC vibration signals are divided two ways. Each signal is fed unprocessed to the A/D multiplexer card for routing in sequence to the FFT card. In addition, each signal is also routed via a low-pass filter to an enveloping card. The enveloped signals are then fed back to the A/D multiplexer card for outputting in sequence to the FFT cards. The

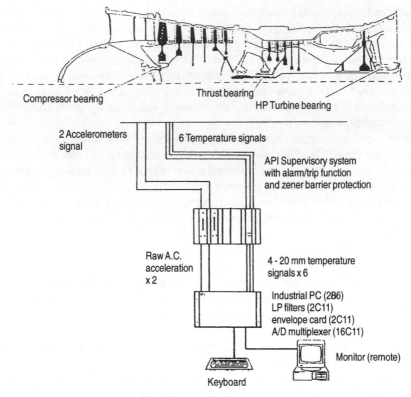

Fig. 14.12   Monitoring system functional layout.

temperature signals are also fed to the A/D multiplexer, and in addition, a tacho signal is provided. The PC software package controls the data acquisition and signal processing such that the temperature values, overall vibration levels, vibration spectra and envelope spectra are all read to the system database.

The diagnostic software develops diagnostic indicators from the envelope spectra. For as bearing faults will exhibit unique patterns in the spectrum dictated by bearing geometry and shaft speed, the frequency ranges of interest can be computed. The only variable is speed, thus the computation is continually updated. 'Windows' are created in the frequency domain around the unique bearing frequencies for the inner race, outer race and rolling elements. When a fault develops with a rolling element bearing, energy will be generated within the appropriate frequency band. For a bearing in good condition, no energy apart from background noise is expected. The software therefore continually 'looks' for significant increases in power spectral density within each frequency 'window'.

The diagnostic indications D1, D2, D3 in each case can then be trended with time. When faults occur the location can be readily identified, for example, oil starvation of the bearing will characteristically give a simultaneous increase in all three diagnostic indicators. In all cases, corroborative evidence can be obtained by viewing the high-frequency acceleration spectrum and temperature trends and spreads. This approach utilizes high-quality vibration signals from accelerometers located very close to the bearing and a well-proven signal processing technique to 'look' for the onset of bearing damage.

The engineering required to fit such sensors is relatively straightforward, and does not threaten the integrity of the machine. It could in the not too distant future be a standard fitment on some gas turbines, and it is envisaged by ATL that this system will be refined and further evolved. This is explained in the next section, where the system outlined above may be used in conjunction with an expert system for automatic fault indication (Hill, 1991).

14.7 EXPERT SYSTEM APPLICATION

Many chapters in this book have discussed various measurement parameters and techniques for monitoring the health of rotating equipment. The basic principle is to monitor parameters which are sensitive to developing machine faults or performance degradation. However, interpreting monitored data and relating changes in value to machine faults invariably requires a significant input from specialist machinery engineers, as variations in parameter values may result solely from routine changes in machine operating states.

Alternatively, changes in certain parameters may be the consequence of any one or more of a number of faults, and as shown above, in this instance, specific fault identification will rely on the experience of the engineer, supported by corroboration using other monitored parameters. When a fault is confirmed the experience of the engineer is required to assess the severity, and advise on continued machine operation plus any associated remedial actions. The time spent by engineers reviewing monitored data adds significantly to the cost of implementing a condition-based maintenance programme.

There is considerable confusion between Artificial Intelligence (AI) and expert systems. Although the two are related, AI is much broader, and is strictly speaking a collective name for a field of academic research. AI encompasses a number of subfields including natural language interfaces, robotics and automatic programming as well as knowledge-based or expert systems. Expert systems are the most practicable application of AI and is that area which to date has seen the most commercial development.

Expert systems use knowledge and inference techniques to analyse and solve problems. By encoding the knowledge and reasoning skills of human experts within an expert system program, it is possible to diagnose problems and make recommendations that would previously have required a human expert's attention. Expert systems differ from conventional programming techiques in three key areas:

1. the representation of knowledge;
2. heuristic search;
3. separation of knowledge from inference and control.

Conventional programming must reduce a problem to mathematical terms which can be manipulated by an algorithm, whereas expert systems, and AI techniques in general, encode knowledge as linguistic expressions, words and sentences, which are manipulated using logical procedures.

Heuristic search enables a computer to produce probable answers to a problem which is not fully defined, in the same way that a human would use 'rules of thumb'. With conventional programs, knowledge about a problem must be embedded within algorithmic procedures, using code which is only understandable by a programmer. With expert systems the 'program' provides the inference capability, but the knowledge that is used is stored separately in a knowledge base, as text which can be understood and modified by non-programmers. AI has also led to the development of new symbolic languages such as LISP and PROLOG, which are able to process 'data' or knowledge expressed in words or sentences.

One of the earliest and best-known expert systems is MYCIN, which was developed at Stanford University in California in the early 1980s. It was developed to assist physicians with the diagnosis of bacteraemia or meningitis, in order to recommend drug treatments. Another well-known expert system is XCON, which is used by Digital Equipment Corporation (DEC) to configure computer systems according to a customer's requirements. XCON has some 6000 'rules' relating to the way that components can be combined to create a computer system. DEC employs eight full-time specialists to keep the system up to date; however, without the support of an expert system, it would not have been possible to offer the flexibility of configuration currently available. XCON now 'knows' far more about how to configure DEC computers than any human expert.

In the application of expert systems to condition monitoring, it is not necessary to resort to fundamental development from scratch. The more cost-effective approach is the use of an expert-system building tool or shell. These are partially developed programs, and include a number of components, which are shown in Figure 14.13. The knowledge base contains all the information that the human expert would use to solve a

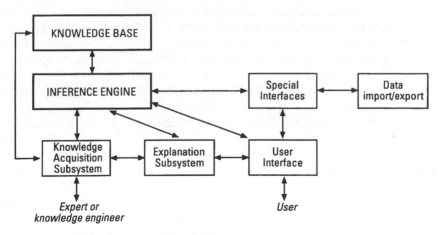

Fig. 14.13 The architecture of an expert system.

problem. This could be based on heuristics, or observational evidence which cannot be rigorously proved, and facts, or knowledge which is provable. The simplest way of representing knowledge is as rules. The general form of a rule is:

IF (condition) THEN (consequence)

If the conditional part of the rule is satisfied then the consequence follows. Other forms of knowledge representation include complex facts such as:

Context–Attribute–Value

These are organized into structured sets and object orientation information.

The inference engine is a program which provides the expert system with its ability to reason and arrive at a diagnosis. The most common inference strategies are termed 'backward' and 'forward' chaining. In backward chaining, the system starts with the goals the user has listed in the knowledge base, and works backwards via rules to determine what initial data are required to determine if that goal can be recommended. Ultimately, the user is asked questions, to provide the system with the initial data it needs to qualify or disqualify a particular recommendation. Backward chaining is therefore used in interactive sessions associated with some 'offline' condition monitoring systems.

Forward chaining systems must be provided with initial data before applying the rules to determine which inferences can be made. A second cycle then begins in which additional rules are examined based on the initial data and the results of the first cycle. This process is repeated until all inferences have been determined, and if possible, a recommendation made. Forward chaining systems are more appropriate for 'online' monitoring systems processing data derived from sensor inputs. Expert system shells normally provide both inference strategies which can be mixed together.

Another important aspect of expert system shells is the development interfaces that are provided. These include a facility for creating and editing the knowledge base, a means for adding explanations and viewing rule structures, and special interfaces which include hooks to custom software modules for data import/export and enhanced graphics capabilities. Originally expert system shells were developed using symbolic languages such as LISP, but more recently there has been a trend towards the C language, which offers better performance, ease of integration with other software modules, and transportability across computer platforms.

Application of an expert system to condition monitoring will require some form of integration with a data acquisition system. The simplest expert systems use the operator to make this link, via an interactive question and answer session, during which the operator inputs machine details and values of various parameters which have been monitored independently. The more integrated approach is to acquire data using offline portable data collectors, a dedicated online system, or a combination of both, into a system database. Ideally the acquisition system should include some 'intelligence' to permit alarm processing, normalization and reduction of data to parameters which are sensitive to machine faults.

This applies particularly to vibration frequency spectra, for which, as shown above, specific components are of most value, rather than the full spectrum. The expert system then processes the data stored in the database to yield current fault diagnoses and remedial operator actions. Trends in regularly stored historical data can also be processed to forecast progress of developing faults and associated maintenance requirements. A typical system configuration is shown in Figure 14.14.

As an expert system can only be as good as the knowledge embodied within it, the key to success is the effective development of the rules in the knowledge base. This process is normally referred to as knowledge engineering. It involves working with human experts to identify and refine the knowledge needed to solve a particular type of problem. In the case of condition monitoring the experts will include operators, maintenance engineers, instrument and process engineers, and rotating machinery specialists. It will also be necessary to consider machine operating and maintenance histories, and which monitored parameters are available. Knowledge-base development is an on-going and evol-

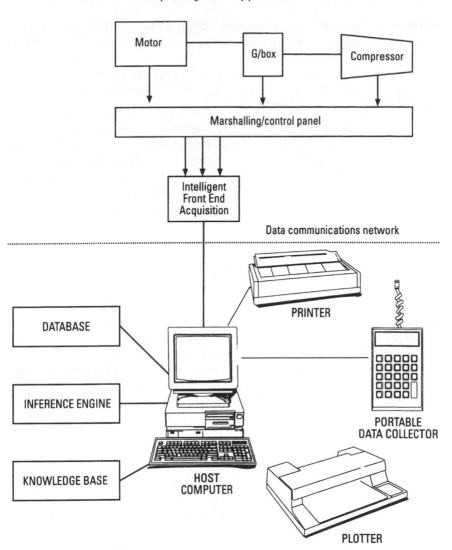

Fig. 14.14 Schematic diagram of a condition monitoring expert system.

utionary process which should start simply, perhaps only considering one aspect of a machine's operation or specific types of fault, and then be expanded and refined with experience.

The major benefits of using expert systems in condition monitoring are as follows:

- automatic diagnosis of faults is achieved from monitored parameters;
- repeatability of diagnosis for a given set of input data;

- the expert system can incorporate the knowledge of more than one human expert, and can be available 24 hours a day, simultaneously at a number of sites;
- the knowledge base is separate from the program, and therefore can be easily understood and modified by non-programmers, using a word processor or similar kind of interface.

The major problems which may be experienced with expert systems are false alarms and false diagnoses. It is therefore essential that great care is taken in the specification of the monitoring system, the selection of appropriate fault specific parameters, and the data processing and normalization procedures as well as development of the knowledge engineering phase, in order to ensure that the operational system can accurately assess machine behaviour.

14.8   CONCLUSIONS

The above project cases illustrate the application of vibration monitoring techniques in a commercial sense. Although using the specific example of vibration monitoring on gas turbine systems, the projects quoted show that the use of vibration techniques can diagnose potential failures and be cost-effective in a commercial setting. Other rotating machinery can in a similar way be monitored just as effectively via the use of these techniques, and with the development of knowledge-based expert systems the cost/benefit of such applications is becoming more favourable on an increasing range of industrial machinery.

14.9   REFERENCES

Anon (1994) *Case histories – machinery acceptance testing, ATL Application Notes,* Southampton, UK.
Hill, J. (1991) Application of expert systems to condition monitoring, Special Feature, reprinted from *Noise and Vibration Worldwide,* **22**(4), Elsevier, The Netherlands, 20–2.
Shrieve, P. (1994) How, when and why to use condition monitoring – is the complexity fully justified, *ATL Application Notes,* Southampton, UK.
Smith, R. (1990) Aeroderivative gas turbines – bearing monitoring using advanced vibration diagnostic techniques, Special Feature, reprinted from *Noise and Vibration Worldwide,* **21**(8), Elsevier, The Netherlands, 17–20.

PART FIVE

Techniques for Wear Debris Analysis

CHAPTER 15

Detection and diagnosis of wear through oil and wear debris analysis

A.L. Price and B.J. Roylance

Tribology Centre, University of Wales College of Swansea, UK

15.1   INTRODUCTION

The earliest recorded perceptions of wear and wear debris were recorded over 2000 years ago as follows:

> a ring is worn thin next to the finger with continual rubbing. Dripping water hollows a stone, a curved plough-shear, iron though it is, dwindles imperceptibly in the furrow, we see cobble stones of the highway worn by the feet of many wayfarers. We see that all these are being diminished since they are worn away. But to perceive what particles drop off at any particular time is a power grudged to us by our ungenerous sense of sight. (Lucretius, 95–55 BC)

Today, the power to observe and determine the characteristics of wear particles is no longer begrudged us, so how do we detect and diagnose wear? In this chapter, we shall direct our attention to three main issues:

1. The generation of debris as a direct consequence of wear and its implications.
2. The techniques available to perform oil and wear debris analysis.
3. How the techniques are implemented in machinery condition and health monitoring programmes in industry.

Figure 15.1 shows a cut away sectional view of a typical industrial gearbox unit. While Figure 15.2 provides evidence of what happens when a deteriorating gear tooth condition is allowed to progress too far,

Handbook of Condition Monitoring
Edited by A. Davies
Published in 1998 by Chapman & Hall, London. ISBN 0 412 61320 4.

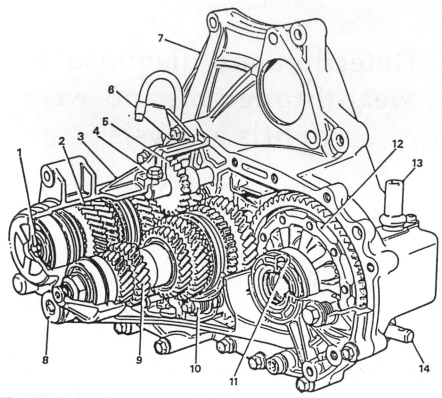

Fig. 15.1   A typical gearbox configuration.

and Figure 15.3 represents an example of the debris generated from a gear-wear situation.

Hence the scope for analysis of used oil, and the wear debris transported within it, can be illustrated schematically in Figure 15.4. Both online and inline monitoring are carried out directly or in conjunction with the machine during its operating condition. This facility provides the opportunity to obtain data in a real-time mode. The Magnetic Debris Plug (MDP) unit for example, is a ferrous debris capture device which is positioned strategically within the lubricating circuit, e.g. the sump. Its periodic removal permits the debris to be analysed offline, that is away from the machine. Likewise, a filter unit traps material above a designated size. This can be subsequently analysed after the filter has reached the end of its useful life and is replaced by a new one.

Taking a sample from a lubricating system at prearranged intervals means that two kinds of analysis can be carried out offline, these are oil and wear debris analysis. Performing oil analysis enables the state of the fluid and any evidence of contamination or corrosion to be determined.

Fig. 15.2 Severe damage on a spur gear.

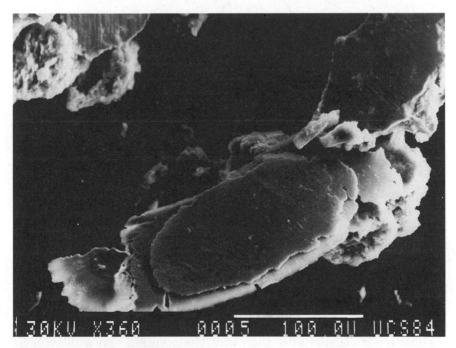

Fig. 15.3 Gear wear particle.

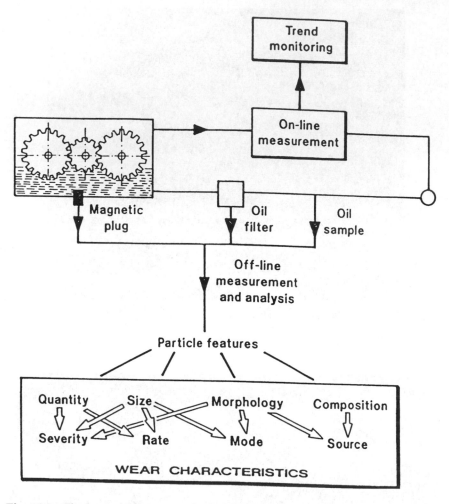

Fig. 15.4 The basic components of wear debris analysis, their characteristics and relationships to wear.

As with oil analysis, wear debris analysis is undertaken in a number of different ways but broadly the particles are categorized in terms of their quantity or concentration, size, morphology and composition. The associated wear characteristics are the severity, rate, mode and source of the wear. In terms of the implications for maintenance practice, it is important to appreciate that whereas oil analysis can be utilized as part of a pro-active maintenance strategy, wear debris analysis, by the very nature of wear itself, can only be used to monitor active primary wear in a reactive mode. That is after the event has occurred and as shown in Figure 15.5.

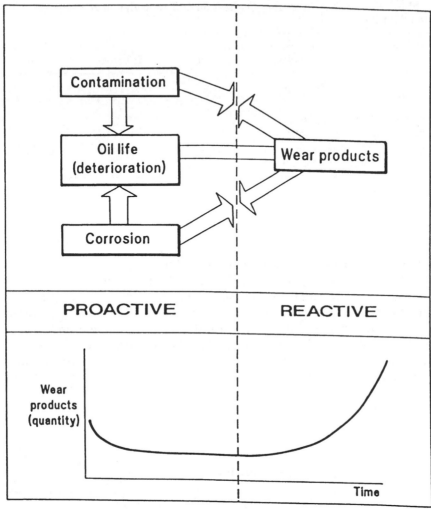

Fig. 15.5 The proactive and reactive components of wear and associated products.

These aspects can best be demonstrated by way of an example. The American company, Computational Systems Incorporated (CSI) Ltd, manufacture and market an oil analysis system known as 'Oilview'. It is an on-site or in-shop system which enables the determination of oil degradation, contamination and machine wear by utilizing a combination of relative dielectric measurements and magnetic differentiation of particles. By using the Oilview device, the preventive maintenance team at the Chrysler stamping plant in Warren, Michigan, identified impending failures in each of the 1000 ton Hamilton presses located at this plant.

Table 15.1 Data obtained from Oilview oil and wear debris analyser for Hamilton Press

Sample date	Oil life index	Corrosion index	Contaminant index	Ferrous debris index	Debris type
21.3.93	26.7	0.1	22.9	272.0	F
03.6.93	29.7	0.0	996.6	1086.0	F & NF
Action: Broken rocker stud removed. New stud installed					
28.6.93	19.4	0.6	15.7	183.0	F & NF
11.8.93	12.7	0.0	10.5	79.0	F

F = Ferrous.
NF = Non-ferrous.

One of the presses suffered a sheared stud supporting a rocker arm. The oil analysis history as reported in Table 15.1 reveals what happened, and Figure 15.6 shows the trend plot for the data collected for this press. The press was repaired within 24 hours, thus avoiding several months of lost production and this was attributed to detecting the problems at an early stage. The estimated savings ranged between $50 000 and $1 000 000, depending on how the return on investment is calculated. Having thus illustrated the scope for performing oil and wear debris analysis, we now examine these aspects in more detail, commencing with a description of the basic wear mechanisms encountered in many practical situations.

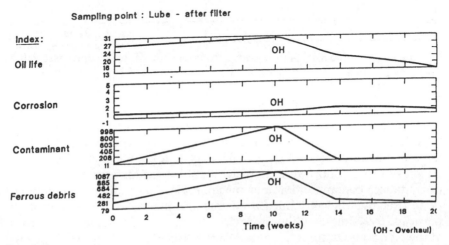

Fig. 15.6 Trends in condition monitoring data.

15.2 WEAR IN LUBRICATED SYSTEMS

Tribology is the study of what happens when surfaces come into rubbing contact with each other. It embodies what in previous times was otherwise known as 'friction, lubrication and wear'. It soon becomes evident that tribology is a multi-disciplinary subject involving in the

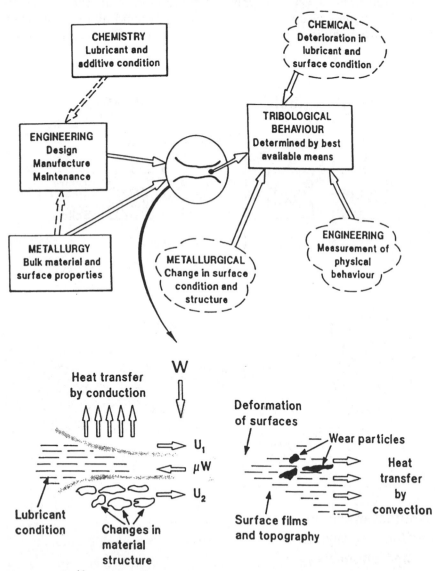

Fig. 15.7 Interacting surfaces in relative motion (the inter-disciplinary implications).

main, chemistry, materials science and mechanical engineering. Figure 15.7 is representative of the principle activities involved, as interactions take place in the contact region of the rubbing surfaces. The extent of the separation of the surfaces is also crucial in determining the type and severity of wear that will take place.

Figure 15.8 outlines the 'Stribeck Curve', in which the three principal 'regimes' of lubrication are specified. When the surfaces are separated by the lubricant, a film thickness much greater than the combined surface roughness of the two surfaces is present and hydrodynamic or elasto-hydrodynamic lubrication conditions apply. The former usually

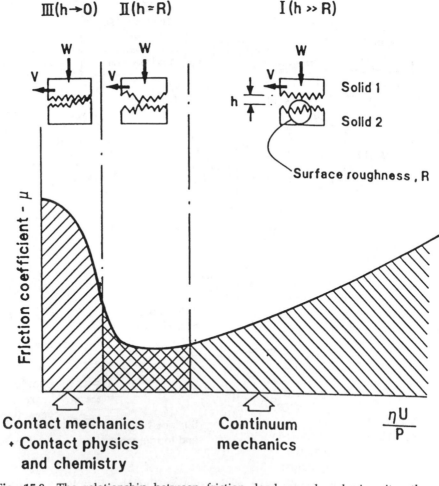

Fig. 15.8 The relationship between friction, load, speed and viscosity – the Stribeck curve.

relates to plain-journal or thrust-bearing operation, while the latter is more indicative of the lubrication of a rolling element bearing or gear tooth contact.

At the other extreme, when the surfaces are only separated by a very thin layer of lubricant, typically a mono-layer, then we say that the surfaces are boundary lubricated. The region in between these two extremes is called the 'mixed regime' and here intermittent contact occurs. This latter regime is often the most critical phase of operation, one in which wear debris analysis has an important part to play in determining the true wear state of the machine. The relationship between wear and the regime of lubrication is epitomized by the behaviour of rolling element bearings and gears, as shown in Figures 15.9 and 15.10.

Wear is defined as the progressive loss of substance resulting from mechanical interaction between two contacting surfaces. It can take various forms depending upon several interconnected factors such as geometry, motion, load, speed, and the state of lubrication. Figure 15.11 demonstrates for different applications the link between the nature of the contact, conformal, sliding, low stress/conformal, sliding, rolling, high stress, the mode, or manifestation of the wear and the underlying mechanism or cause. This approach can be further extended to show the

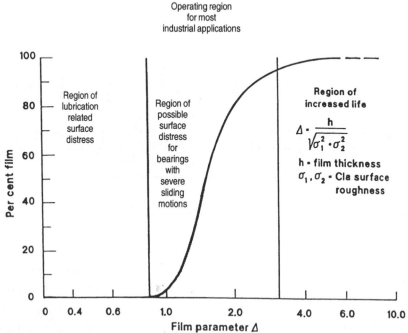

Fig. 15.9 Surface distress versus film thickness ratio.

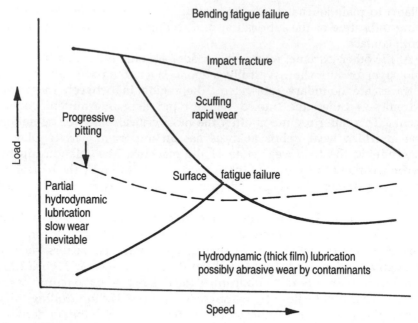

Fig. 15.10 Load/speed conditions characteristic of principle sources of gear tooth failure.

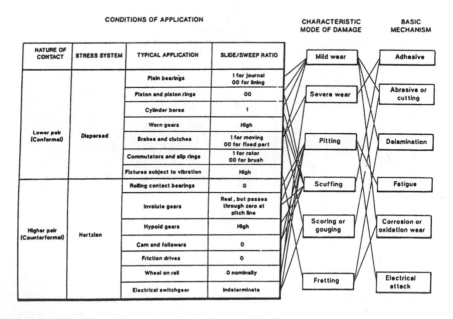

Fig. 15.11 Relationship between modes and mechanisms of wear for different applications.

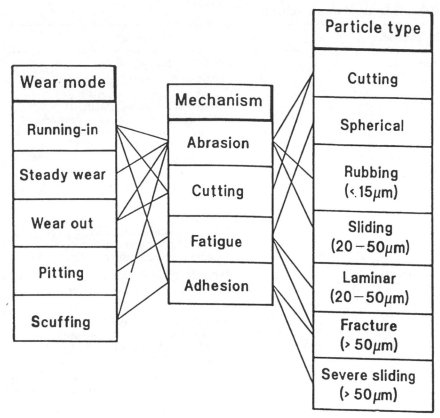

Fig. 15.12 Wear modes, mechanisms and particle types encountered in gear operation.

relationship between wear mode, mechanism and the type of ferrous particles which are generated and subsequently captured in the lubricant.

The particular descriptions shown in Figure 15.12 are those normally used to describe the morphology of the particles analysed, when using the analytical ferrography method. Table 15.2 summarizes the underlying factors involved in defining the five principal mechanical wear mechanisms encountered in practice, while Table 15.3 provides further information, which includes other wear manifestations such as those due to erosion and corrosion. Table 15.4 further relates component type to the contact situation, and the associated class of debris. The wear particle descriptions associated with analytical ferrography being summarized in Table 15.5.

In addition to describing wear in terms of its mode and mechanism it is also necessary to be able to specify its extent, rate and severity. For

Table 15.2 Basic wear mechanisms

Mechanism	Description
Adhesion	Occurs in sliding.
	Plastic deformation → junction growth.
	Failure at junction of contact or elsewhere → metal transfer occurs during initial stages of operation (running-in) and when thermal effects due to sliding lead to local 'welding' under high friction forces.
Abrasion	*Two-body*
	'Hard' asperities abrade softer contacting surface.
	Three-body
	'Foreign' bodies or hardened particles abrade both parent surfaces in passing through the contact.
Delamination	Asperity–Asperity contact
	Fracture process caused by repeated loading–unloading.
	Softer surface becomes smoother and cracks nucleate in sub-surface leading to formation of thin sheets–delamination.
	Harder surface experiences cyclic loading inducing plastic shear deformation.
Pitting fatigue	Occurs under rolling action.
	Repeated cyclic loading–unloading.
	Development of cracks at maximum shear stress locations, surface flaws and material impurities.
	Sub-surface fatigue → spalling (large chunks).
	Surface-initiated fatigue → pitting or micro-pitting.
Fretting	Surface damage arising from small amplitude oscillatory slip under normal load.
	Initially, adhesive asperity contacts are formed and junctions fracture. Loose particles become oxidized which cause wear by abrasion.
Corrosion wear	Process in which chemical or electrochemical reactions with the environment predominate. Likely to encompass corrosion of worn surfaces and wear of corroded surfaces.
	The effect of temperature is such that the reaction rate approximately doubles for a 10°C rise in temperature.

Table 15.3 Worn surface appearance and associated debris

	Description	
Type of wear	*Worn surface*	*Wear debris*
Adhesive	Tearing and transfer of material from one surface to another.	Large irregular particles >10 μm, unoxidized.
Mild wear	Fairly uniform loss of material; slight surface roughening.	Fine particles <10 μm, fully oxidized.
Two-body abrasion	Harder surface little or no damage. Softer surface exhibits scores, grooves or scratches.	Consists mainly of softer material (fine swarf), unoxidized material.
Two-body abrasion	Softer surface contains embedded hard wear particles–maybe scored or wiped. Harder surface scored/ grooved corresponding with hard embedded particles in counter surface.	Contains material from harder surface in form of swarf containing unoxidized material. May also contain softer material in lumps.
Fretting	Surfaces heavily pitted; pits may be small or larger, or overlapping to produce rough area. Surface of pits rough on macroscopic scale with oxidized appearance.	Fine, fully oxidized. If from ferrous metal it will contain Ferric Oxide (Fe_2O_3) with cocoa (rust) appearance. Some spherical particles likely.
Three-body abrasion	Surfaces exhibit deep scratches or grooves deeper on softer surface, similar on both surfaces if equally hard.	Fine, may contain some unoxidized metal among loose abrasive material.
Erosion	Surface very unevenly worn with smooth, matt appearance. Some embedded fine particles likely.	Difficult to detect among loose abrasive material.
Corrosive wear	May show signs of oxidation, corrosion or general chemical reactions.	Always highly reacted, i.e. no free metal present. Usually fine and amorphous.

example, we can distinguish mild wear from severe wear, as shown in Figure 15.13a, as a function of some operating parameter such as load or speed. Likewise, the rate at which it is occurring as a function of time, may also be important; this is illustrated in Figure 15.13b, and has important implications for wear debris monitoring. Figure 15.14 shows the same trend but depicted this time in terms of debris production rate. This idealized representation of wear progression with time forms a

Table 15.4 Examples of typical wear debris produced by specific machine components

Component	Contact situation	Type of debris
Rolling bearing Gear teeth Cam/tappets	Non-conformal rolling–sliding	Ferrous particles Varying shape Size range 10–1000 μm
Piston rings and cylinder	Load concentrated in small area	Ferrous particles
Splines Gear coupling		Iron oxide particles Size < 150 μm
Plain bearings	Load distributed over a large area	Small ferrous and/or non-ferrous particles
Piston Cylinder		

Table 15.5 Wear particle morphology–ferrography descriptors

Particle type	Description
Rubbing	Particles < 20 μm chord dimension and $\approx$ 1 μm thick. Results from flaking of pieces from mixed shear layer – mainly benign.
Cutting	Swarf-like chips of fine wire coils. Caused by abrasive (two-body or three-body) cutting tool action.
Laminar	Thin, bright, free-metal particles, typically 1 μm thick, 20–50 μm chord width with holes in surface and uneven edge profile – emanates from mainly gear/rolling element bearing wear.
Fatigue	Chunky, several microns thick from, e.g. gear wear, 20–50 μm chord width.
Spheres	Typically ferrous, 1 to > 10 μm diameter generated when microcracks occur under rolling contact fatigue condition.
Severe sliding	Large > 50 μm chord width, several microns thick. Surfaces heavily striated with long straight edges. Typically found in gear wear.

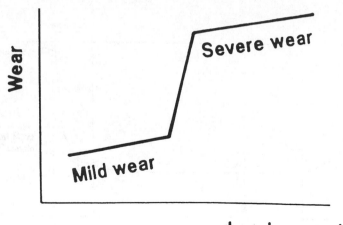

(a) Wear transition

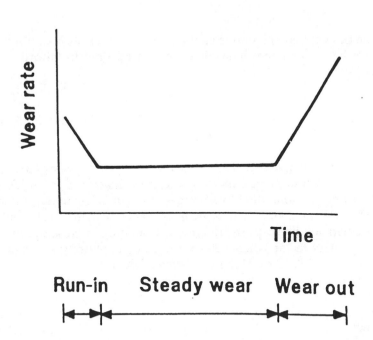

(b) Representative wear history characteristic

Fig. 15.13 Wear characteristics.

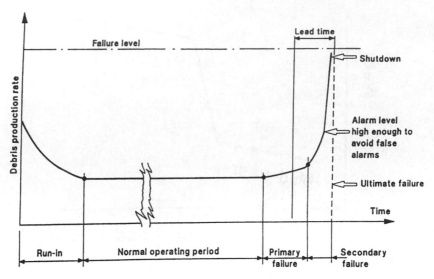

Fig. 15.14   Bath tub curve.

baseline of expected behaviour against which factors such as the setting of alarm levels, and the determination of lead time to failure, can be established.

15.3   WEAR DEBRIS – TRANSPORT, MONITORING EFFICIENCY AND ANALYSIS

Figure 15.15 illustrates diagrammatically the various ways in which debris, as it is transported away from the contact region, is likely to become trapped within the lubricating system, or is lost to the surroundings. The efficiency with which wear debris are carried to the point of measurement, on-line capture, magnetic drain plug or oil sampling point is an important factor when determining the monitoring strategy for a particular machine. The overall efficiency of the process is, however, the cumulative effect of three distinct phases in the whole monitoring operation. These are the lubricant transport efficiency, the efficiency of the debris capture device and, ultimately, the measurement or analysis; efficiency of the monitoring instrument. The range of overall efficiency, recorded under systematically controlled conditions of assessment, shows that it can vary from 10% to 90%.

Having obtained a sample of the wear debris, the scope for performing analysis resides with determining their quantity, size, composition and morphology. Table 15.6 shows the relationships between wear characteristics and wear particle feature. Quantity – or more correctly the concen-

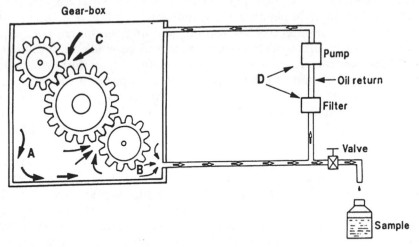

A - Wear particles lost to walls of sump/pipework
B - Oil to filter, etc.
C - Secondary degradation through gears
D - Wear particles lost to filter and pump

Fig. 15.15 Parent fluid-loss mechanisms.

Table 15.6 Relationships between wear characteristics and particle features

Wear characteristic	Wear particle feature
Severity	Quantity (concentration)
	Size
	Morphology
Rate	Quantity
	Size
	Morphology
Type	Size
	Morphology
Source	Composition
	Morphology

tration of particles per unit volume of lubricant sample–and size are determined quantitatively, i.e. by measurement. Composition can also be determined quantitatively by elemental analysis or by other, more subjective means. Morphological analysis is performed in the first instance by viewing the particles through an optical microscope, the debris having first been separated from the carrier fluid medium. In determining the first three features, the debris do not necessarily have to be separated from the fluid in order to perform the analysis.

There are numerous methods available to measure concentration. This value is the most direct, simple and cost-effective method for obtaining a first indication of impending wear and deterioration. It is, therefore, the foremost 'front-line' tool for use in plant and machinery condition monitoring. Measurements of concentration can be made online, on-site or in the laboratory. It is particularly useful for performing trend analysis over a period of time. Methods used to perform the measurements are in the main either optical, light blockage, or electrically inductive. The units used to define the measurements are directly or indirectly related to the concentration of particulate, e.g. weight per unit volume of lubricant sample.

In a number of instances, however, the particular instrument used can provide a non-dimensional number or index, which is proportional to the concentration. For surveillance purposes this is often quite acceptable, on the grounds that, in carrying out trend monitoring, it is only the relative values that matter and absolute determinations are therefore not essential. Most of the methods used to measure concentration are size-dependent and hence are usually size-limited. The total size range of interest for most particulates divide into two distinct size ranges: 1–100 microns and 100–1000 microns. The former are those particles which are contained within an oil sample, the latter are those which are captured, either magnetically, or by filters.

Size distributions are important in wear particle analysis because an indication of an increase in size often signals the transition from mild benign to active severe wear. Automatic methods for determining size distribution have existed for many years, notably in the area of monitoring hydraulic system contamination. However, for identification of particle type, the debris must also be visible, and this requires the use of image analysing systems. Recent work using these techniques has shown that, by classifying the type of size distribution, the type of wear, its rate and severity can also be confirmed.

When carrying out visual inspection of debris under the optical microscope, it is usually possible to assess the composition. This is especially true if the particles have been deposited onto the substrate using magnetic force. For more precise information, however, it is necessary to use a scanning electron microscope (SEM). Alternatively, by taking the original oil sample and carrying out spectrometric analysis, the elements present in the sample can be identified. Further details of these and other methods are presented later in the chapter. Morphological determinations are the most subjective procedures used in the entire spectrum of particle analysis. It is based almost entirely on a pattern recognition approach, as exemplified by reference to Table 15.5. Further developments in the use of image analysis procedures have led quite recently to the development of specific shape descriptors, and these can be utilized for distinguishing different types of wear.

15.4 LUBRICANT PROPERTIES AND OIL ANALYSIS METHODS

So far, we have made reference to the functions of the lubricant in terms of its general lubricating requirements and also as a means for transporting wear debris within the system. We now examine briefly the properties that constitute the substance of what a lubricant is and expecially how, by measuring the changes occurring in those properties, the results can be utilized as part of an overall oil and wear debris monitoring strategy. In practice, there are numerous factors which must be taken into account to ensure that the lubricant will function correctly throughout its working life. Some of the key factors involved are listed in Table 15.7 and the principle properties are recorded in Table 15.8. Among the properties listed, the most important single factor is viscosity, with thermal effects being important in terms of stability, and deterioration due to corrosive elements, can well point to problems occurring in the system.

Table 15.7 Key factors in determining lubricant properties

Factor	Comment
Thermal stability	Lubricant must not breakdown as a consequence of increased operating temperature.
Chemical stability	Lubricant can be chemically attacked by oxygen, (from air) water or other substances related to thermal stability.
Compatibility	Ability to function with other materials.
Corrosiveness	Lubricant must not attack other material and resist corroding itself during use.
Thermal conductivity	Ability to conduct heat away.
Heat capacity	Related to the amount of heat which can be carried away by the lubricant.
Flammability	Must not be prone to catching fire.
Toxicity	Must not affect the environment and hence, the health of people.
Availability	Lubricant escapes easily and has a finite life. Hence, it needs to be replaced.
Price	Cost vs quality.

Table 15.8 Properties of lubricants

Property	Description/comment
Viscosity	A measure of a lubricant's resistance to flow. Helps to keep surfaces apart but is sensitive to its molecular weight and temperature. Also modified by presence of contaminants.
Extreme pressure and anti-wear	Aim is to prevent welding of metal surfaces at point of contact and, hence, reduce friction.
Oxidation stability	Hydrocarbon oils react with oxygen (in air) to form acid or sludge products. Time taken for oil to oxidize is function of operating conditions but is highly temperature dependent. Numerous laboratory tests exist to determine oxidation stability.
Corrosion inhibitors	Moisture present in oil or condensing from the atmosphere can lead to corrosion in engines and circulating systems. Rust inhibitors in the oil afford protection through the formation of a thin protective film on metals. Combustion products can form acids which corrode bearings and engines; also, from the products of oil degradation – oxidation inhibitors prevent this and detergent dispersant additives exhibit neutralizing properties.
Pour point	The temperature at which oil stops flowing. De-waxing oils or using pour point depressants are used to reduce the level.
Demulsibility	The ability of the oil to separate readily from water which gets into the lubricant. Additives designed to give stable emulsions are used in soluble cutting oils, fire-resistant hydraulic fluids, and also in some diesel engine lubricants.
Flash point	The temperature at which it gives off a flammable vapour in specified conditions. Provides a useful check on possible contamination in a low flash product such as petrol.
Acidity/alkalinity	Acidity is conventionally measured in terms of the amount of potassium hydroxide (KOH) needed to neutralize the acidity. Alkalinity is the amount of KOH that is equivalent to the acid needed to neutralize alkaline constituents in a fixed quantity (1 gram) of sample.
Detergencing	Detergent properties are required to keep combustion and oil-degradation products in suspension and prevent their depositing and baking in high-temperature zones. Low-temperature deposits are also a problem, notably in petrol engines.

The ingress of either solid or liquid contaminants is often the first stage of a serious unplanned deterioration in a system's life-cycle. By detecting the presence of contamination in the lubricant at a sufficiently early stage, avoiding action can be taken which will save considerable expenditure. If not detected until severely damaging wear has taken place, the result can be very expensive indeed. Accordingly, the basic techniques employed to monitor a lubricant's condition are described below, as preparation for demonstrating the setting up of a programme to conduct routine surveillance of industrial oil samples. Normally, this programme is coupled to the operation of a systematic and computerized procedure for reporting the results of the analyses.

Oil analysis methods are spectrometric techniques used for the examination of oils, and for indicating the elemental chemical composition of the debris present, as well as the concentrations of the individual elements detected. The major elements present in oil samples result from the wear of ferrous and non-ferrous metallic components, the ingress of extraneous matter such as dust etc. and the lubricant with its associated additive package. A typical commercial spectrometric analysis would be carried out to include the following elements:

- **wear elements**   iron, chromium, aluminium, copper, lead, tin, nickel, manganese, titanium, silver and molybdenum;
- **additive elements**   zinc, phosphorus, calcium, barium, magnesium, sodium and boron;
- **contaminants**   silicon, sodium, boron and vanadium.

The results of the analysis are normally reported as parts per million (ppm) of the individual elements, and the size of the wear debris particles contained within the lubricant sample often determines the type of spectrometric technique used. There are a range of commercial spectrometers available to the oil and particle analyst, and those which are most widely used in oil and wear debris analysis are summarized in Table 15.9 together with their advantages and limitations. These techniques can be briefly described as follows.

- **Atomic Absorption (AAS)**   This technique is used for wear debris analysis and is a method whereby the various elemental atoms absorb light energy of specific wavelengths. The oil sample containing the debris is mixed with a diluent and sprayed into a high-temperature atmosphere. This is achieved by means of a nebulizer, mixing chamber and an acetylene–type flame. An expansion of the technique is where the oil sample is heated in an inert gas furnace instead of a flame. The sample first dries, then becomes decomposed and finally atomized. Using the furnace method increases the sensitivity and detection limits of the technique. However, it also greatly extends the process analysis time, which is an important factor when considering commercial applications.

Table 15.9 Established spectrometric techniques available for oil/debris analysis

Basic technique	System identification	Manufacturer	Cost		Max particle size	Analysis			Particle destroyed	Comments
			Instrument (£K)	Analysis (£)		Quantity range	Time/ element	Time for 6 elements		
Atomic absorption	Atomic Absorption Spectrometry AAS	Perkin Elmer Varian GBC Hitachi	10–40	very low	1 μm	ppm (flame) ppb (furnace)	40s (flame) 3m (furnace)	4m (flame) 18m (furnace)	Yes	Accurate and free of interference. Two systems – flame and furnace. Flame (slow), furnace (very slow). Limited particle size.
Atomic absorption	Infrared IR Fourier Transformation FT-IR	Perkin Elmer	10–20	very low	n/a	n/a	3–5s	3–5s	No	Used for oil analysis only. Comparative method against standards. Very rapid method.
Atomic emission	Atomic Emission Spectrometry AES	Perkin Elmer Baird Spectro Analytical	10–30	very low	5 μm	ppm	40s	4m	Yes	Slow, but with an extended range of elements able to be analysed. Larger particle size range than AA.
Atomic emission	Rotating Disc Electrode RDE (Rotrode)	Spectro Inc. Baird	40–60	low	8 μm	ppm	30s	30s	Yes	No sample preparation required. Quick method. Better accuracy with smaller particles but can be used up to 8 μm and possibly larger with different source.

Atomic emission	Inductively coupled Plasma ICP	ARL, GBC Baird, Varian Perkin Elmer Spectro Spectroil	50–80	medium	5 μm	ppb-ppm	60s	60s	Yes	Quick method with high sample throughput. Analytically sensitive to debris and oil additives. Large dynamic range.
Atomic emission	Direct Current Plasma DCP	ARL–Applied Research Labs	20–50	low	5 μm	ppb-ppm	60s	60s	Yes	Analytically similar to ICP but with smaller dynamic range and running costs.
Atomic fluorescence	X-Ray Fluoresence XRF (EDXRF) (WDXRF)	ARL, Kevex Oxford Analytical Seiko, Outokumpa	20–100	very low	independent of size	ppm-100%	5s	60s	No	Analyse solid or liquid samples. Large dynamic range. Non-destructive. Quick method. Specialized identification possible.
Atomic fluorescence	Energy Dispensive X-Ray Analysis EDS (with SEM)	Link, Kevex	30–65	medium	independent of size	%	100s	100s	No	Analysis of dry debris. Fairly slow. Used with scanning electron microscope. Large dynamic range.

- **Infrared** The infrared spectroanalysis technique detects and measures molecular compounds. Instruments generally use a scanning optical wedge to produce an inter-ferrogram, which is resolved by Fourier transformation into a spectral plot of infrared absorbence versus wavelength. Differences between the spectrum of the oil and that of a reference sample of the clean oil measure contamination or chemical changes in the used oil. Contaminants normally measured by infrared analysis include water, 'blow-by' products, ethylene glycol, unburnt fuel and refrigerant gases.

  Degradation of the oil through oxidation and nitration is measured directly, as is the depletion of the additives which provide the oil's detergency, dispersancy, alkalinity and antiwear characteristics. The advent of specialist procedures, such as Attenuated Total Reflection (ATR) cells, and the automation of sample throughout using transmission cells has made this technique an invaluable analysis procedure for used oils. Due to its versatility, this technique has become an essential analysis method for monitoring the degradation of lubricants.

- **Atomic Emission (AES)** This method is also used for debris analysis, it is a technique whereby a spectrum is formed by excited atoms or ions producing wavelengths relating to the elements present. Excitation is achieved by the use of a high-temperature flame, and the intensity of the spectral lines produced varies according to the quantity of each element present.

- **Rotating Disc Electrode (RDE)** This is a development of AES which gives a faster throughput and increased accuracy. Again, light excitation of the atoms or ions is achieved by a high-temperature source. In this case, a discharging arc, thereby producing the various wave lengths associated with the elements present. A small sample of the oil containing debris is placed in a tray, and a portion of the sample is lifted from the bulk by means of a rotating vertical graphite wheel. This is burned at the top of the lift by the discharging arc.

  The technique is commonly termed 'Rotrode', and usually used for the smaller-sized wear debris of less than 8 microns. Further innovations have been introduced to extend the size range of the debris which can be analysed. One such innovation is the use of a high-repetition rate AC pulsed arc, together with a large particle detection system. Another method is to use an ashing technique, in which the base oil is burned away to leave an ash; this is excited to provide a spectra for analysis by atomic emission. The technique is known as the Ashing Rotating Disc Electrode (ARDE).

- **Inductively Coupled Plasma (ICP)** This is also used for the analysis of wear debris and oil additive elements. A high-temperature source, i.e. a plasma, is produced by the transfer of power from a radio frequency source into an electrically conducting ionized gas by means of inductive heating. The oil sample is introduced into the high-temperature source as a spray by means of a nebulizer. Dilution of the sample is frequently required due to variations in viscosities and nebulization. Burning of the sample takes place at the high temperatures of the plasma torch, giving an atomic emission spectra according to the elements present. This is a rapid technique with wide range of analysis.

- **Direct current plasma (DCP)** Very similar to the ICP method. However, the plasma temperature is very much lower and the geometrical configuration of the plasma source is also different, using two graphite anodes and a tungsten cathode. The oil sample is again introduced as a spray via a nebulizer. This is a low-cost technique with a rapid throughput and wide analysis range.

- **X-Ray Fluorescence (XRF)** Again used for wear debris analysis, this is a technique in which the oil sample is bombarded by X-rays. The irradiation causes changes in energy levels to occur and the elements present to emit X-rays, or fluoresce. The intensities of the emitted X-rays enable the detection and quantification of the elements present. The sample being analysed may be solid or liquid, and the elements detected will depend upon the type of X-ray source(s) used.

- **Energy Dispersive X-ray Analysis (EDX)** This is a technique used in conjunction with a scanning electron microscope for the analysis of dry wear debris. The wear debris is first extracted from its fluid carrier to provide the dry sample. This operation can be carried out in a number of ways, e.g. magnetic separation, filtration etc. The sample is then usually coated with a thin layer of a conductive element, such as carbon, to improve the overall conductivity and hence the resolution, when acted upon by the electron beam of the SEM. The electron beam, on striking the sample, causes the emission of X-rays from the elements within the debris. These emissions can be used to give qualitative information (mapping) or quantitative data when compared with standard elements.

15.5 WEAR DEBRIS ANALYSIS METHODS

A wide variety of basic techniques are used in the detection, and/or evaluation, of the wear debris present in a lubrication system. The monitoring instruments which have evolved from these techniques may be used inline, online or offline. Those techniques and instruments most widely used are summarized in Table 15.10 for on/inline, and Table 15.11 for offline. The various basic techniques can be briefly described as follows:

- **Optical (light obscuration)** This technique uses the change in light intensity which occurs when particulate debris pass through a light beam. The intensity change is detected using a photodiode and the

Table 15.10 On/in-line wear debris monitoring techniques

Basic technique	Method	Manufacturer and model	Prime cost	Particulate size range	Comments
Optical	Light obscuration	UCC International Ltd CM20	£6.5K	>5, >15, >25, >50 μm	On-line method particle sizer/counter over the various size ranges. Used for hydraulic fluids with viscosities up to 500 cSt. Results to NAS and ISO standards. Storage and display facilities.
	Light obscuration (light blockage)	Partikel-Messtechnik PMT System	£10K–30K	1–2000 μm	Can be used in- and on-line (also off-line). Particle sizer/counter. Size range can be extended using special sensors. Used for contamination control of hydraulic fluids. Display of size/number plus memory.
	Light obscuration (light absorption)	Hydac S.a.r.l. Parity Controller RC 1000	£7K	5–100 μm	On-line contamination control of hydraulic fluid systems. NAS and ISO contamination classes displayed. Can be used with oil viscosities from 15–1000 cSt at flow rates of 50–150 ml/min.
	Time of transition (dynamic shape analysis)	Galai Production Ltd CIS-1000	£25K	2–3500 μm	On/in-line method suitable for some engine oils. Sizer/counter. No calibration necessary. Requires pressurized system (maximum 11 bar). Image analysis also available. (Can be used off-line).

Filter blockage		Product	Price	Size range	Notes
	Time of transition (scanning laser)	Laser Sensor Technology PART-TEC	£25K	1–4000 μm	On/in-line method suitable for clear or semi-clear fluids. Size distribution calculated. Can cope with high particle concentrations. No calibration required. (Can also be used for off-line).
Filter blockage	Pore blockage	Diagnetics CONTAM ALERT	£8K	5 μm, 10 μm, 15 μm screens	On-line device suitable for hydraulic and lubricant systems. Uses three screen sizes producing count/size distribution to ISO and NAS contamination ranges. Viscosity range 8–150 cSt. (Also used in off-line mode).
		Lindley Flowtech Ltd FLUID CONDITION MONITOR (FCM)	£7K	5 μm, 15 μm meshes	On-line mesh device suitable for a range of lubrication systems provided contamination classes to ISO and NAS standards as well as count greater than mesh sizes. Viscosity range 5–350 cSt which can be extended. (Can also be used off-line).
	Mesh obscuration	Coulter Electronics Ltd LIQUID CONTAMINATION MONITOR (LCMII)	£7K	5 μm, 15 μm, 25 μm meshes	On-line mesh system suitable for a wide range of lubrication and hydraulic systems. Measurements of size and count range as well as lubricant viscosity assessment (10–500 cSt). Standard ISO display result available.

Table 15.10 (Continued)

Basic technique	Method	Manufacturer and model	Prime cost	Particulate size range	Comments
Wear	Thin film sensor	Fulmer Systems Ltd FULMER WEAR DEBRIS MONITOR (WDM)	£4K	35–150 μm	On-line method can detect very small quantities of debris, typically <1 ppm. Must be calibrated for viscosity and particle type. Ferrous debris 35–150 μm. Trend or instant readings. Can also be used off-line.
Ultrasound		Monitek Technologies Inc. MPS-D	£9K	0.8–3000 μm	On-/in-line sensor suitable for hydraulic fluids and lube oils. Independent of flow rate. Particulate given as total concentration in ppm up to percent values. Able to multiple size to provide a distribution.
Radioactivity	Thin layer activation (TLA)	Cormon Ltd/AEA Technology (Harwell) ACTIPROBE	Variable £1K plus	n/a	On/in-line detection of loss of material from an irradiated component. Depth of wear monitored in 25–300 μm. Sensitive method suitable for lube systems.
	Surface layer activation	Spire Corp SPI-WEAR	Variable £6K plus	n/a	In-line method measuring loss of material thickness from irradiated component. Detection up to about 10 μm loss possible. Can also use a tracer technique with the lubricant which washes the component.

Magnetic attraction	Magnetic flux path change	Ranco Control Ltd CONTINUOUS DEBRIS MONITOR CDM-TECALERT	£1.5K	n/a	In-line (low pressure) or on-line method suitable for lube systems. Continuous monitoring of debris build up with time reported in terms of weight. For maximum efficiency fluid flow rate should be 0.1–0.5 m/s. Debris returned to system at fixed intervals or when weight reaches certain value.
	Magnetic capture	Gas Tops Ltd FERRO SCAN	£3K	1–1000 μm	Real-time on-line method used in flowing or pressurized lube systems. Accuracy ± 1 ppm up to concentration levels of 100 ppm with flow rate of 0.5–5 m/s. Sensitivity can be impaired by changes in viscosity, temperature and vibration. Uses repeated cycles of debris build-up; the debris being returned to the system after each cycle.
		Muirhead Vatric Components Ltd MAGNETIC CHIP COLLECTORS	Variable £50K plus	All sizes	Probe collects ferrous debris and the whole removed for examination. Self-closing valve system prevents lubricant loss on removal. Low cost, simple method.
		Vickers Inc. QUANTITATIVE DEBRIS MONITOR (QDM)	£8K	$> 500\ \mu m, > 1000\ \mu m$	In-line system used mainly in aviation applications. Electronics produce signal relating to particle (chip) size. Further signal processing gives trending and rate of debris production.

Table 15.11 Off-line wear debris monitoring techniques

Basic technique	Method	Manufacturer and model	Prime cost	Particulate size range	Comments
Optical	Light obscuration	Climet Instruments Co. CLIMET C1-1000	£15K	1–5000 μm	Used for debris monitoring in oils, fuels and hydraulic fluids. Dynamic range of 1:50, viscosity working range of 1–100 cSt and maximum flow rate of 700 L/m. Six size bands displayed.
	Light obscuration (light extinction)	HIAC/ROYCO (Pacific Scientific) HIAC/ROYCO 8000	£10K–20K	0.5–9000 μm	Particle sizer/counter used for lube and hydraulic fluids monitoring up to concentrations of 20 000 particles per ml. Can be used with viscosities up to 2500 cSt. Results to NAS and ISO standards. Filtration tests.
	Light obscuration (light exteinction)	Malvern Instruments Ltd MALVERN AUTOCOUNTER	£12K	2–150 μm	Particle sizer/counter for oil systems. Dynamic range of 50. High particle concentrations of up to 47 000/ml can be tolerated. Viscosity range 1–300 cSt. Output to NAS and ISO standards. Includes graphic and numerical displays.
	Light obscuration	Particle Measuring Systems PMS LBS-100	£10K	0.5–60 μm 2–150 μm 5–300 μm	Used for hydraulic fluid and oil analysis. Particle counter/sizer. Three range sensors with a maximum particle concentration of 10 000/ml. Viscosity range 1–50 cSt. Sampling capacity up to 1 litre.

Principle	Instrument	Price	Size range	Description
Light obscuration, and light scattering	Particle Measuring Systems PMS MICRO LASER PARTICLE SPECTROMETER	£9K	0.05–0.2 μm (1) 0.2—3 μm (2) 0.3–5 μm (3) 0.5–60 μm (4)	Particle contamination and fine debris in oils and hydraulic fluids. Range used depends upon type of sensor and sampler–volumetric (2) and (4) with concentrations up to 15 000/ml, in-situ sampling ranges (1) and (3) with concentrations of up to 10000/ml.
Light obscuration (light blockage)	Partikel Messtechnik PMT SYSTEM	£10K–30K	1–2000 μm	Contamination control of hydraulic oil. Dynamic range of 100+. Particle range can be extended up to 15000 μm using special sensors. Wide range of fluid viscosities can be used. Displays size/number.
Light obscuration	Rion Co. Ltd RION KL-01	£7K	5–150 μm	Used for debris in hydraulic and lube oils. Maximum particle concentration 800/ml. Other liquids can be monitored using different sensors. Automatic instrument after setting calibration.
Forward reflectance (forward light scattering)	Met One Inc. MET ONE LIQUID PARTICLE COUNTER	£10K	0.5–600 μm	Method for contamination control or clean lubrication. Large size range covered by six sensor bands. Maximum counts up to 10 500 particles/ml. Multichannel, multifunction instrument operating up to pressures of 140 bar.

Table 15.11 (Continued)

Basic technique	Method	Manufacturer and model	Prime cost	Particulate size range	Comments
Optical	Forward reflectance (near angle light scattering)	Spectrex Corp. SPECTREX SPC-510 LASER PCS	£12K	0.5–100 μm	Particle counting in lube oils, fuels and hydraulic fluids while sample is still within container (bottle). Reusable standards. Maximum particle concentration 999/ml. Dark fluids should be avoided.
	Time of transition (dynamic shape analysis)	Galai Production Ltd GALAI C1S-1	£15K	0.5–1200 μm	Particle size with shape capabilities using image analysis. Suitable for some engine oils (also on-line version).
	Time of transition (scanning laser microscope)	Laser Sensor Technology Inc. LAB-TEC	£14K	1–1000 μm	Suitable for high particle concentration –up to 200 000 particles per second can be counted. (Also on-line version available.)
Filter blockage		Same methods/instruments as for on/off-line – *see Table 15.10*			
Visual appearance		Hydrotechnik UK Ltd CCK4	£1K	1 $\mu m \to$ upwards	Suitable for oil or hydraulic fluids. Uses membrane filters with small hand-held microscope. Comparison against standards. Portable.
		United Air Specialists UK Ltd CHECKER-KIT	£1K	1 $\mu m \to$ upwards	Suitable for oil or hydraulic fluids. Portable. Uses filter membrane and small hand-held microscope.

Howden Wade Ltd Thermal Control Products CONPAR PARTICLE COMPARISON KIT	£1.5K	$>1->100\,\mu m$	Suitable for general hydraulics and fuels. Large contaminant range covered. Portable. Nine classes of cleanliness available and correlate to NAS, ISO and DEFSTAN standards. Membranes examined using microscope and compared with master slides.
Fairey Arlon Ltd FAS-CC100	£2.5K	$1\,\mu m \rightarrow$ upwards	Dedicated to contamination control of lube and hydraulic fluids. Uses membrane filters and a microscope with build in standards to 150 specification. Portable.
Millipore Corp. MILLIPORE PATCH TEST	£1.5K	$5\,\mu m \rightarrow$ upwards	Used for lube oils and hydraulic fluids. Uses membrane filters and can go down in particle size detected. Compares filter (patches) with standards of varying colour according to density, compensation being made for colouring due to oil. Portable.
Oilab Lubrication Ltd OILAB PORTABLE LUBE AND FUEL ANALYSIS KIT	£2.5K	$1\,\mu m \rightarrow$ upwards	Multipurpose oil and hydraulic fluid analysis measuring physical properties of oil as well as contamination levels. Uses membrane filters together with Wear Particle Atlas and illuminated microscope. Portable. Range of 'add-on' extras.

Table 15.11 (Continued)

Basic technique	Method	Manufacturer and model	Prime cost	Particulate size range	Comments
		Parker Hannifin (UK) Ltd PARKER PATCH TEST KIT	£1K	$1\ \mu m \rightarrow$ upwards	Measures contamination levels of oil or hydraulic fluids. Uses membrane filters and hand-held microscope. Filters compared to supplied patches of varying contamination classes.
Magnetic attraction	Ferrography	Predict Technologies FERROGRAPH		$1-100\ \mu m +$	Used primarily for lubricating oils containing ferrous wear debris although some non-ferrous particles are also deposited. Debris contained on substrate which is observed using microscope or measured for density. Indicates type and quantity of wear taking place. Fairly slow and expensive with respect to consumables.
	AC magnetometry	Swansea Tribology Centre PARTICLE QUANTIFIER PQ90	£4K	$<1-2000\ \mu m$	Can be used for all fluids containing ferrous debris. Quick, versatile method measuring patches (filters), substrates (slides), plastic pots and bottled. The quantity of debris detected is displayed as a PQ index. Instrument has in-built memories, printer, display and PC interface connection. Portable with battery as well as mains operation. Requires calibration against standards.

Method	Product	Price	Size range	Description
Rotational magnetic field	Swansea Tribology Centre ROTARY PARTICLE DEPOSITOR RPD	£4K	<1–2000 μm	Uses a rotating magnet assembly which produces an excellent dispersion of wear debris extracted from the carrier fluid. Debris is deposited on a glass or plastic substrate in a form suitable for microscope examination or image analysis. Both ferrous and non-ferrous debris are deposited. Quick method with low running costs.
High gradient magnetic separation	Tribometrics Inc. WEAR PARTICLE ANALYSER MODEL 56	£4K	0.5 μm → upwards	Special magnetic filter (FIRON) used which captures both ferrous and non-ferrous particles. Variety of filter grades available. Instrument can be in analogue or digital mode. Particles removed for microscopic analysis. Portable.
Inductance	Staveley NDT Technologies DEBRIS TESTER	£3K	<1 μm → upwards	Extracted magnetic debris on a slide is placed within a detector. AC balanced bridge which causes an unbalance and quantifies the debris in terms of Debris Test Units (DTU), a mass equivalent of up to 200 mg. Storage and downloading facilities. Requires calibration against standards.
Gravimetric	Millipore Corp. MILLIPORE FLUID CONTAMINATION ANALYSIS KIT	£1.5K	0.45 μm → upwards 0.8 μm → upwards	Suitable for a wide range of oils and fluids. Uses membrane filters in matched pairs so that the weight of debris in a quantity of fluid can be measured without pre-weighing. The debris collected can be examined using a microscope or spectrometrically

output is calibrated to give the particle size for the flow conditions. Any change in flow conditions and/or particle properties requires a recalibration of the instrument. The instruments can be of the static or flow type. The former uses a comparison between a standard screen and a membrane filter containing the debris, offline. The latter utilizes an adjustable flow system to regulate particle travel through the light beam, giving an online system. These instruments also provide a particle size count.

- **Optical (time of transition)** This technique uses a scanning laser beam and is based on the principle that the time of interaction of a particle within the beam depends directly on the particle size. The technique is independent of the type of fluid used and therefore does not require calibration. Instruments can be used either online or offline and provide particle size distribution data.

- **Optical (forward reflectance)** This technique is based upon the reflectance of light at a very shallow angle of incidence. The light is reflected forward in a narrow angle band, and occurs at an intensity depending upon the surface area which is impinged by the light beam. This in turn depends upon the particle size. Several offline instruments are available which provide a particle count using this technique.

- **Filter blockage** This technique depends upon the change in the pressure/flow characteristics, which occur when an orifice is blocked by debris within the liquid passing through that orifice. In practice, screen or mesh is used which consists of a large number of same-sized orifices. Any particles in the fluid which are larger than the orifice size will cause a blockage, thereby decreasing the flow rate through the mesh. The technique can either utilize a constant flow rate and measure pressure changes, or maintain a constant upstream pressure and measure the change in flow. These instruments can provide a particle count and size distribution, and may be used either on- or offline.

- **Magnetic attraction** This technique uses the magnetic susceptibility of ferrous contaminants to separate the debris from the carrying fluid. The separation is brought about by a variety of methods such as the use of a permanent magnet or a magnetic filter. Those instruments which separate the debris in a manner suitable for further examination and analysis are generally offline monitors. Magnetic plug services, which provide this same facility, are online debris collectors. Inline techniques collect the ferrous debris from the passing fluid by using magnetized sensing heads or fields. The debris are allowed to build up over a specified period of time, and the wear rate is calculated from the weight collected or the change in magnetic flux. The debris are released back into the system at the end of each measuring cycle by demagnetizing the collecting zone.

- **Wear** This is a technique whereby the electrical resistance of an electrically conducting thin film sensor is continuously monitored. The wear debris contained within the flowing fluid are allowed to impinge upon the sensor, causing a wearing away of the sensor material, and hence increasing its electrical resistance. The change in resistance depends upon the rate of change in sensor wear and therefore upon the hardness, sharpness and frequency of the particles striking the sensor. The technique may be used online or inline and compensation for changes in fluid temperature are made by placing a recording sensor downstream of the monitor.

- **Ultrasound** This technique uses a pulsed ultrasonic acoustical beam focused into a fluid, such that it will sense the presence of particulate matter within that liquid. Maximum sensitivity is achieved at the focus of the acoustic beam. Hence, any passing debris will interrupt the beam and cause a change in the strength of the reflected pulse signal, as seen by the receiver. The rate of change in the strength of the reflected pulse can be used to quantify and/or size the particles present in the fluid stream. The technique can be used in- or online.
- **Radioactivity** This technique involves the monitoring of irradiated wear particles which have resulted from the wear of an irradiated component. The method is carried out by either, monitoring the particles using Gamma ray detection units within the vicinity of the irradiated workpart, or by monitoring the decrease in radioactivity of the component itself. The technique can again be used either in-, or online.
- **Inductance** This method uses the change in magnetic flux which occurs when wear debris are present in a magnetic field. The debris are extracted from the carrier fluid and placed within a sensor area, causing an imbalance in the inductive measuring coil. This imbalance is translated into a displayed value relating to the quantity of debris present. The technique is normally used in an offline mode and is suitable only for ferro-magnetic debris.
- **Gravimetric** This technique assesses the weight of debris contained in a specified volume of liquid by passing the fluid through a fixed size, chosen, membrane filter. The increase in filter weight, after compensating for fluid effects, gives the amount of debris present. The method is suitable for offline monitoring only.
- **Electrical conductance** A technique which depends upon the electrical conductivity of the debris within a non-conducting fluid. The capture of conductive particles between two electrical terminals results in bridging the gap between them, and this causes a short circuit which indicates their presence. The capture is brought about by using a magnetic plug arrangement and therefore the debris must be ferro-magnetic. The technique may be used either in-, or online.
- **Visual appearance** This technique involves viewing the debris, usually with a microscope, and making an assessment of the type, size and number of particles present. Normally, a fixed volume of debris containing fluid is passed through a filter and then washed clean of oil. The residue is then examined and compared to standards or charts. The technique is for offline use only.

- **Image analysis** Basically a technique involving the computer analysis of video camera images of dried and cleaned particles extracted from a carrier fluid. The particles are first extracted using filtration or magnetic separation etc. onto a substrate for viewing by means of an optical microscope. The microscope image is translated into an electronic digital picture through a video camera and an image processing system. The electronic image is then analysed using computer software techniques to produce information with respect to size, shape, texture, colour, etc.

15.6 PHYSICAL TESTING OF LUBRICANTS

- **Viscosity** The viscosity of an oil may, in simple terms, be considered as a measurement of the fluid's resistance to flow. Industrial oils are classified in terms of their International Standards Organisation (ISO) rating. That is the kinematic viscosity in Centistokes, measured at 40°C. Automotive lubricants, however, have various international specifications which require the additional measurement of kinematic viscosity at −30, −20, −10, and 100°C. For example, SAE rating. Used automotive lubricants are, however, only tested at 40 and 100°C.

  The changes in viscosity found in engine drain oils are complex and depend upon a number of effects, which separately can either increase or decrease the drain oil viscosity. An increase in viscosity may indicate oxidation, contamination with dirt or water, or the addition of a higher viscosity oil to the system. The presence of lacquers and other oxidized substances in the oil cause an increase in the oil viscosity, as will the presence of any particulate solid material in an engine oil.

  Insoluble levels, in excess of 5% weight in diesel engines, may cause the oil to become extremely viscous and result in difficulty in starting, block filters, or give rise to oil starvation leading to mechanical failure. It is seldom that the viscosity of an industrial oil decreases in use. If this occurs, it suggests contamination with a solvent or lower viscosity oil. Fuel dilution in engine oils will also cause a noticeable reduction of viscosity. While shear breakdown of multigrade lubricants in heavy-duty applications, will similarly cause a reduction in the viscosity.

- **Acid and base numbers** Acid and base-number determinations are carried out in non-aqueous solutions and therefore are not directly related to absolute levels of acidity or alkalinity. Total Acid Number (TAN or Neutralization value) (ASTM D974 D664, IP139, 177) is a measure of the total amount of both weak and strong organic acidity present in the oil. Compounds which form in the early stages of oil oxidation are not in themselves harmful. Further oxidation will convert initial oxidation products into acids which attack and corrode metals. Hydraulic oils have a residual total acid number. With service, this number will increase, and twice the original value is an actionable level. Strong Acid Number (SAN) (ASTM D974 D664, IP139, 177) is a measure of the strong inorganic acidity in the oil. The strong acid number should be monitored for synthetic hydraulic oils and may rapidly increase during service. The presence of strong acids in an engine drain oil is undesirable and is indicative of strong inorganic sulphur acids, which may lead to copper/lead bearing corrosion and rust formation.

Total Base Number, (TBN) (ASTM D664 D2896, IP177, 276) is a measure of the reserves of alkalinity present in the oil. Crankcase oils are continuously monitored for TBN, particularly in marine and residual gas engine applications, where quality of the fuel is suspect. High sulphur fuels quickly deplete the reserve alkalinity of the crankcase oil which is then unable to neutralize the harmful acids formed as a result of the combustion process. In the case of crankcase oils, both total base and total acid numbers exist in the new oil. During service, the base number will stabilize at a lower level than that of the new oil. A further decrease from this stabilized level and an increase in acid number will then determine the appropriate oil drain.

- **Insolubles** Several methods are used for the determination of insolubles in used oils. In centrifuging, the insoluble material is separated from a weighted oil sample which is mixed with *n*-pentane or *n*-heptane solvent, by centrifugal force. The insolubles collected are then dried and weighed. In filtration, a weighed amount of used oil is diluted with *n*-pentane and filtered through a membrane filter of known porosity. The insolubles collected are then dried and weighed as before. The blotter spot method involves depositing an oil spot on special paper and measuring its opacity in several areas. Thermogravimetry involves using approximately 50 mg of used oil and heating it in a stream of nitrogen up to 650°C. After 5 minutes at 650°C, 10% air is introduced into the stream of nitrogen to oxidize the carbon, further air is then added up to 100%, with the sample weight being continuously recorded. In the photometric method, a small measured amount of used oil is mixed with *n*-pentane or *n*-heptane in a glass cell, placed in a photometer and compared against calibrated standard filters.

15.7 IMPLEMENTATION OF WEAR DEBRIS MONITORING AND OIL ANALYSIS PROGRAMMES

Numerous programmes have been developed to present condition monitoring results in a form suitable for inclusion in a Maintenance Management System (MMS). Information may then be constantly updated, providing data which is required for repair action, and benchmarks to help in measuring the effectiveness of maintenance operations.

Trends can be established, and evaluated data for projections are an unlimited source of management information. At a primary level, 'unit condition reports' may be itemized to indicate problems and to recommend corrective action to maintenance personnel.

The current condition of equipment may be indicated in terms such as satisfactory, marginal or critical. Any discrepancies or deficiencies, in either mechanical or lubricant condition, can then be identified and appropriate corrective action specified.

A secondary level of information may summarize the condition reports into a management oriented overview of the mechanical integrity of the operation and also the effectiveness of various maintenance procedures. Other levels of information may also define long-term management requirements from the data generated by these programmes. Comparative evaluations can also be made of equipment, lubricants, components, maintenance activities and operational procedures.

A typical analysis programme may include the following, spectrographic wear debris analysis, infrared spectroscopy, viscometry, total acid and base number evaluation, plus wear particle analysis, that is particle quantifier or ferrography. The following case-study examples show how the combination of these techniques may be complied to suit specific applications. It is not always necessary to carry out all the available analytical procedures. Thus experience becomes paramount in matching these techniques so as to reduce the number of tests, and hence cost, without significantly reducing the effectiveness of the monitoring programme.

The first example, shown in Fig. 15.16, illustrates a programme developed to monitor gas turbine compressors used in natural gas-pumping stations. These turbines have large plain white metal journal bearings in their assembly. It should be noted that the spectrographic analysis showed increasing levels of both lead and copper. Wear of the white metal bearing, which is symptomatic of lead detection, would not, in its initial stages, be expected to result in damage of the rotor shaft. However, as further damage occurs and the presence of copper in the oil is detected, there is an indication that wear has progressed through the white metal into the backing material, which in this case is bronze.

If wear was allowed to progress, bronze being a harder material than white metal would cause significant damage to the rotor shaft. In this case, borescope inspection revealed slivers of white metal protruding from the journal bearing. Subsequent dismantling of the bearing showed that wear had progressed through the white metal into the hard backing material and that no damage had occurred to the shaft. The white metal bearing was repaired at minimal cost and thus the equipment state prediction avoided expensive subsequent damage to the rotor shaft.

The second example concerns a six-cylinder diesel engine in a combined heat and power system. Here, the inclusion of infrared spectroscopy in this case, enhanced the monitoring programme enabling the

Mechanical Integrity Analysis

Swansea Laboratory Services, University College of Swansea, Swansea SA2 8PP. Tel 0792-295219 Fax 0792-295701

File Name: ex.004.01

Date: 23 Feb 95

Customer: Illustrative Example.
Operation:

Unit No:
Manufacturer:
Lubricant:

Component: Gas Turbine
Model:

Sample Date		09 Feb 89	19 Mar 89	10 Apr 89	02 May 89	19 May 89
Sample No						
Lubricant Hrs/Mileage		n/p	n/p	n/p	n/p	n/p
Equipment Hrs/Mileage		4433	4895	5255	5600	n/p
Makeup Lubricant		n/p	n/p	n/p	n/p	n/p
Wear	Iron	0	2	2	2	4
Elements	Chromium	0	0	0	0	0
ppm	Aluminium	0	2	0	0	2
	Copper	4	5	11	15*	15*
	Lead	10	24*	59**	82***	114***
	Tin	0	2	2	0	1
	Nickel	0	1	0	0	0
	Manganese	0	0	0	0	0
	Titanium	0	0	0	0	0
	Silver	0	0	0	0	0
	Molybdenum	0	1	0	0	0
Additive	Zinc	99	48	42	33	55
Elements	Phosphorus	72	72	79	66	72
ppm	Calcium	7	7	1	1	2
	Barium	23	3	1	1	1
	Magnesium	0	8	5	4	6
Contaminant	Silicon	0	2	0	0	3
Elements	Sodium	0	4	6	5	12
ppm	Boron	0	0	0	0	1
	Vanadium	0	1	0	0	1
Physical	Fuel	-	-	-	-	-
Tests	Water	Neg.	Neg.	Neg.	Neg.	Neg.
	Dispersancy	-	-	-	-	
	TBN/TAN	0.19	0.24	0.13	0.12	0.11
Ferrography	Dl	2	24	21	8	8
	Ds	1	16	18	4	5
	Is	3.0E0	3.2E2	1.2E2	4.8E1	3.9E1
Viscosity (cst at 40°C)		45	42	44	44	43
Result (Internal Ref No)		D1	D1	D19	D19	D19

* - Reportable , ** - Unacceptable , *** - Urgent

Sample Date - 19 May 89

Condition - Critical - This Unit should receive immediate attention
Discrepancy - Abnormal wear trends suspected.
Recommendation - Inspect unit for bearing damage (boroscope)
Report results of visual inspection.

Fig. 15.16 Oil analysis data for gas turbine.

degradation of the lubricant to be trended. The high levels of oxidation and nitration found were the result of lubricant degradation, which correlated with high values of kinematic viscosity. The oil change interval of between 800 and 1000 hours was therefore reduced.

15.8 CONCLUSIONS

Diagnosis, subsequent to analysis, requires considerable experience, and knowledge of the interaction of the lubricant, mechanical equipment and the working environment. A few samples of this interaction have been detailed above and are further listed in Table 15.12. This list is intended as an illustration of the permutations and combinations of results, which have to be considered for reliable diagnostics to be obtained from an oil analysis monitoring programme. Accumulated results from a large database enable interpretative decisions to be refined and improved.

Aside from helping to establish normal wear patterns, concentration profiles are also useful in comparative evaluations of equipment or

Table 15.12 Typical permutations and combinations of results from oil analysis

Symptom	Diagnosis
Diesel engine	
Silicon, iron, chromium, aluminium	Damage to the air filter, cracks or absence of clips on the air manifold system allowing the ingress of abrasive silicon dust and subsequent damage to the cylinder liner/piston/piston ring.
Sodium, copper, lead	Coolant leak with damage to the main bearings or the ingress of salt through the air manifold.
Chromium	Bore polishing and damage to the piston rings.
Copper	Material leeching from the oil cooler.
Copper and lead	Main bearing damage.
High viscosity	Degradation of the lubricant through oxidation and nitration, or high soot loading, or incorrect lubricant top up.
Low viscosity	Fuel dilution (low flash point) or incorrect lubricant
Low total base number	Lubricant degradation
Low total acid number	Lubricant degradation
High oxidation (infrared)	Lubricant degradation
High nitration (infrared)	Lubricant degradation
Gearbox	
Silicon, iron	Abrasive wear resulting from ingress of dust
Iron, chromium, nickel	Wear of bearing material (rolling element)
Low viscosity	Wrong lubricant
High viscosity	Lubricant degradation or wrong lubricant
Hydraulic system	
Silicon, iron	Abrasive wear resulting from ingress of dust
Increasing total acid number	Lubricant degradation
Low or high viscosity	Wrong lubricant

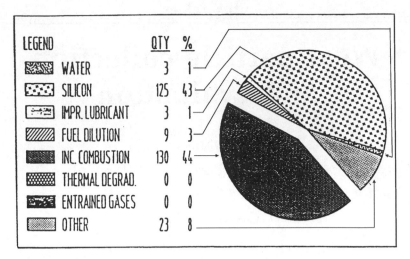

Fig. 15.17 Annual evaluation report: City Bus Company.

operating conditions, lubricants, components and maintenance procedures. Close supervisory control is essential to a viable monitoring programme. The computer system can assist in this responsibility by keeping track of sampling schedules and reporting on equipment due or overdue for monitoring. These checks also make sure that recommendations are followed. The annual evaluation report as shown in Figure 15.17, then enable the monitoring scheme to become part of an effective overall maintenance management program.

15.9 BIBLIOGRAPHY

Hunt, T.M. (1993) *Handbook of Wear Debris Analysis and Particle Detection in Liquids*, Elsevier, Amsterdam, The Netherlands.

Hutchings, I.M. (1992) *Tribology: Friction and Wear of Engineering Materials*, Edward Arnold, London, UK.

Jones, M.H. (ed.) (1984) *Condition Monitoring '84, International Conference Proceedings*, Pineridge Press, Swansea, UK.

Jones, M.H. (ed.) (1987) *Condition Monitoring '87, International Conference Proceedings*, Pineridge Press, Swansea, UK.

Jones, M.H. (ed.) (1991) *Condition Monitoring '91, International Conference Proceedings*, Pineridge Press, Swansea, UK.

Jones, M.H. (ed.) (1994) *Condition Monitoring '94, International Conference Proceedings*, Pineridge Press, Swansea, UK.

Neale, M.J. (1979) *A Guide to the Condition Monitoring of Machinery*, HMSO, London, UK.

CHAPTER 16

Wear particle collection and evaluation

A. Davies

Systems Division, School of Engineering, University of Wales Cardiff (UWC), PO Box 688, Queens Buildings, Newport Road, Cardiff, CF2 3TE, UK

16.1 INTRODUCTION

Fluid monitoring is performed for three major reasons; these are:

1. to ensure the minimization of contamination in lubrication oils and hydraulic fluids, by particulate matter or other contaminants, which if unobserved may result in machinery failure.
2. to ensure a minimization of the degradation of the physical condition of the fluid, which again if unobserved can also lead to equipment failure.
3. to ensure that any wear debris, contained in a fluid as a result of component damage, is detected, collected and subsequently analysed, thus allowing its source to be identified for repair.

By undertaking such monitoring from the initial commissioning of plant and machinery, it may be possible to predict component failures before they occur, or at the very least, identify conditions which cause increased equipment wear and thereby reduced operating life. It is important to note that, as with other condition monitoring schemes, an initial quantifiable operating signature or baseline reference is required in the form of data to which any subsequent sample may be compared. If this is not done then prediction via trend analysis is not possible, and at some time in the future when abnormal conditions are detected, all that can be done is to attempt damage limitation, fault diagnosis and rectification.

Handbook of Condition Monitoring
Edited by A. Davies
Published in 1998 by Chapman & Hall, London. ISBN 0 412 61320 4.

In this chapter we are mainly concerned with the first and third reasons given above for fluid monitoring, that is the detection, collection and identification of particulate wear debris, the contamination of liquids by solid particles having been for some time identified as a major cause of failure in fluid power machinery. This relationship was established beyond doubt by a government sponsored study into the problem (Anon, 1980) and is accentuated by modern high-pressure systems, where running clearances are usually kept as small as possible to achieve maximum efficiency and to reduce leakage. This obviously calls for increased levels of fluid cleanliness and the eradication if possible of particulate contamination.

As stringent limits in respect of particle contamination can only be achieved and maintained by appropriate monitoring and filtration, there is a great deal of interest in the design and development of wear debris collection and analysis systems. Initially, it should be noted that particle contamination of fluids is the result of three major factors which are:

1. contamination via the working environment through seals, reservoirs or fresh supplies of pre-contaminated fluid;
2. via debris generated within the system and due to normal equipment operation;
3. through design, manufacture, assembly, repair or other maintenance activity, where particles are retained in the system and subsequently contaminate the working fluid.

Accordingly we should note that particulate level monitoring, collection and analysis are most suitable for high value, repairable, and recirculatory fluid systems. In this type of equipment, they have the advantage of providing both good fault detection and identification of failures in those components washed by the working fluid. Both online and offline monitoring methods can be used, with the choice depending on application factors such as information required, timeliness, cost, usability, expertise etc. It should also be noted that some methods only detect, collect and quantify magnetic particles, while others can provide a more detailed analysis, outlining the presence of other elements in the fluid.

In addition, recognition must be made of the fact that different methods may have dissimilar detection capabilities in respect of particle size. Accordingly, care should be taken in this respect over method selection for a particular application. It should also be remembered that as wear/fatigue are continuous processes while machinery is in operation, the particles size and quantity are not likely to be stable values. Indeed it is on this fact that trend detection is based, having first obtained at the outset a normal operating signature for the equipment. In the sections which follow, debris characteristics and methods of wear particle collection and analysis are outlined, together with some guidance on their most appropriate use.

16.2 MONITORING FLUIDS

As already outlined above, one reason why fluids are monitored is because they are likely to contain evidence of damage to various components within a machine. On the one hand they provide a ready-made communication path to the operating parts within a system, while on the other they can cause confusion by arbitrarily transmitting debris from a variety of locations within the machine. Precise fault identification is further obstructed by the fluid carrying particles which may have been already in the system as a result of manufacture, maintenance or ingested by the machine via seals, reservoirs or filling points etc.

However, in mitigation of the above disadvantages fluid monitoring has two saving graces and these are:

1. the use of a 'designed-in' machine feature, the fluid circuit, which makes the technique a relatively low-cost monitoring system.
2. the high probability that if a fault exists it will be detected, although perhaps not accurately located without some effort.

The last point of accurate fault location can be somewhat assuaged by sample point positioning. That is by monitoring both above and below critical components and filters in the fluid circuit, and by expert interpretation of the sample analysis. The latter, can of course, only come from experience in debris analysis.

Fluid condition can be altered by three groups of possible contaminants; these are as follows.

- Other liquids, which mix with and contaminate the base fluid being monitored. Note that this does not include chemical 'additives' which are supposed to be present in the base fluid formulation for highly specific reasons, and are designed to confer desirable properties on the liquid. It is important, however, that care should be taken to ensure that none of these 'additives' attack components or seals within the machine, otherwise damage will occur, and as a consequence, debris generated. The presence of acidic or alkaline 'foreign' chemicals in a monitored fluid is however a cause for concern. Their presence may be indicative of worn seals, pistons or other actuators or they may have been formed due to high temperature oxidation.

    Other types of contamination in this area include biological growth and gummy precipitates. For example, oil in water emulsions such as machine tool coolant suffer from this type of problem, which tends to be caused by atmospheric conditions. Hence the need for micro-biological additives to control emulsion stability and the fluid lubricating properties. Filter blockage, corrosiveness and an unpleasant smell arising from the machine are all features which indicate this sort of problem is present. Monitoring for biological growth can be undertaken by means of culture tests via pre-treated dip-slides.

Other difficulties can arise when the water content of an oil increases to the point where it falls out of solution and forms dispersed droplets in the fluid. A loss of performance and oxidation may well be the result of this undesirable condition, particularly in hydraulic fluid. The presence of water in lubricating oils is always a cause for concern, if only because of the corrosive effect which can ensue. As indicated in other chapters of this book, liquid contamination is specifically tested for when monitoring fluids, and the results often taken as a measure of the machinery condition/operation.

- Gas absorption in the fluid. Plain ordinary air is normally the main offender in this particular area, as all liquids have to a certain degree, a value of natural absorption dictated by atmospheric pressure. Concern is warranted for two reasons: first the increased risk of corrosion taking place within the machine, due to the presence of air in the fluid, and secondly the possibility of a change in the liquid compressibility value.

Air contamination of a fluid can result from a number of possible causes such as low liquid levels, poor tank design, circuit leaks or turbulent flow. Tests for air contamination are given in other chapters of this book, as are those for other gases which may be present due to some chemical reaction or unusual process. In the same way as with the problem of biological growth, these are normally revealed by a most unpleasant smell emanating from the equipment.

- Solid matter in the form of particulate debris. This type of contamination can appear in a variety of different guises and for several different reasons. Figure 16.1 indicates some of the possible forms of this type of contamination, and some of its likely characteristics and causes. The manufacture of components, plus their assembly, packing and transportation, is an obvious area to suspect as a prime source of debris in fluids. Processes such as machining, extrusion, pressworking, shotblasting and welding contributing by their very nature to particulate generation.

Despite careful and in some cases meticulous cleaning, particles from this source can still be 'built-in' to equipment, along with debris resulting from finishing operations such as plating and painting. Almost all hydraulic systems are cleaned thoroughly during assembly, and on completion prior to operational use. This is done by flushing with a lower viscosity oil than would be normally used in the system, and by passing this through the circuit at a higher velocity than is used in practice. Even so, and despite careful monitoring, it should be noted here that not all debris are removed by this procedure.

Another obvious source of solid debris is that generated by the wearing process taking place within the machinery during its normal operation. Initially, this material is to be expected as part of the normal 'running-in' process for new equipment, and as such it may not be detrimental to machine operation. The particles should however be

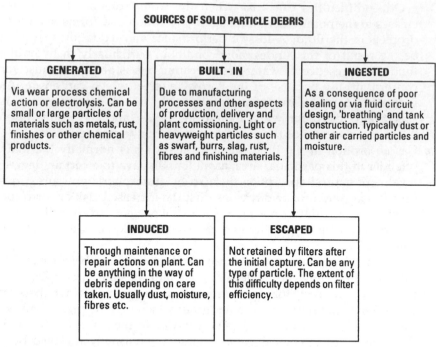

Fig. 16.1 Origin and characteristics of solid particle debris.

trapped, removed from the system, and analysed to ensure that their presence in the fluid is normal, and as one would expect.

Components damaged by high levels of stress or fatigue can also be sources of solid debris. These particles tend to be large, usually above five microns in size, and should be traced back to the damaged machine part when detected. Low melting point alloys such as those used in bearings can also shed material into the fluid as globules, if the temperature rises above their melting point in localized hot spots. Further sources of contamination may involve the surface treatment given to the various materials which make up the equipment, or be traced to the operational process undertaken by the machinery in question.

Because fluid circuits cannot be completely isolated from the surrounding environment, debris suspended in the atmosphere can also be ingested into the system. Accordingly this must be recognized as a possible source of any contamination found in a fluid. With damaged fluid seals, reservoirs, filling points and the filling process itself all being possible sources of debris ingestion. Finally, poor quality maintenance action and the inefficiency of filtration can also result in the presence of unrequired debris.

This chapter mainly confines itself to discussing the last category of contamination outlined above, that is the detection, collection and analysis of solid particles in fluids. To that end, we now turn our attention to the characteristics of the material in question.

16.3 TYPES OF DEBRIS

Given the number and variety of sources from which debris in fluids can originate, it must now come as no surprise to learn that the nature of these particles can be almost infinite. To aid the analyst, however, several excellent wear particle catalogues exist which contain photographs, plus a wealth of detail in respect of such debris. These publications, also provide a great deal of information on the likely origin of particles within many typical machine systems (Anderson, 1982). Although the absolute categorization of particles is difficult, several standard groupings exist based on debris appearance, and to a certain extent these are independent of size. They are generally recognized as spheres, pebbles, chunks, slabs, platelets, flakes, curls, spirals, slivers, rolls, strands and fibres.

Spheres have been identified in many cases of debris analysis with their sources being attributed to grinding, welding, wear processes and normal machine operation. Informed opinion is that spheres tend to be produced as a consequence of the melting and subsequent rapid cooling of microscopic material, most particularly, metals, within the machine system (Jin and Wang, 1989). However, fatigue has also been proposed as the generation mechanism for spheres, with some evidence being available in respect of bearing failures (Anderson, 1982). Non-metallic spheres can also be found during debris analysis, such as glass beads associated with peening, polymer particles made up from oil constituents, and perhaps spheres associated with particular industrial processes.

Pebbles are smooth surfaced, roughly shaped ovoids of quartz or silica, and are normally attributable to ingestion from dusty working environments. Chunks tend to be rough, equally proportioned lumps of metallic material, often produced by fatigue processes. While slabs are hard, thickish flat particles, usually with sharp or rough brittle edges, and generated via sliding fatigue wear. They are often temperature marked and have similarities with machine shop swarf. Platelets and flakes are also very similar to each other, and tend to originate from normal machine operation. Rubbing marks can sometimes be seen on this type of debris, which also may be coloured and have random outline. The colouring of these particles may well identify their source as being from paint applied to the machine, or rust generated within the system due to condensation.

Curls, spirals and slivers have been linked in shape to the debris left over from machining. They are usually brittle and their shape is as a consequence of temperature distortion during production. Wear action by moving parts in the machine can also be a source of this type of debris shape. If that is the case, the particles will not show any temperature marks and are generally softer. Rolls can be produced from debris which were originally complete spirals or undistorted plates. Accordingly their source is quite likely to be the same as that for those particular shapes. Strands and fibres are needle- or string-like shapes, which are usually non-metallic in nature and are linked to ingested debris. Typical materials can be wood or artificial clothing fibres, cotton and other elements present in atmospheric dust. If metallic, their source may well be due to finishing operations during manufacture.

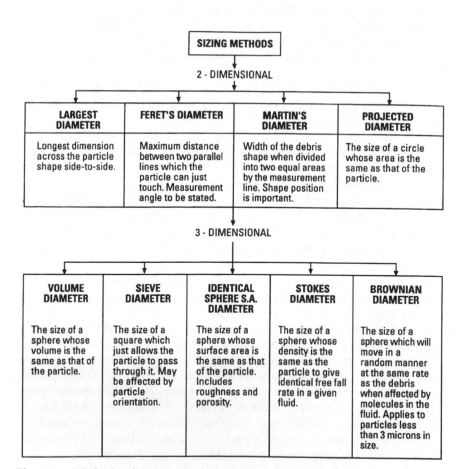

Fig. 16.2 Methods of sizing wear debris.

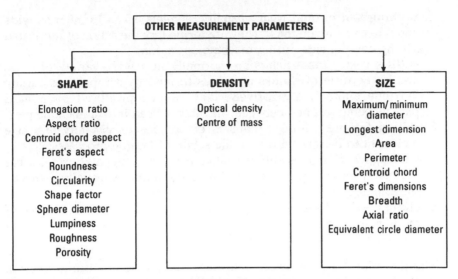

Fig. 16.3 Alternative measurements for wear debris.

Although difficult in an absolute sense, shape classification as shown above is reasonably logical, and can give an indication of origin to aid fault location. The size of debris on the other hand occupies a very wide spectrum, and the methods of particle classification are numerous. A good explanation of the various methods of 'sizing', as outlined in Figures 16.2 and 16.3 is provided in Hunt (1993). As can be seen from the figures, the range of debris descriptors is quite formidable, and hence an overall size classification based on some composite of these parameters is not really feasible. Thus two pragmatic approaches are commonly used:

1. the employment of the characteristics most applicable to the expected debris;
2. to relate the debris found to an ideal sphere via the use of a shape factor.

Note that there is no correct 'size' as such, simply that value which is produced by the method used, and this should be stated as part of the analysis report. It should also be noted that each of the different 'sizing' methods are not comparable and also that extensive research is being done in this field, with for example, fractal image analysis (Kirk and Stachowiak, 1993). For further reading in this area, and an interesting discussion on debris size/instrument calibration, the reader is referred to Hunt (1993).

Debris may also be classsified by origin, and in this sense five groups are normally recognized as sources of wear debris. They are:

1. **rubbing and normal wear** usually platelets 0.5 to 15 microns wide and 0.15 to 1 micron thick. They are formed by the action of lubricated sliding components during machine operation;
2. **cutting wear** these debris are normally in a wide variety of sizes, and in the form of chunks or spirals formed by components cutting into one another; alternatively, they can be produced by a hard particle being trapped between, and then forced into, a machine part;
3. **rolling fatigue** chunks, platelets or spheres of virtually any size material can be formed due to the action of fatigue processes;
4. **sliding wear** this is differentiated from rubbing or normal wear actions by its severity, it can produce large particles, usually platelets, showing signs of high stress;
5. **sliding and rolling wear** typical of gear tooth 'scuffing', this type of mechanism produces platelets or chunks with smooth surfaces and greater than 10 microns in size.

This grouping of debris by origin is a useful aid to fault location, and the method was originally developed through military research (Beerbower, 1975; Bowen and Westcott, 1976).

It should be noted, however, that the way in which any of the classification systems above are used in practice very much depends on the analyst, and also on the application being investigated at the time. Each system, be it shape, size or origin, has its own advantages, and none is necessarily mutually exclusive to the others. Indeed, the key use of debris classification is to provide understandable evidence for fault location, and in this context all methods are surely permissible. Accordingly, how such debris are collected for analysis is now the area to which we must turn our attention.

16.4   DEBRIS COLLECTION

In this exercise, ensuring that a representative sample of debris is analysed is of prime importance. In other words:

> Howsoever the sample is obtained, care must be taken to ensure that it is an accurate representation of the state of the system, and that in the transfer from machine to analyser, it must remain unchanged.

This statement should be adhered to in both online and offline monitoring situations. Thus the sample must contain the actual debris evidence only, and no contamination via the sampling system or from the atmosphere. As debris are not part of the fluid and may not mix with it easily, samples are usually taken a period of time after the machine has commenced operation. This is to allow for system stabilization, and the extraction of hopefully a representative sample. In this context, the type of flow experienced by the fluid is a major factor which affects mixing.

Turbulent rather than laminar flow is preferred, as this aids particle mixing and the provision of a representative sample. However, most fluid circuits are designed to be laminar, thus improving efficiency, by reducing the flow resistance and the pressure losses associated with turbulent flow. Sample point positioning is, therefore, a quite important exercise when designing a monitoring system. Fortunately, turbulence does occur naturally in fluid systems at bends, edges, rough surfaces and changes in pipework diameter. In addition, 'mixers' can be fitted to create turbulent flow, if no suitable feature is present prior to a sampling point. Care should also be taken to sample only from pipework that experiences a regular flow of fluid and, in addition, the extraction line should be flushed clear of any 'standing' liquid.

This, again, is a precaution designed to ensure that a representative fluid sample is acquired, for the way particles disperse in a fluid is quite complicated. The non-homogeneous nature of the liquid/debris mix can be affected by viscosity, density ratio, a non-liquid state, particle porosity, surface tension, electrostatic potential and dimensional size. In addition, for sub-micron particles, their movement in the liquid can be influenced by fluid molecular bombardment, or 'Brownian' motion. All this is aside from the actual shape of the particle, which may in itself make its motion somewhat unpredictable, along with other factors such as fluid velocity profiles, agglomeration and pipewall boundary effects. For an interesting discussion on particle dispersion in fluids the reader is again referred to Hunt (1993).

To capture particles for analysis, four methods are usually employed to extract debris from a machine system.

1. **Magnetic debris collectors** These systems are well known and have been extensively used over the years particularly in aircraft and motor vehicles. Normally they have a self-sealing arrangement whereby the magnetic head can be removed for inspection without necessitating fluid loss. Particle assessment can take place local to the machine, or if required in a laboratory, by either removing a sample of debris or by utilizing a spare magnetic head to replace that sent for analysis. If a sample is taken, the head should first be 'dried' using a solvent to remove any fluid, and the sample debris then transferred to clear adhesive tape, for storage and subsequent analysis.

   The main disadvantage of this type of 'plug' arrangement is fairly obvious; that is they only detect/collect particles attracted by a magnetic field. Other problems are that some debris may be lost during removal, or escape due to 'build-up', while seals and heads can stick if left unchecked, and 'catch' efficiency can be variable depending on positioning, fluid velocity and magnetic strength. They are, however, a relatively cheap form of condition monitoring, easily retrofitted to machinery, although better when 'designed-in'. In addition, they can also be designed to indicate/count debris for information analysis or alarm activation (Neale, 1979).

2. **Existing filtration arrangements** This is a fairly obvious way of checking particulate contamination in fluids, especially as filtration is widely used to eliminate debris contamination. The regular examination of the contents of machine filters would therefore seem to be an easy, logical and low-cost method of condition monitoring. There are some disadvantages to doing this, however, such as ease of filter removal and replacement, fluid drainage and replenishment, particle extraction for analysis and machine stoppage during inspection. Some types of filters such as centrifuges do not necessarily suffer from these problems and accordingly each installation should be assessed individually for the feasibility of employing this approach.

3. **Special filters** These systems are normally designed to collect all debris down to the mesh size of the filter. They overcome many of the disadvantages of ordinary filters by being designed for removal while the machine is operating and without loss of fluid. Such filters are often used in conjunction with magnetic plugs for the detection of non-ferrous debris and where high rates of contamination are expected. Accordingly, they need to be inspected and cleaned regularly to ensure a blockage does not take place and that the filter is operating correctly.

   Another arrangement which can be used in this area is equipment designed to cause particle migration, rather than trappage in a standard mesh type filter. These systems are designed to move debris from the fluid by centrifugal force into a holding area where they are captured by magnetic attraction. In essence, the design forces the fluid to rotate at high speed, throwing the particles out to the periphery, along with any air in the fluid. The equipment has no moving parts and the debris drop down the outer wall of the device into the holding area.

   Other particle migration devices include centrifuges, electrostatic and ultrasonic filters. In the electrostatic system, the fluid passes adjacent to a vertical filter arrangement and though an electrical field. This charges the particles such that they move towards either the positive or negative poles of the device and are then subsequently picked up on the filter. In the ultrasonic system, a frequency of around 1 to 2 megaHertz (mHz) is used to agglomerate the particles which then fall under gravity for subsequent collection.

4. **Oil sample** Self-sealing valves can be fitted to fluid lines where samples are to be taken under pressure, and these can then be used

with appropriate sampling equipment. The positioning of sampling points is important, and relates to the information demanded. If it is the current debris level that is required, a vertical pipe, or the side/top of a horizontal pipe is to be preferred. A total debris measurement would dictate a position on the underside of a horizontal pipe, as this would act as a 'dead end' or settlement point for all particles in between samples. The container used to hold the sample must be clean, in the sense of not being contaminated by solid or biological matter. Commercial 'bottles' normally come pre-cleaned to a very high standard, and this in practice is the best solution to adopt for sample acquisition.

This type of sampling, where the system has achieved operating and temperature stability, is often referred to as dynamic sampling. It can be considered as 'best practice' because it is repeatable and useful for trend prediction. Two other sampling methods in this category are, however, often used; these are reservoir and drip feed sampling. Both are prone to error due to mispositioning during sampling acquisition and may also cause more debris to enter the system via a lack of cleanliness during the sampling procedure. Static sampling is also possible, and if this technique is used, care should be taken to ensure the fluid is well mixed prior to a sample being taken.

Having looked at the way in which fluids are monitored, the type of debris likely to be encountered, and the way in which debris are collected, we now turn our attention to the way in which these particles are analysed.

16.5 DEBRIS ASSESSMENT

At the start of this chapter, the main reasons for fluid monitoring were outlined, and thus far we have looked quite closely at the characteristics of the liquid/debris mix. However, it is important to remember that fault identification/location in machinery is the essential reason for the use of this technique. Accordingly, the assessment of debris to provide clues to their origin is a vital aspect of this work. Particle contamination of fluids can have three major consequences which may ultimately lead to failure, these are:

1. **Machine component wear** This is typified by particles acting as abrasive material when trapped between two sliding metal surfaces. Mid-range particles, 3–13 microns, appear to cause the most trouble, being slightly larger than the lubricating fluid film. In several bearing studies, fine filtration has been found to extend life, while heavy particle contamination significantly shortened it. A similar situation has also been found to occur in fluid power systems, with spool valve

wear due to particle contamination, reducing system efficiency quite significantly.

2. **Component seizure** This is not a common occurrence due to the presence of particulate wear debris alone, unless severe neglect of the equipment has taken place, and the particles are of sufficient size to jam the moving parts. Accordingly, this type of failure in machinery is more likely to be caused by some other mechanism inducing an initial component breakup, the debris from which then inflict the damage. It should be noted, however, that spool valves in hydraulic fluid systems are sensitive to particulate presence and, because of the close tolerances involved, can seize given the right conditions.

3. **Flow reduction** This may be a problem in fluid power circuits where close tolerances are involved, although in lubrication systems, where the flow of oil tends to keep the liquid pathways clear, it is not so significant. However, silting may occur in some systems, especially if the flow rate is low or non-existent.

To avoid these problems, particle detection, identification and assessment is an essential task. This does not just mean elemental analysis but also, as we have seen, the classification of size, shape and quantity. In many applications, the identification of size and quantity via trend analysis is enough to allow the instigation of corrective measures, such as fluid replacement or an improvement in system filtration. Shape and elemental analysis are normally used for fault location.

It should be noted, however, that simple precautions can help to reduce particle contamination in fluids and thus reduce the requirement for sophisticated debris assessment techniques. For example, the flushing of equipment after manufacture with an appropriate fluid and then sealing the circuit is one useful method. The use of a clean area for final assembly and the improvement of fluid transfer procedures, via the use of filters and sensible drum handling, are others.

In critical systems, however, where there is a risk of contamination, suitable arrangements for the detection of particles in the machine should be in place. Although debris detection, counting and chemical analysis systems are discussed elsewhere in this book, a word about particle quantification may not be out of place at this point. There are many systems available for determining the quantity of debris in a given volume of liquid, and they should not be confused with particle counters. Each will normally have a lower limit on debris size, due to the detection and capture method used, together with an estimation of the dimension distribution in the fluid.

It should be noted that it is the lower size limit which is important, and indicates if excessive wear is occurring. Three measurements are commonly used in instruments of this sort and each is linked to a lower limit of debris size. They are:

1. **The weight of debris per volume of fluid** This can be achieved via the use of a double membrane filtration method as discussed in Hunt (1993).
2. **The intensity of the debris per volume of fluid** This can be achieved via a variety of different techniques, each of which tries to assess a quantitative change in the amount of debris; typical instruments may be based on the use of filters, photometric dispersion, nephelometric, piezoelectric, and other electrical principles (Hunt, 1993).
3. **The shape of the size distribution** This can also be achieved in a number of different ways, although most instruments use one of four possible methods. These are Fraunhofer diffraction, sedimentation, sieving and time of translation, with the distribution being displayed either as a cumulative plot or as a histogram (Hunt, 1993).

Chemical analysis of the elements present in debris, usually provide good indicators in respect of the source of the trouble. However, an analysis of the particle's physical characteristics can also aid in fault location. As we already seen, size, shape and quantity are useful parameters to monitor, while the visual aspect, hardness and bulk of particles can also provide some evidence as to their source. Careful separation of the debris from the fluid carrier is quite important in this type of analysis, so that none of the sample is lost or contaminated by poor preparation procedures. Three main analysis techniques are used:

1. **Microscope analysis** Or the visual inspection of debris, via the use of a microscope which may have some additional features to help in particle identification. Typical extra features are top, bottom or oblique illumination, polarising filters and a magnetic stage for identifying ferrous particles. The technique should be used in conjunction with a suitable debris atlas to assist in particle identification.
2. **Scanning Electron Microscope (SEM) analysis** This is a much more costly system designed to examine particles at very high magnification. Such systems are normally fully computerized and use a parallel beam of electrons to obtain the image. Sample resolution down to 2 nanometres can be achieved with these systems, allowing debris of micron size to be analysed for shape size and chemical content.
3. **Image analysis** Essentially the electronic visualization and examination of debris. These systems are again expensive and computerized, but have the very great advantage of being able to provide dimensional sizes and to record analyses. Typically they consist of a microscope, television camera and image processing hardware/software. Colour imaging is also possible with these systems together with automatic counting, differentiation and shape estimation. Automated identification of particles from a debris sample, via the use of expert system software linked to this type of system is also under investigation (Albidewi et al., 1991).

It should be noted here that there are many other items of equipment which are used in the particle identification area. These are referred to in other chapters of this book and may be used for specific classification purposes, such as magnetic/non-magnetic or metallic/non-metallic debris separation and evaluation. Accordingly, these will not be discussed here, although it should be remembered that all such equipment is designed primarily to help in the examination, trending and comparison of debris with known standards.

16.6 CONCLUSIONS

This chapter has attempted to review both particle collection and evaluation within the area of wear debris analysis. Emphasis has been placed in this discussion on the debris particles themselves and their role as information carriers in respect of fault location. As indicated in the review, there is a very large amount of highly technical published information in respect of the methods, techniques and equipment used in this area. Accordingly, it should be noted that successful fault location via debris analysis is really the province of those with expertise and experience. Thus the interested reader/potential practitioner is referred to the reference list below as an initial entry into what is a fairly complex subject area.

16.7 REFERENCES

Albidewi, A, Luxmoor, A.R., Roylance, B.J., Wang, G. and Price, A.L. (1991) Determination of Particle Shape by Image Analysis, *Proceedings of the Condition Monitoring '91 International Conference*, Erding, Germany, 14–16 May 1991, 411–22.

Anderson, D.P. (1982) *Wear Particle Atlas*, 4th edn, 1991, Spectro Incorporated, Littleton, MA, USA.

Anon (1980) Saving money on machine repairs with condition monitoring, *The Production Engineer*, **59**(3), 42–3.

Beerbower, A. (1975) *Mechanical Failure Prognosis through Oil Debris Monitoring*, Exxon Research/USAAMRDL report No TR-74-100, USA.

Bowen, E.R. and Westcott, V.C. (1976) *Wear Particle Atlas*, Naval Air Engineering Centre, Lakehurst, NJ, USA.

Hunt, T.M. (1993) *Handbook of Wear Debris Analysis and Particle Detection in Liquids*, Elsevier, London, UK.

Jin, Y. and Wang, C. (1989) Spherical particles generated during the running-in period of a diesel engine, *Wear*, **131**, 315–28.

Kirk, T.B. and Stachowiak, G.W. (1993) Fractal image analysis of wear particle boundaries and surfaces for machine condition monitoring, *Proceedings of COMADEM '93*, University of the West of England, Bristol, UK, 121–6.

Neale, M. and Associates (1979) *A Guide to the Condition Monitoring of Machinery*, HMSO, London, UK.

CHAPTER 17

Lubricant analysis as a condition monitoring technique

N.H. Riley

Fuchs Lubricants (UK) Plc, New Century Street,
Hanley, Stoke-on-Trent, ST1 5HU, UK

17.1   INTRODUCTION

Oil analysis is a technique which has been traditionally used in the industrial market place to determine the condition of those lubricants commonly used in plant and equipment. This analysis has, however, also been used on automotive equipment for many years to determine the condition not only of the oil but also of the equipment as well. When correctly applied, the analysis techniques involved can be used on oil samples from industrial plant and equipment, to establish the condition of the machine as well as the state of the lubricant. Whatever technique used, however, it should be noted that condition monitoring relies on taking measurements on a regular basis and upon the subsequent trend analysis to establish equipment condition. In this case it involves the taking of oil samples on a regular basis.

In addition to regular sampling, there are four essential prerequisites to ensure the effectiveness of this technique.

1. accuracy and representativeness of the sample;
2. use of a sophisticated and repeatable range of analytical techniques;
3. precise interpretation of the results;
4. information feedback to confirm the interpretation of the analysis.

Accordingly, this chapter outlines in detail the above criteria, and demonstrates how oil analysis can be used as an effective condition monitoring tool.

Handbook of Condition Monitoring
Edited by A. Davies
Published in 1998 by Chapman & Hall, London. ISBN 0 412 61320 4.

17.2   LUBRICANT SAMPLING

Initially, it is essential to establish a suitable sampling point on the equipment to be monitored, such that consistent samples can be taken as easily as possible, preferably without having to remove guards or other parts of the machinery. Depending upon the equipment type and design, the following may be considered as examples of suitable sampling points:

- **for gearboxes** dipstick tube into sump, level plugs, and pressure feed lines if pump fed;
- **for hydraulic systems**   reservoir, pressure and return lines and pressure test points;
- **for engines** dipstick tube into sump and sample tap fitted to oil feed pipes;
- **for compressors** dipstick tube into sump and sample tap fitted to oil feed pipes.

Ideally, the sample should be taken when the machine has reached its normal operating temperature, and the equipment has been working in the normal way. The lubricant should be extracted, preferably while the fluid is in turbulent flow, with the machine still running. If this is not possible, then it should be done as soon as possible after the lubricant has stopped circulating. Any dead areas of fluid circulation, such as an infrequently used section in a hydraulic circuit should not be used. If there is any doubt that the sample is not suitably representative, a number of samples should be taken from different points on the machine to establish which is most suitable.

Having established the best place to take the sample, some consideration should then be given to the way in which it is extracted from the equipment. Wherever possible, it is advisable to fit a tap-off point. The tap is situated so that there is not a dead area created, where debris can collect in the tap-off line. If this is not avoidable, sufficient oil should be run off to clear any residual debris prior to taking the sample. Tap-off sample points are by far the quickest, easiest and most consistent way to take samples. Reservoir drain plugs should be avoided unless there is no alternative, as they can give artificially high readings in terms of sedimentary dirt, dust and wear metals etc. If a drain plug is to be used, special consideration of this fact should be made during the interpretation of the results.

Samples can also be extracted using a tube with a vacuum pump or syringe if no suitable tap-off point exists. If a vacuum pump or syringe is used, it is recommended that the tubing or syringe are used only once, to prevent cross-contamination which would result in incorrect interpretation of the analysis. Where a vacuum pump is used, care should be taken to ensure that oil is not drawn into the pump body itself, as this will reduce pump life and efficiency, and may cause cross-contamination. When taking the sample using a tube, it is important to ensure that the sample is drawn from the same point in the system, i.e. the tube is not pushed to the bottom of the sump where residual debris could be drawn up.

Once the best method of sampling is established, it is important that it is adhered to, as variations in sampling techniques can lead to inconsistent results. This is also true of the next thing that must be determined, i.e. the frequency at which the samples should be taken. Table 17.1 depicts typical sampling frequencies for differing applications. These frequencies are, however, only a suggested guide, as the frequency is dependent upon the environmental conditions, the duty cycle and the severity of operation of the equipment. It is also suggested that sample frequencies are increased on specific equipment when the results of the analysis start to give cause for concern. Rapid sample turn round time is also recommended. This, if possible, should be less than 48 hours from receipt of sample, which means that the sampling frequencies can be increased to weekly or more, when plant conditions indicate.

If sampling frequencies are too long, i.e. every six months or even yearly, the results of any analysis cannot be used as a condition monitoring technique. It is then just a simple oil analysis, which can only

Table 17.1 Typical sample frequency

Applications	Conditions	Frequency
Gearbox – automotive and industrial	Heavy loading/high temperature ($>65°C$)/wet, dirty conditions/continuous operation	Monthly
	Light loads/normal temperature ($<50°C$)/clean operating conditions	2–3 monthly
Hydraulic – industrial	Heavy loading/high temperature ($>65°C$)/wet, dirty conditions/continuous operation	Monthly
Hydraulic – machine tools, light applications	Light loads/normal temperature ($<50°C$)/clean operating conditions	3–4 monthly
Hydraulic automotive	Wet, dirty conditions/continuous operation	Monthly or 250 hrs
Air compressors	Continuous running ($>60\%$ duty)	Monthly or 500 hrs
	Intermittent running	2 monthly
Engines – mobile plant	All conditions	10 000Km or 250 hrs
Engines – static plant	All conditions	500–1000 hrs

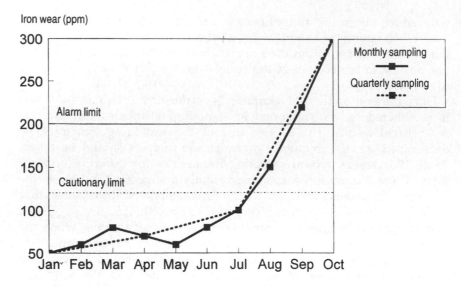

Fig. 17.1 Sampling frequency.

comment on the condition of the oil at that point in time and consequently, any further potential problems may not be seen. With reference
to Figure 17.1, the solid curve represents the variations in iron level for
samples taken on a monthly basis from an item of industrial plant. The
dotted curve shows the trend if the samples had been taken on a
three-monthly or quarterly basis. From this example, it can be seen that
the cautionary alarm level would have been reached with the August
sample. However, had the sampling frequency been quarterly, an alarm
level would not have been reached until October. Thus the sample is the
most important part of this form of condition monitoring. Unless it is
consistent, representative and taken at the correct frequency, the analysis
of the oil and its subsequent interpretation are meaningless.

17.3 SAMPLE ANALYSIS

Once a suitable sampling point and sample frequency have been established, the samples need to be analysed in such a way that the condition
of both the lubricant and the machine can be determined. The tests
carried out on the sample must not only be accurate and repeatable, but
also must be rapid enough to provide an acceptable turn round time.
Ideally, turn-round time should be between 24 and 48 hours from receipt
of the sample. The tests carried out on a sample will depend upon the
equipment being monitored and the lubricant in use. Table 17.2 outlines

Table 17.2 Selection of test methods

	Viscosity	Insolubles	Water	TAN	Flash point	TBN	Optical	PQ	ICP	ISO count
Gearbox	X	X	X				O	X	X	
Hyd system	X	X	X	O				X	X	O
Engine	X	X	X		X	O		X		
Compressor	X	X	X	O			O	X		
Turbine	X	X	X	O			O	X		O

those tests which should be carried out on some typical applications. The tests normally undertaken are as follows.

- **Kinematic viscosity at 40 and 100 °C** Industrial machines usually require lubricants of a specific viscosity range to reduce friction and prevent excessive wear of components. This test is used to detect contamination of the lubricant by some other oil, severe oxidation, or by another fluid such as petrol, diesel or a solvent etc. The effects produced by lubricant viscosity being too high or too low can be extremely detrimental, such as premature or accelerated wear, high-temperature operation, sluggish movement and reduced lubricant life. From the two viscosity results produced by this test, the viscosity index of the lubricant can be calculated. Note that the viscosity index is a number which relates to the viscosity/temperature relationship of the lubricant. This calculation also assists in establishing that the lubricant being used is correct and meets the required specification. It is essential therefore to know the specification of the lubricant in use.
- **Water content** This is an important test, for any water present in the lubricant could promote corrosion, increase wear on gears, bearings, hydraulic components or cause foaming of turbine oils. The presence of water in the sample suggests ingress via seals, gaskets or possibly from condensation. This test will, therefore, depend upon knowledge of the equipment to indicate seal failures, cylinder head gasket damage or water cooler leaks. This test may also be used to check on the condition of special water-containing fluids, e.g. water glycols, or HFA and HFB fire resistant hydraulic fluids.
- **Insolubles** In this test, the sample is filtered through typically a 1 or 5 micron filter. The weight of the debris left on the filter paper can then be expressed as a percentage. This test gives a good indication of filter efficiency or seal effectiveness. It may also be used to give an

indication of poor fuel combustion, in the form of fuel soot within an engine crankcase.

- **Total Acid Number (TAN)** This parameter is normally used as an indication of prolonged high-temperature use or the oxidation of a lubricant. High exposure to oxygen or aeration, by a lubricant, may accelerate the oxidation process. In advanced stages, the viscosity of the lubricant will begin to rise significantly, which may induce corrosion and lubrication problems. Typical indications of oxidation, are a characteristic discoloration of the lubricant and a burnt odour.

- **Total Base Number (TBN)** This is used to determine the alkalinity of a lubricant, normally for crankcase applications. When such a crankcase lubricant is formulated, reserves of alkalinity are built-in. These are gradually depleted by the acids formed during the combustion process of a sulphur containing fuel, i.e. diesel. Once the TBN has been depleted to 50% of its original value, the lubricant should be changed to prevent corrosion of components within the engine.

- **Flashpoint** This is normally associated with the detection of a fuel in the lubricant, i.e. petrol or diesel. This is generally due to fuel injection system faults or to intermittent driving, stop/start or short journeys. It may also be used to detect the presence of certain dissolved solvents or gases in the lubricant.

- **Particle Quantification (PQ)** This is a number which when quoted indicates the amount of ferro-magnetic material present in a lubricant sample. The test reacts to all size ranges of ferrous material. However, it reacts best to large debris, i.e. greater than 20 microns. It is designed to capture potential failures such as fatigue pitting, scuffing and spalling, i.e. the failure modes which normally disassociate themselves from abrasive wear, and where the majority of particles are in the less than 10 micron range.

- **Optical microscopy** Using the filter patch prepared for the insolubles test, particle shapes, sizes and colours can be examined. Bright ferrous, or rusty material, yellowed metals, flat flakes and spirals etc. can all be identified from the sample. It is also possible to give an indication of their source, plus the nature of the wear taking place within the machine, e.g. scuffing etc.

- **Elemental analysis by Inductively Coupled Plasma (ICP)** By the use of ICP techniques, wear elements can be detected and quantified in parts per million (ppm). The combination of elements detected can be combined to isolate such materials as steel alloys, brasses, bronze, white metals etc. Possible other contaminants from the process or environment, in the form of minerals and many chemical compounds, can also be identified. In addition to checking for the internal wear of a machine, or possible contamination from the environment, the ICP test can also detect additive elements in the lubricant sample. Additives such as anti-wear, extreme pressure, detergent and dispersants can be quantified by this test to ensure that they are present in sufficient quantities to fulfil their duty. It should be noted, however, that the measurement of additive levels does require an accurate knowledge of the lubricant in use.

- **Particle counting** Certain hydraulic and turbine systems may require 'super clean' lubricants to quantify the cleanliness of a system by particle counting. Should a particle count be required, the results may be expressed in several ways, i.e. the actual numbers of particles counted within certain size ranges, or by generally accepted methods such as ISO and NAS cleanliness codes. Particle counting is an essential test in the majority of large volume hydraulic and turbine applications. Regular testing for cleanliness by this method allows for the effectiveness of seals and filters to be monitored.

17.4 INTERPRETATION OF RESULTS

To enable the correct and meaningful interpretation of lubricant sample data, there are six main requirements which should be considered.

1. **Knowledge of the equipment being monitored** Before any item of plant can be monitored it is essential to establish its design function, operating parameters and history of recent failures. The person interpreting the data should have sufficient information to be able to understand how the equipment works, the layout of the components within the system, and whether ancillary equipment such as coolers, filters and/or lubrication pumps are fitted. It is also advisable to know the duty cycle of the equipment, e.g. 24 hours per day, six days per week or eight hours per day, five days per week? Service histories when available also provide an excellent insight into the expected types of failure.

2. **Metallurgy of components** The wear elements found in a lubricant can, when interpreted correctly, enable a component within an item of plant to be identified as the source of the wear. Therefore, a thorough knowledge of the metallurgy of all the component parts is necessary. By knowing the metallurgy, it is then possible to look at the ratios of wear elements in the sample and compare them to the ratios of elements found in the components, and thus identify the likely source of wear. For example, consider the set of data for an oil sample, taken from the gearbox of an underground coal-cutting machine and shown in Table 17.3. Examination of the results shows that wear is occurring at more than one source. By examining the table of metallurgical compositions, as given in Table 17.4, it can be seen that the most likely source of wear is the gears and the gear selectors. The level of tin (SN) allied to the copper (CU), indicates gear selector wear and

Table 17.3   ICP analysis results

Element	Fe	Cu	Zn	Pb	Sn	Ni	Cr	Mo	Mn
ppm	440	110	3	4	10	17	3	1	2

Table 17.4   Metallurgy of components

Component	Fe %	Cr %	N %	Mo %	Mn %	Pb %	Cu %	Zn %	Sn %
Bearing cages						0.1	60.5	39.5	
Rolling element bearings	96.8	1.36			0.52				
Gears	94.72	0.85	3.4	0.17	0.45				
Gear selectors			1.0			3.5	87.0	2.0	7.5
Shafts	95.74	1.15	1.55	0.27	0.57				
Gearbox casing	93.37				0.3				

the nickel/chrome/molybdenum ratio suggests gear wear as opposed to that of shaft or casing. It must, however, be appreciated that there are also instances where it is possible to identify the exact component which is becoming worn. A typical example would be an engine where the metallurgy is such that identification of individual components is possible.

3. **Contamination** One major advantage of oil analysis over other forms of condition monitoring is its ability to identify and monitor levels of contamination within the lubricant. In our experience, contamination is one of the major causes of failure in a wide variety of industries, e.g. mining, process plant, quarrying, component manufacture and building products etc. Typical examples of contamination are as shown in Table 17.5. Having identified the potential sources of contamination, it is necessary to analyse them to establish which elements may be present and in what proportions. For example, as

Table 17.5   Examples of contamination

Industry	Contamination
Mining	Stone dust, ore, water
Process plant	Process material, dust, water, chemicals
Quarrying	Stone dust, water
Component manufacture	Cutting fluids, dust, water, swarf
Steel making	Mill scale, dust, water
Ceramic	Clays, chemicals, dust, water

Table 17.6 Typical contamination elements

Contaminant	Elements detected
Gypsum	Calcium, Sulphur
Soluble cutting fluids	Sodium, Boron, Sulphur, Potassium
Stone dust	Silicon, Aluminium, Calcium, Magnesium

shown in Table 17.6, and once this has been established, it is possible to identify the process contaminants in the lubricant.

4. **Understanding the lubricant in use** To derive the maximum benefit from condition monitoring by oil analysis, not only is a thorough understanding of the equipment being monitored required, but also a comprehensive knowledge of lubricants, their additive packages and applications. The analysis of the lubricant should provide information on not only wear and contamination but also on the lubricant itself. Types of tests which show the condition of the lubricant are, for example, viscosity, total base number, total acid number and the spectroscopic analysis for additive elements. From these tests, one can determine whether the lubricant in use is as specified for the equipment in question, or whether it is contaminated with another grade of oil. In the latter case, further tests to those listed above may be required to identify the contaminant. For example, gas liquid chromatography or infrared spectroscopy. In either event, the analysis carried out on the sample should be sophisticated enough to determine whether the lubricant has exceeded its useful life or not.

For example, let us consider an automotive transmission fluid used in a semi-automatic gearbox of a heavy duty dump truck. The transmission required a lubricant to Allison C-3 specification, and this had to operate under high load conditions in a low gear for abnormally long periods of time. The lubricant in use contained a viscosity index improver which, when operated under high shear conditions, is liable to deterioration and thus give a reduction in the viscosity of the fluid. The transmission manufacturers recommended an oil-change period of 1000 hours, but regular monitoring produced the trend shown in Figure 17.2. While the wear metal results all showed acceptable levels of wear up to 500 hours, the viscosity of the lubricant beyond this point fell below the Allison C-3 requirement. The consequence of this was an unacceptably high level of wear. There were two possible answers to this problem. First, due to the severity of the application, the transmission oil could be changed at 500 hours and not the 1000 hours recommended. Secondly, by having a thorough understanding of lubricants and their applications, an alternative transmission fluid could be recommended. The latter solution was

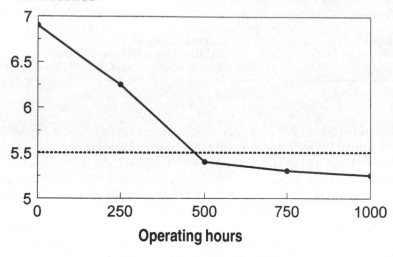

Fig. 17.2   Kinematic viscosity @ 100°C.

used and an alternative fluid was selected. Monitoring showed that it exceeded the Allison C-3 requirements for the recommended change period of 1000 hours, producing acceptable levels of wear during that period. This solution therefore provided the user with reduced servicing time and lubricant costs, increased plant availability and optimum transmission service life.

5. **Monitoring of trends**   To monitor an item of plant accurately, it is essential to look for trends in the lubricant sample analysis results. It has been found that all equipment wears in a different manner. Even two identical pieces of plant driven by identical motors, carrying the same loads and operating on the same duty cycle, will show different initial levels of wear and different wear rates. Therefore, each item to be monitored has to be considered unique, and residual debris levels and wear rates will be specific only to that individual item of equipment. As can be seen from Figure 17.3, samples 1 to 5 show a gearbox generating iron wear debris at one rate, and then from samples 5 to 8, the wear rate has increased. This of course assumes that the samples are taken at regular intervals during the operating cycle of the gearbox.

6. **Establishing alarm levels**   When interpreting sample results it is possible to identify process contaminants, lubricant condition and wear trends. For condition monitoring by oil analysis to be effective, however, it is essential to establish alarm or out of control levels,

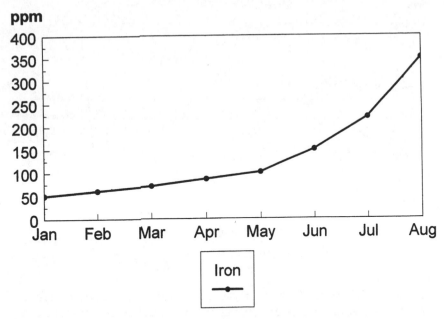

Fig. 17.3 Spur gearbox – monthly samples.

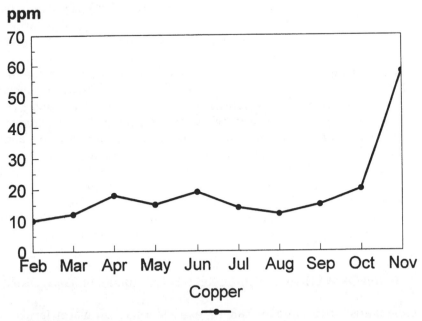

Fig. 17.4 Hydraulic system – monthly samples.

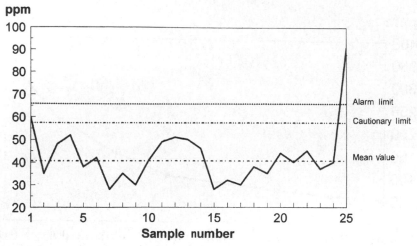

Fig. 17.5 Wear metal alarm limits.

above which some form of remedial action is required to prevent machine breakdown. Alarm levels in the initial stages of monitoring will have to be set from experience. However, the system used to trend the results, via the use of a suitable computer software package, must be flexible enough to allow for alarm levels to be tailored to suit an individual item of plant. One method of establishing alarm levels is through the use of Statistical Process Control (SPC) techniques. SPC is used by many industries to monitor manufacturing and to improve the quality of products constantly. First, as can be seen in Figure 17.4, the wear debris levels remain constant over a series of samples, until there is a rapid increase in wear debris from one sample to the next. It can be seen that the copper levels have remained relatively constant for nine samples with a sudden increase indicating the onset of a failure. The second scenario when monitoring trends is where the item wears at a relatively constant rate, and as the onset of failure occurs, the rate of wear will increase as shown in Figure 17.3. A typical control chart is presented in Figure 17.5. This shows that with a suitable number of results, the suggested minimum being 12, alarm levels can be established statistically. That is by calculating the mean and then adding two standard deviations to the mean for a cautionary limit, and three standard deviations for an alarm limit.

17.5 INFORMATION FEEDBACK AND COMBINED MONITORING

While statistical techniques can be used to establish alarm levels, no monitoring system can be effective unless there is a feedback of information from the plant operator. Therefore, there must be a dialogue

Table 17.7 Benefits of oil and vibration monitoring

Oil monitoring	Vibration monitoring
• Recirculating oil systems	• Grease and oil systems
• Detects low levels of abrasive wear	• Detects fatigue cracks and pitting
• Identify process contamination	• Mechanical looseness
• Water contamination	• Out-of-balance rotating components
• Incorrect grade of oil	• Misalignment of rotating components
• Monitor wear trends	
• Ideal for reciprocating equipment, i.e. engines	• Not ideal on reciprocating equipment, i.e. engines
• Identify component(s)	• Identification of component

between the person interpreting the sample results and the personnel operating the plant. This is to ensure that when mathematically derived alarm levels, or those set by experience, are exceeded then the information is both meaningful and accurate. The generation of false alarms is acceptable in the early stages of monitoring, as one is becoming familiar with an item of plant, but once an accurate database has been established, false alarms will only increase maintenance costs and plant downtime. The result of this is a lack of confidence in the condition monitoring technique.

Thus, it is well recognized that no single condition monitoring technique is the universal answer to the prediction of equipment failure (Riley and Mann, 1988). To this end, the lubricant is an excellent carrier of information; it can be and is used in conjunction with other monitoring techniques such as vibration analysis in a variety of applications. In terms of complementary monitoring techniques, oil and vibration analyses are an extremely powerful combination when monitoring complex drives. The benefits of using both oil and vibration monitoring are shown in Table 17.7. Sampling data relating to Figures 17.1 to 17.5 is given in Table 17.8

17.6 CONCLUSIONS

Correctly applied oil analysis can be used as an effective method of condition monitoring an item of plant. The main strength of this technique is its ability to identify both wear and contamination at an early stage, thus allowing for remedial action to be taken to prevent equipment failure and/or optimize the scheduling of maintenance. Oil analysis has in many industries proved to be an effective method of controlling contamination, reducing wear and improving plant reliability, which subsequently reduces both maintenance and production costs. With the increasing use of computerized planned maintenance schemes

Table 17.8   Time: Data relating to the figures

(a) Figure 17.1

Month	Monthly sampling	Quarterly sampling
Jan	50	50
Feb	60	
Mar	80	
Apr	70	70
May	60	
Jun	80	
Jul	100	100
Aug	150	
Sep	220	
Oct	300	300

(b) Figure 17.2

Hours	Viscosity	Min viscosity for Allison C-3
0	6.9	5.5
250	6.25	5.5
500	5.4	5.5
750	5.3	5.5
1000	5.25	5.5

(c) Figure 17.5

Sample	Metal ppm	Mean	Caution	Action
1	60	40.58	57.4	65.81
2	35	40.58	57.4	65.81
3	48	40.58	57.4	65.81
4	52	40.58	57.4	65.81
5	38	40.58	57.4	65.81
6	42	40.58	57.4	65.81
7	28	40.58	57.4	65.81
8	35	40.58	57.4	65.81
9	30	40.58	57.4	65.81
10	41	40.58	57.4	65.81
11	49	40.58	57.4	65.81
12	51	40.58	57.4	65.81
13	50	40.58	57.4	65.81
14	46	40.58	57.4	65.81
15	28	40.58	57.4	65.81
16	32	40.58	57.4	65.81
17	30	40.58	57.4	65.81
18	38	40.58	57.4	65.81
19	35	40.58	57.4	65.81
20	44	40.58	57.4	65.81
21	40	40.58	57.4	65.81
22	45	40.58	57.4	65.81
23	37	40.58	57.4	65.81
24	40	40.58	57.4	65.81
25	91	40.58	57.4	65.81

(d) Figure 17.3

Month	Iron ppm
Jan	50
Feb	60
Mar	70
Apr	85
May	100
Jun	150
Jul	220
Aug	350

(e) Figure 17.4

Month	Copper ppm
Feb	10
Mar	12
Apr	18
May	15
Jun	19
Jul	14
Aug	12
Sep	15
Oct	20
Nov	58

in industry today, there will be an increased need for all forms of condition monitoring to link directly into such systems. The technology for this exists, and the development of totally integrated systems will be the next phase in the advancement of maintenance techniques.

17.7 REFERENCES

Riley, N.H. and Mann, M.D. (1988) *Condition Monitoring – A Basic Guide to Techniques and their Applications*, I.Prod.E, London, UK.

Riley, N.H. and Mann, M.D. (undated) *The Role of Lubricant as a Condition Monitoring Tool*, Century Oils Application Note.

CHAPTER 18

Commercial applications of wear debris analysis

A. Davies and P.R. Drake

Systems Division, School of Engineering, University of Wales Cardiff (UWC), PO Box 688, Queens Buildings, Newport Road, Cardiff, CF2 3TE, UK

18.1 INTRODUCTION

In essence, as we have seen in earlier chapters, the technique of wear debris analysis is useful where it is necessary to monitor components in contact with a fluid, and which are likely to be afflicted by the process of progressive wear. Periodic analysis of the fluid can reveal the quantity and nature of the wear debris, thereby allowing the rate, type and extent of wear on critical machine components to be determined. The degradation of the physical condition of the fluid, such as for example its viscosity, can also lead to either component or system failure, and as a consequence result in a reduced equipment life-cycle. Typical items which can be monitored by this technique include bearings, gears, cams, tappets, piston rings, cylinders and seals. In addition, complete equipment systems can be surveyed for wear, a problem which is unavoidably associated with plant operation and aging.

It is possible to argue that only by monitoring the physical condition of a fluid and by seeking to identify the presence or absence of contamination can effective fault prediction be implemented on industrial machinery. This takes place via the detection of the conditions which cause an increase in wear, and thereby reduce an equipment's life-cycle. If excessive wear debris are found in a fluid, then a fault or abnormal condition is already present, and all that can be done is to take action which will avoid or minimize further damage. This includes diagnosis, and the correction of both the fault and its cause. Thus, it is

Handbook of Condition Monitoring
Edited by A. Davies
Published in 1998 by Chapman & Hall, London. ISBN 0 412 61320 4.

important to remember that one of the main advantages of wear debris analysis is that it can provide a clear indication of the cause of a fault. For in a fault situation, only the unhealthy machine components contribute an excessive amount of particles, and the path of the debris from source to sampling point is much clearer.

Accordingly, the method of wear debris analysis may be sub-divided into the three main areas outlined below (Neale, 1979). These are based on sample acquisition, and it should be noted that, by definition, the process of wear debris analysis is in the main restricted to circulatory fluid arrangements. However, the hydraulic, lubrication and cooling systems of industrial plant and machinery can, for example, be monitored via this methodology. It also has the added potential advantage of, in some cases, being independent of machine operation or speed. The three main methods are:

1. direct detection, which involves arranging for the fluid to flow through a device sensitive to the presence of debris;
2. particle collection, which requires a means by which it is convenient to remove a sample of debris from the system for 'offline' examination;
3. fluid analysis, which similarly requires a sample to be removed, and subsequently analysed to determine its condition and the extent of any contamination.

The means by which these monitoring methods may be implemented include, as we have seen in previous chapters, inductive or capacitative sensors, filters, optics, magnetic plugs, centrifuges, particle counters, ferrography and spectrographic oil analysis. Each scheme of implementation has its advantages or drawbacks, and their individual use depends upon the requirements in any particular application. In general, these techniques usually monitor ferrous debris, but some can identify non-ferrous and non-metallic particles in sizes which range from around 0.5 microns upwards. The type of surveillance offered may be periodic or continuous with a reasonable to good detection rate for rapid or gradual failures. Several systems have the ability to work 'online' and to activate alarms, together with producing a quantitative rather than qualitative output should this be required.

In modern fluid power systems, the running clearances of components within both plant and equipment are constantly reducing. This is occurring in an attempt to achieve maximum efficiency or increased power output, and also to reduce the leakage caused by higher system pressures. As a consequence of this reduction in running clearances, the requirements in respect of fluid cleanliness have increased. Accordingly, there is a growing need to monitor for particle contamination in such systems to ensure that the high levels of cleanliness required are maintained. Thus stringent maximum permissible contamination levels are now being applied to many industrial fluid systems, in order that they may achieve greater in service reliability. These levels are currently being obtained by the use of appropriate filtration technology, and via the use of post filter debris analysis, which links them to fluid condition monitoring.

It is also important to note that the three main causes of particle contamination are (Hunt, 1993):

1. ingress via the atmosphere and industrial environment, through seals, reservoirs and replenishment fluid;
2. generation from within the system, by its operation, and the unavoidable contact between moving parts;
3. retention due to the design, manufacture or maintenance of the system.

In the sections which follow, the commercial implementation of wear debris analysis systems is considered, along with selection criteria and some practical examples of their use in different industrial applications.

The criteria by which a condition monitoring method is selected obviously includes cost, ease of use, degree of technical expertise required, accuracy and repeatability. However, there are two other criteria which are particularly important in the selection of wear debris analysis methods (Williams *et al.*, 1994). The first of these is the speed of turnround in respect of the analysis, and the second being: what information is provided by the method adopted? The fastest methods will operate in an 'online' mode, although it may be difficult to achieve a detailed fault diagnosis when using the currently available methods. Some of the more sophisticated and well-established techniques require 'offline', or possibly 'off-site' analysis at specialized laboratories.

In general, therefore, there is a trade off between the analysis turnround time and the quantity/quality of the information provided by any particular technique. The provision of 'on-site' analysis will require greater 'in-house' skills and also the purchase of appropriate analysis equipment, thereby pushing up the cost of an effective monitoring programme. The use of electronic computerized analysis systems may help to improve this tradeoff dilemma, by allowing the implementation of more sophisticated analysis techniques on smaller portable equipment. This was touched upon in Chapter 3, and will now be further outlined in the sections on commercial implementation which follow.

18.2   DIRECT DETECTION SYSTEMS

This approach essentially provides for some form of sensor arrangement to monitor the fluid stream, and thus give a direct indication of the amount of wear debris being generated by the equipment under surveillance. Typical low-cost sensing methods involve inductive or capacitative units, which have the ability to detect the presence of conducting debris in the fluid flow. A representative inductance sensor consists of a cylindrical coil of wire wound around a non-metallic pipe through which the fluid passes. This obviously has the advantage of not disrupting the fluid flow, while any conductive particles in the fluid which passes through the coil will cause small changes in the inductance of the sensor. Ferrous particles will increase the inductance while non-ferrous debris will cause a decrease in the flux and hence a lowering of inductance (Whittington and Flynn, 1992).

One useful advantage of this type of non-contacting inductive sensor is that it can differentiate between ferrous and non-ferrous debris in an 'online' mode. However, it does not capture particles for further detailed diagnostic analysis and is limited to the detection of relatively large particles. This may imply that these units are insensitive to irregular wear conditions, and also that they are not suitable for monitoring the general cleanliness of a fluid. It should further be noted that the method may experience problems if fluid aeration is encountered, electrical interference or high levels of vibration (Yarrow, 1991). The capacitative sensor relies on the fact that metallic debris lying between the plates of a capacitor will alter its alternating current (AC) capacitative coupling. Two capacitors are normally used, with one compensating for temperature effects while the other records the amount of debris in the fluid flow (Neale, 1979).

Optical direct detection systems are also available, in which a light beam is passed through the fluid in order to measure its turbidity. These sensors work on the principle of luminosity interruption as a beam of light is shone across the fluid from a source on one side to a detector on the other. If the light blockage or extinction technique is employed, a detector is placed on the far side of the fluid, such that it experiences a reduction in the sensed intensity as particles in the fluid pass between it and the source. If the light-scattering principle is to be used, then the amount of scattering produced by the particles in the fluid rather than its intensity is sensed by the detector, which is again placed on the opposite side of the fluid to the illumination source (Williams *et al.*, 1994).

Optical systems can detect ferrous/non-ferrous and non-metallic particles, although problems may arise in respect of other debris and fluid opacity shrouding critical readings. Another advantage is that these systems can be made automatic, providing an output linked to flow rate and designed for self-calibration over all likely operating conditions. Most optical transducer systems are virtually maintenance-free and can be applied to a wide range of fluids. Their main disadvantage is cost at around £5000 per 'online' transducer. A typical unit is the Climet liquid quality transducer which uses laser diode light scattering to detect particles as low as 0.01 microns (Hunt, 1993).

A further technique, which can be used for the direct detection of wear debris, is that of electrically conducting filters. This arrangement senses the presence of abnormal wear debris in a fluid stream by using an electrically conductive mesh filter. The large particles are trapped in the filter and segments of the mesh are bridged by the metal debris. This allows the quantity of debris to be sensed by the electrical system and output to a meter or automatic alarm. The method is insensitive to normal wear particles and the presence of non-conducting debris can be detected by the differential pressure across the mesh. Chemical changes in the fluid are not detected but temperature measurement is possible using separate sensors on the same circuitry. Mesh sizes can vary and are removable for close inspection of the debris (Neale, 1979).

As outlined in Table 18.1, these systems can be used to maintain a continuous surveillance of plant or equipment and, via a quantitative output, can give adequate warning of rapid or gradual failure. Aside from those methods mentioned above, other techniques include the use of radiation, abrasivity and ultrasonics. In most cases, the systems outlined have a reasonable ability to withstand operation in a high-temperature environment, but have a tendency towards poor performance when excessive vibration is encountered. They may be used on large-volume circulation systems, but not on equipment which uses grease as a lubrication medium.

For the most part, they are insensitive to the addition of fluid to the system being monitored, or to the machinery/equipment operation. They do not usually require any operator involvement, and can be retro-fitted to items of plant with minimal or no difficulty. Trained personnel, or operation via a centralized plant monitoring system, is not necessarily required in all cases. Thus, they provide reasonably cost-effective monitoring, particularly the electrical sensors where costs are relatively low, and only routine maintenance is required.

18.3 PARTICLE COLLECTION AND ANALYSIS SYSTEMS

This approach involves inserting some type of acquisition arrangement into the fluid flow, to trap the wear debris, and to permit their removal for fluid cleaning/diagnostic examination. By performing this type of inspection, the amount of debris being generated can be determined along with their shape and material composition. Three main methods are in common use, these are magnetic plugs, filters and centrifuges. Particle collection for 'offline' analysis is commonly carried out by the use of removable magnetic plugs which are usually of the quick release and self-sealing type to prevent loss of fluid. Typical examples are manufactured by Muirhead Vactric Components Ltd (Hunt, 1993).

Table 18.1 Typical 'online' direct detection particle counting systems

System	Comments
1. Thin Layer Activation AEA Technology	This system is based on the tracking of radioactive particles, eroded from a thin irradiated layer of atoms in the component. Cost is approximately £1000/radioactive item, which must be irradiated at the supplier's site.
2. Climet CI-1500 Climet Instruments	Liquid quality transducer based on optical–nephelometry/turbidity principle. Laser diode light scattering on or 'inline' monitor, which displays continuous turbidity value. Cost about £5000/unit. Detection range 0.01 microns up.
3. CM 20 UCC International	Portable, 'online' optical – light obscuration system, weighing only 10 kg and costing about £7000/unit. Detection range 5–200 microns. PC compatible for downloading of results.
4. Fulmer Wear Debris Monitor Fulmer Systems	A thin electrically resistive metallic film of 0.1 microns thickness, deposited on a ceramic substrate and subject to erosion by abrasive particles is the basis of this unit. Cost about £4000/system with £200/replacement sensor. Detection range ferrous debris 35–150 microns.
5. Micropure Monitek Technologies	An ultrasound 'online' particle contamination monitoring system, which uses an 'inline' pipe wall mounted sensor. Cost ranges from £8500–£30 000 depending on application. Detection range 0.2 microns to 3 millimetres. Useful for monitoring corrosive fluids and others where flow restriction cannot be tolerated.
6. Rion KL — 20/22 Rion Co Ltd	Optical sideways light scattering unit using a laser diode and compatible with a range of fluids. Capable of 'online' use at pressures up to 3 bar. Particle detection range 0.2–2 microns. Cost depends on application.

Obviously, this approach to monitoring will only detect magnetic material in a fluid, and as such, it is particularly suitable for monitoring ferrous components which produce debris in large pieces, e.g. rolling element bearings and gears. The trapping of particles in excess of 5 microns is normally quoted for these units, and it should be noted that they are not suitable for monitoring the general cleanliness of a fluid, which may depend upon the presence of a non-magnetic material such as silicon. Magnetic debris collectors are, however, a low-cost monitoring arrangement, with typical plugs costing between £60 and £200 pounds per unit, depending on the size and complexity of the design.

It is possible for the magnetic plug principle to be automated. In one design, for example, a magnetic plug is placed behind a fluid tight non-magnetic membrane on which it collects the wear debris. The amount collected is automatically measured, and when a fixed value has been achieved the magnet is withdrawn for a short time to release the debris. The monitored variable is the frequency of the withdrawal, which will of course depend upon the amount of wear debris generated. When this frequency reaches an appropriate threshold an alarm can be triggered and a sample of debris retained for analysis. The release of debris back into the fluid system is a disadvantage in this design as it reduces fluid cleanliness and encourages faster debris pickup.

Another approach often used with these systems is to place several magnetic plugs close together, so that if a particle is sufficiently large enough to bridge the gap then an electrical circuit is completed and this sets off an alarm (Veinot and Fisher, 1989). Such a system has been developed in the Vickers TEDECO quantitative debris monitor, which both counts and gives an indication of the size of trapped particles. It is possible with this system for a build-up of fine debris to result in the gap being bridged and a false alarm triggered. To overcome this problem, some commercial systems apply an electrical pulse to burn off the fine debris without removing the larger more critical particles. A burn-off count can then be used to monitor the fine debris generation rate. Typically these units cost around £7000 depending on the application (Hunt, 1993).

The deposits left on filter elements may also be used as the basis of particle analysis. A major disadvantage with this approach is, however, that filter removal and replacement may be difficult and usually requires the machine under surveillance to be stopped. While special filter designs are available to overcome these problems, another disadvantage of this approach is that the filters will not capture wear debris which are smaller than their pore size. However, an advantage of the method is that there is a 100% capture efficiency for debris which are larger than the pore size. Several commercial systems are available which utilize this principle. The Diagnetics CONTAM-ALERT being fairly typical at around £7000 per unit (Hunt, 1993).

The centrifuging of a fluid, or a sample of it, removes all the debris from a liquid. When this process is complete, either the sludge, or the particles contained in a sample, may be assessed and possibly related to some aspect of equipment wear. A drawback to sample centrifuging in the laboratory is that it does not necessarily arrange the debris according to individual size. It may, however, be able to put them into bands, or display the particle distribution depending upon the equipment design. In oxidized oil the carbon particles may obscure the wear fragments which is another drawback to this technique.

However, laboratory centrifuges have the advantage of being relatively inexpensive when compared with other equipment, especially on a like-for-like information output basis. Typical modern laboratory centrifuging systems are the applied imaging disc centrifuge DCF4

Table 18.2a Typical particle collection and sizing systems

System	Comments
1. Magnetic chip collectors Muirhead Vactric	Designed to capture ferrous debris contained within lubricating oil. Can be used up to 14 bar pressure and a temperature range of −70 °C–200 °C. Detection range all sizes of ferrous debris. Cost £60–£200 depending on application.
2. Continuous Debris Monitor Ranco Controls	This system detects the build up of ferrous particles over time. It uses magnetic attraction and flux path change to detect particle quantity. Debris can be released at a set value, or an alarm activated. Cost about £1300. Debris weight range varies, 100–600 mg typical. An 'inline' low pressure sensing unit.
3. Contam Alert Diagnetics	Portable monitor which can be used 'online' or with bottled samples. The system uses a photoetched screen filter/blockage/flow decay method to determine particle size/distribution. Cost about £7000 with a detection range up to 15 microns.
4. Fluid Condition Monitor Lindley Flowtech	This system uses the rise in pressure drop across a filter, to determine the number of particles trapped. Automatic reversal and backflushing takes place, with 4 counts being averaged. Cost about £7000. Detection range up to 15 microns.
5. AI Disc Centrifuge Applied Imaging	This system uses the sedimentation principle via a disc centrifuge. It is a PC-based desktop system which determines particle size and distribution. Cost about £15 000. Detection range 0.01–60 microns.
6. Microparticle Classifier George Fischer	Classification of fine metal powders and combustion dusts by sieving via an air centrifuge elutriator. Test time up to 2 hours for 8 fraction sizes. Cost about £12 000. Detection range covers particles below 60 microns. Dry sample used.

Table 18.2b   Typical particle/debris image analysis systems

System	Comments
1. CUE systems Olympus Optical	Desktop PC-based laboratory system with microscope and CCD camera. Black and white or colour analysis possible with high resolution. Several advanced functions and up to 35 different particle parameter measurements available. Total system cost £11 000–£42 000.
2. Hamamatsu systems Hamamatsu photonics	Real-time video system which can count up to 30 000 particles. An average of 8 counts always displayed. Area of count and size selectable, with pseudo colour display. Total system cost around £20 000.
3. Keyance Microscope Keyance Corp	Portable microscope probe, controller module and optional monitor. × 1000 magnification and × 30 depth field over optical microscope. Hand-held analysis possible. 6 lens. Total system cost around £10 000.
4. Magiscan Applied Imaging	Complete system includes process modules, monitors, microscope controller, high resolution CCD camera, printer and extensive software. Particle sizing/metallurgy applications. 72 000 images may be stored. Total system cost around £20 000.
5. Microeye/Microscale Digithurst	PC-based system with image capture, compression, analysis and output. Colour video images, 3D profiles and surface visualization. Many particle features can be analysed. Total system cost about £10 000.
6. Quantimet Leica Cambridge	Powerful processing and analysis systems are available for use with a microscope. PC-based and Windows-compatible with a wide range of analysis functions. Fully automatic, true colour and chemical analysis, plus high resolution available. Total system cost depends on requirements.

manufactured by Applied Imaging International, and the BAHCO micro-particle classifier produced by the Dietert division of George Fischer Inc. Both cost around £15 000, see Table 18.2a (Hunt, 1993).

As most particle collection systems operate on a periodic rather than continuous basis, frequent sampling and analysis is necessary for good warning of failure. They can detect both rapid and gradual wear, but as most involve 'offline' optical or image analysis techniques for quantitative assessment, see Tables 18.2b and 18.2c, their ability to automatically trigger alarm systems is limited. The magnetic plugs and filter arrange-

Table 18.2c Typical offline optical/other particle sizers

System	Comments
1. Coulter LS series Coulter Corp	In this instrument, light is diffracted around the particle at angles inversely proportional to its size. Multiple optical trains are used with special binocular technology. Polarisation Intensity Differential Scattering gives a detection range of 0.1–800 microns. Dry or liquid samples may be tested. Desktop PC-based. Cost on application.
2. Debris Tester Stavely NDT	The unbalanced AC bridge principle is used in this equipment. Sample debris being placed on a slide after transfer from magnetic plug. Detects ferrous particles only with results capable of downloading into a PC for trend analysis. Portable unit, mass range 0.02 mg–200 mg, weight 6 kg, cost £3000.
3. Elzone Particle Data Inc	Particles are caused to flow through a small sensing orifice almost individually. A current is passed across the sensor to determine particle size by change of impedance during transit. Limited to use on electrically conductive fluids. Detection range 0.4–1200 microns. Cost around £15 000.
4. Fritsch Analysette 22 Fritsch GmbH	PC-based particle sizer designed to measure debris wet or dry in suspension. Laser system with convergent beam. Built in ultrasonic bath for dispersion before measurement. Detection range 0.16–1250 microns. Cost depends on features required around £20 000–£45 000.
5. Malvern Mastersizer Malvern Instruments	This instrument uses Fraunhofer diffraction as the basic principle. It has a wide detection range 0.1–2000 microns and both wet and dry dispersions analysed. Desktop and PC-driven cost is high.
6. Polytec Particle Sizer Polytech GmbH	An optical light scattering instrument which uses a white light source. Particle detection is possible on or 'offline'. Detection range 1–125 microns and distributions can be displayed. Cost £25 000.

ments for sampling work satisfactorily in most environments, including large volume circulation systems, and they are relatively impervious to heat and vibration. Magnetic plugs can also be used in non-circulatory systems but not where grease is involved as the fluid medium. Both filters and plugs are also relatively insensitive to fluid system top-up, and it is possible to obtain data while the machine is in operation.

The sampling points for both plug and filter systems need to be in accessible locations. Accordingly, they should be placed in an optimum position during the design of the equipment to be monitored, and not if ossible retro-fitted after manufacture. Trained personnel are required to obtain good performance from particle collectors usually via part of a centralized plant monitoring system. In general the cost is low per sampling point, but significant in terms of say optical or image analysis equipment, personnel for effective interpretation, and on-going system administration. Tables 18.2b and 18.2c outline the features of available particle imaging and optical analysis systems.

18.4   FLUID ANALYSIS SYSTEMS

Fluid sampling is a two-step process, involving firstly the extraction of a representative amount of liquid from the machine being monitored, and secondly undertaking a detailed laboratory analysis of the sample. The second step is required to determine the sample condition and the materials in it. This hopefully will then give an indication of the equipment's condition. Two methods of analysis are commonly performed, viz:

- **elemental analysis**   which determines the chemical elements present in the sample and their quantity;
- **particle analysis**   which determines the quantity, size and shape of any particles present in the sample.

Spectrometric oil analysis is typical of the first type of test, while ferrography and particle sizing/counting are used in the second.

In any form of fluid analysis, the sampling mechanism and subsequent testing procedure is critical to the success of this 'offline' monitoring technique. The type of particles being monitored can be microscopically small, and this means that the corresponding sampling containers must be scrupulously clean. Reputable fluid analysis laboratories which provide an analysis service, will supply their own clean and capped sample bottles, along with training in the correct procedure for taking samples. The sampling point on the monitored system must be chosen carefully, to provide a sample which is representative of the current state of the fluid in the machine. If sampling is carried out downstream from a filter for example, then large-sized wear debris will not be detected. However, such a sampling point may be useful in determining the effectiveness of the filter where very high levels of cleanliness are required in the fluid.

Fluid samples should be taken when the normal system operating temperature has been reached, with the machine running according to its designed duty cycle and the liquid is in turbulent flow. This mitigates against the settlement of larger wear debris within the fluid system, and helps to ensure that a representative sample is obtained for analysis. When designing a fluid sampling arrangement, it is important to consider the anticipated size of the wear debris involved, and also the ability of the liquid circulation to transport them to a sampling point which is not to be located in a 'dead' area. The best method of fluid extraction is via specially fitted tap-off points sited such as not to induce 'dead' areas in the liquid circulation.

These sampling taps provide a reliable, repeatable, easy and quick way of obtaining representative samples, and avoid the possibility of artificially high debris readings, which can occur due to particle accumulation if fluid is taken from reservoir drains. In the absence of tap-off points, syringes or vacuum pumps can be used to extract fluid, usually via the dipstick tubes which enter the sump of the systems to be monitored. When syringes are used, it is imperative that a new clean syringe is used for each sample. This avoids cross-contamination by microscopic particles in successive samples, and the same care should be exercised with the tubing used for vacuum pumps.

Spectroscopy is a well-established and widely used method of elemental fluid analysis. It is conducted in an 'offline' mode with samples either being analysed on-site or at a remote laboratory. The technique has the advantage that it can provide data on the chemical composition and concentration of the debris in the fluid sample, with the added advantage that these levels can be used as a trend indicator for the equipment under surveillance. The disadvantage of spectroscopy is that it only detects small particles up to about 8 microns in size, this being due to a number of factors, including an inability to vaporize larger debris. This inability to detect larger particles can mean that some severe mechanical wear can go undetected. There is no lower limit on the size of particles that spectroscopy can detect, as it is possible to identify elements in solution with this technique.

The method is used extensively throughout industry, with British Rail (BR) reporting a successful application in monitoring the condition of engines for high-speed trains. The technique being used to monitor the condition of the lubrication oil in this application, as well as the engine components which are washed by the fluid. It is also reported to have produced significant financial savings by extending the period between oil changes as well as engine overhauls. The processing of samples has been improved by BR, through investing in automatically controlled laboratory analysis and computerized communications networks, to handle rapid data interpretation and dissemination. The United States Air Force (USAF) has also implemented an extensive spectrometric oil analysis programme to increase the reliability, safety and flight readiness of its aircraft (Williams *et al.*, 1994).

Several different types of equipment can be used in spectroscopy, see Table 18.3a. Only three will be discussed here as representative

Table 18.3a Typical spectrometric instruments

System	Comments
1. Varian Spectrometer Varian Pty Ltd	This instrument uses the atomic absorption technique. A slow method taking up to 3 mins for each element. Detection range up to 1 micron. Cost depends on requirements, £7000–£30 000. Particle is destroyed during the test.
2. Baird Multi Element Oil Analyser Baird Europe	A rotating disc electrode system. Up to 30 elements can be analysed in less than 60 secs. Particle size for analysis up to 8 microns possibly higher. Cost around £40 000. Test is destructive of particle but gives elemental breakdown plus concentration.
3. Varian Liberty Varian Pty Ltd	An inductively coupled plasma unit this system links analysis speed with effectiveness. Commercially available since the 1970s the technique is good for large samples and can be automated. It has good sensitivity and a mechanized capacity of 100 samples per hour. Particle size for analysis up to 5 microns. Cost £50 000–£80 000.
4. ARL 8410 XRF Applied Research Labs	This system uses X-ray fluorescence to determine the elemental composition of the sample and the quantity. Any sample can be tested with up to 80 elements identified. Very fast test, around 5 secs to identify one element and not destructive of particle. Cost in the range £20 000–£100 000.
5. Outokumpu Courier 40 Outokumpu	An X-ray diffraction system, this instrument is used to detect mineral content in a sample, typically slurry. It can analyse any particle size with a fast detection time, about 5 secs for one element. Cost is around £100 000 and the test does not destroy sample.
6. Hitachi Hitachi Scientific Instruments	This system uses mass spectrometry to detect elements and organic products in oil. Simultaneous analysis is possible with the instrument analysing a solution not an individual particle. Very fast test, one element in 1 sec, but destructive of particle. Cost depends on requirements, range £20 000–£15 000 000.

examples, and the interested reader should refer to (Hunt, 1993) for a complete review of techniques and equipment. In the atomic absorption spectrophotometer a sample is diluted and vaporized in an acetylene flame. Light with a wavelength corresponding to the element to be detected is passed through the flame and absorbed if the element is

Table 18.3b   Typical particle analysis systems

System	Comments
1. Ferrograph VD AMOS	Used offline to check lubricating oil samples. Ferrous debris only. The equipment also includes a density measuring arrangement. Detection range 1 micron upwards using bottle samples. Cost around £2500 complete.
2. Particle quantifier Swansea Tribology Centre	This instrument is designed to assess the amount of ferrous debris in an oil sample. Fast test around 10 sec per sample. Carbon material not sensed. Sample can be tested as taken, or with debris separated. Main use on hydraulic and lubrication oils. Particle detection range 1 micron–2 mm. Cost around £4000.
3. Rotary particle depositor Swansea Tribology Centre	Particles are deposited on different striations according to size, by use of a rotating substrate, on which a sample of oil is deposited. Magnetic attraction separates the ferrous particles. 1–2000 microns. Cost around £4000.
4. Wear Particle Analyser Tribometrics	A special magnetic filter made from steel fibres is used in this equipment. An oil sample is drawn through the filter and particles captured by filter action plus magnetic attraction. Analogue or digital output possible with debris available for subsequent microscopic analysis. Detection range depends on filter size used, up to 10 microns non-magnetic debris. Cost about £3000.

present in the sample. Comparison with a known reference standard provides a direct measure of the amount of each element in the sample. The instrument is accurate, compact and relatively simple to operate. It is, however, limited to detecting particles up to 1 micron in size and one element at a time (Hunt, 1993).

The emission spectrometer measures the characteristic wavelengths of light which are emitted when the elements in a fluid are excited by a high-energy source such as an electrical discharge. Elements are then identified by comparing the measured spectra with reference spectra produced by known elements. A direct reading emission spectrometer measures the wavelengths by using photomultipliers linked to spectral lines and consequently several elements can be detected simultaneously. The equipment is large, expensive and requires precise calibration to produce good repeatability. However, it has the benefit of being able to detect bigger particles than the spectrophotometer. The use of emission spectroscopy for the analysis of wear debris in grease has also been reported (Jones, 1989).

In the X-rays fluorescence method, the fluid sample is irradiated causing energy level changes in the atoms, which when reversed emit secondary X-rays. The energy in these X-rays indicate the elements in the fluid sample and the intensity of the emission determines the concentration of the elements. This is a fairly sophisticated method which requires a degree of skill in the interpretation of the emitted X-rays. The equipment can be very expensive but is also reasonably portable. This method has several advantages over the others, in that samples may be in solid or liquid form, the dynamic range is such that the simultaneous analysis of a wide spread of concentrations is possible on the same sample, and the technique is non-destructive (Hunt, 1993).

In the case of particle analysis, there are, as with elemental analysis, several different types of equipment which are commercially available to undertake this type of test. Typical instruments are outlined in Tables 18.2c and 18.3b. The interested reader should refer to (Hunt, 1993) for a complete review of the techniques and equipment available in this area. However, as ferrography is a popular and representative example of the methods used in this area, it is worth briefly outlining here. In this approach, a sample of fluid is brought into contact with a magnetic field, and it is this which causes the separation of the ferromagnetic and paramagnetic particles from the fluid. The debris size range which can be detected using this technique varies from about 1 micron to 100 microns, and non-magnetic particles can sometimes be captured. This does depend, however, on any non-ferrous particle having first had contact with the ferrous components in the system (Desjardins and Williams, 1991). Accordingly, this can be seen as a major disadvantage, which results in the method not being particularly efficient at detecting non-magnetic particles.

Direct reading ferrography precipitates the debris under the influence of an inclined high-intensity magnetic field inside a fine-bore glass tube. The debris are precipitated at two selected points down the tube, with the first of these being close to the entry and monitors the smaller debris. The larger material, typically greater than 5 microns, is detected at the second point. Optical density readouts, obtained at these two points from light sensors, can then be trend-plotted and used to detect debris which require a more detailed analysis. In analytical ferrography, the fluid is caused to flow over a specially prepared substrate, and across a magnetic field that separates out both the ferromagnetic and paramagnetic particles. The larger particles are deposited first, with progressively smaller debris being deposited as the fluid flows across the magnetic field. The morphology of the particles separated from the fluid can be determined by the use of optical microscopes, a scanning electron microscope or image analysing computers. The separation patterns may also provide useful information since they are related to the particle sizes and their distribution.

Fluid analysis techniques do not adequately warn of rapid failure. This is because of their sampling requirements and offline nature. However, in respect of gradual failure, they can perform adequately by producing a precise quantitative analysis of the fluid sample. They are not generally useful in activating automatic alarms and may have problems in dealing with fluid top-up or high-volume systems. Equipment temperature and vibration have no effect due to the sampling procedure used, and the techniques involved in this method can be applied effectively to grease or non-circulatory fluid systems. Fluid can usually be taken while the machine is in operation, although this does require safe access to a designed-in or retrofitted sampling point. Trained staff are required to conduct this type of analysis, usually in a centralized system, and the equipment cost can be high.

18.5 TECHNIQUE COMPARISON

The first attribute which needs to be noted about wear debris analysis methods is the different size of particles that they can detect. Spectroscopy will detect elements in solution but will only detect the presence of particles up to about 8 microns in size, depending on the specific technique used. Severe mechanical wear produces particles larger than 10 microns, with those larger than 20 microns typically providing a reliable indication of extreme wear. Since spectroscopy cannot detect these particles it can result in non-detection of severe wear conditions or, at best, late detection once the debris has become ground down into smaller particles. The ability of ferrography and magnetic plugs to detect the larger particles has led some authors to argue against the use of spectroscopy for the early detection of severe wear (Lewis, 1989).

Methods for enabling spectroscopy to analyse larger particles have been proposed (Lukas and Anderson, 1991). One such method being the use of strong acids to dissolve the large particles. This, however, can be time-consuming and adds greatly to the relatively cheap spectrographic process. Furthermore, it can be argued that even if the larger wear debris are dissolved, their effect on the overall iron concentration levels in the simple fluid will be very small, and iron concentration monitoring is, in any case, not the same as large particle detection (Desjardins and Williams, 1991). Nevertheless, spectroscopy is still a very popular technique, having the advantage over ferrography and magnetic plugs of being able to detect a wide range of magnetic and non-magnetic elements.

These features make spectroscopy particularly useful for monitoring oil additives and the ingress of contaminants such as siliceous material from the atmosphere and general airborne dust. The more comprehensive analysis offered by spectroscopy may give information which can pinpoint the exact location of an impending failure, especially if the metallurgy of the various components in the monitored system is known. If this is so, then the ratio of elements in the wear debris may indicate the source of wear, that is by way of a form of signature analysis. This diagnostic arrangement can be performed, using signatures which define the quantity of elements such as iron, nickel, chromium, molybdenum and magnesium that make up particular machine components. Another positive characteristic of spectroscopy is that it is an extremely sensitive and qualitative technique which suits it to the monitoring of fluid cleanliness.

It may well be that in a particular application spectroscopy will be used in addition to one or more of the other wear debris techniques. With the other method being employed to give an early warning of severe wear conditions, possibly in an 'online' mode. As an example, the case for spectrometric oil analysis being a complement to the ferrographic technique is often made (Evans, 1991). This is also true of other technique combinations which provide comprehensive wear debris coverage. Ferrography will detect particles from about 1 to 100 microns in size, and magnetic plugs will pick up particles from about 20 to 1000+ microns. This means that magnetic plugs are more advantageous when monitoring ferrous components, like roller bearings and gears which produce wear debris in large pieces. However, it should be remembered that large particles may not be picked up in oil samples as their weight precludes their remaining in suspension.

The rotary particle depositor has been developed to overcome conventional ferrography's inability to deal with large sample volumes. Conventional ferrography techniques can normally only cope with a sample volume of about 1 millilitre (ml). Larger sample volumes will obviously give an analysis with tighter confidence limits and they are less likely to exclude large pieces of wear debris which occur in small numbers. In general, the currently available 'online' techniques do not give detailed fluid analysis and fault diagnosis information. Instead, they are used to detect abnormally high concentrations of particles or contamination and large pieces of wear debris. Sensors can also be used downstream from filters to monitor 'online' the effectiveness of the filtration system, and the cleanliness of fluid. One important advantage of 'online' systems, from the practical point of view, is that they are not dependent on oil sampling which, as described above, must be carried out with the utmost care.

18.6 CONCLUSIONS

The key advantage of wear debris analysis is that, by monitoring contamination and the fluid state, it can detect the conditions which cause wear, or malfunction of machinery, before the effects become too severe and failure occurs. Thus the methods of wear debris analysis should be seen as complementary to the other techniques involved in condition monitoring. This provides both wider fault coverage, and corroboratory information for failure prediction and diagnostics.

18.7 REFERENCES

Desjardins, J.B. and Williams, W.S. (1991) Comparative effectiveness of measuring iron particles using spectrometric and ferrographic techniques, *Proceedings of Condition Monitoring '91*, Erding, Germany, 364–71.

Evans, W.A. (1991) Spectrometric oil analysis: a complement to ferrographic analysis, *Lubrication Engineering*, **47**(6), 437–9.

Hunt, T.M. (1993) *Handbook of Wear Debris Analysis and Particle Detection in Liquids*, Elsevier Applied Science, Barking, UK.

Jones, M.H. (1989) Wear debris analysis of grease, *Proceedings of Condition Monitoring and Preventive Maintenance*, Atlanta, Georgia, USA, 132–9.

Lewis, R.T. (1989) The mechanical component of wear, what spectroscopy doesn't tell you, *Proceedings of Condition Monitoring and Preventive Maintenance Conference*, Atlanta, Georgia, USA, 78–82.

Lukas, M. and Anderson, D.P. (1991) Techniques to improve the ability of spectroscopy to detect large wear particles in lubricating oils, *Proceedings of Condition Monitoring '91*, Erding, Germany, 372–98.

Neale, M. (1979) *A Guide to the Condition Monitoring of Machinery*, HMSO, London, UK.

Veinot, D.E. and Fisher, G.C. (1989) Wear debris as a condition monitoring technique for the Sikorsky Sea King Helicopter main gearbox, *Proceedings of Condition Monitoring and Preventive Maintenance*, Atlanta, Georgia, USA, 1–4 May, 119–31.

Whittington, H.W. and Flynn, B.W. (1992) Wear debris monitoring in a lubricant, *Condition Monitoring and Diagnostic Technology*, **3**(2), 61–4.

Williams, J.H. *et al.* (1994) *Condition Based Maintenance and Machine Diagnostics*, Chapman & Hall, London, UK.

Yarrow, A.S. (1991) Condition monitoring by wear debris analysis, *Noise and Vibration Worldwide*, **22**(5), 24–8.

PART SIX

Adopting a Condition-Based Maintenance Strategy

CHAPTER 19

Financial implications and cost justification

G. Eade

Asset Management Centre Ltd, Ludworth Trout Farm,

Marple Bridge, Cheshire, SK6 5NS, UK

19.1   INTRODUCTION

The simple process of financial justification for an investment project, would normally be to compare the initial and on-going expenditure with the expected benefits, translated into cost savings and increased profits. If the capital can be paid off in a reasonable time, and concurrently earn more than an equivalent investment in secure stocks, then the project is probably a good financial investment.

The case for buying a new machine tool, or setting up an extra production line, can be assessed in this way, and is the normal basis on which a business is set up or expanded. The purchase price plus installation, recruitment and training costs, must be paid off within a limited number of years, and continue to show a substantial profit after deducting the amount of borrowed capital, operating cost and so on.

However, the benefits from an investment in a condition monitoring (CM) system are more difficult to assess, especially as a simple cost/benefit exercise. This is because, to put it simply, the variables are much more intuitive and less measurable than pure machine performance characteristics (Eade, 1989).

The ultimate justification for condition monitoring is where a bottleneck machine is totally dependent on a single component such as a bearing or gearbox, and failure of this component would create a prolonged unscheduled stoppage affecting large areas of the plant. The cost of such an event could well be in the six-figure bracket, and the effect on sales and customer satisfaction beyond quantification.

Handbook of Condition Monitoring
Edited by A. Davies
Published in 1998 by Chapman & Hall, London. ISBN 0 412 61320 4.

Yet a convincing financial case is highly dependent on knowing how often this sort of disaster is likely to happen, and also a precise knowledge of the unquantifiable factors referred to above. At best, one can only say that, whatever the cost, if it is likely to happen, it would be foolish not to install some method of predicting it, so that the appropriate preventive action could be taken.

19.2   ASSESSING THE NEED FOR CONDITION MONITORING

Any maintenance engineer's assessment of plant condition is influenced by a variety of practical observations and analyses of machine performance data, such as:

- the frequency of breakdowns;
- the randomness of breakdowns;
- the need for repetitive repairs;
- the number of defective products produced;
- the potential dangers linked to poor performance;
- any excessive fuel consumption during operation;
- any reduced throughput during operation.

These, and many more pointers, may suggest that a particular item of plant requires either careful monitoring, routine planned preventive maintenance, better emergency repair procedures, or some combination of all these approaches in order to ensure a reasonable level of operational availability. The engineering symptoms can, however, rarely be quantified accurately in terms of financial loss. Very few companies can put an accurate figure on the cost of downtime per hour. Indeed, many have no reliable records of their aggregate downtime at all, even if they could put a value per hour on it!

Thus, although a maintenance engineer may decide that a particular machine with a history of random bearing failures requires condition monitoring, if problems are to be anticipated, and the plant taken out of use before a catastrophic in-service failure occurs, how can he or she justify the expenditure of say £10 000 on the appropriate monitoring equipment, when plant and production records may be too vague to show what time and expense could be saved, and what this saving represents in terms of profit and loss to the company? This is a dilemma which can be a daily occurrence, and one which does face engineering and maintenance staff in large and small companies throughout the country.

As if the practical problems of quantifying both the potential losses and gains were not difficult enough, then the status of maintenance engineering in many organizations is such that any financial justification, however accurate, can be meaningless. Usually the maintenance department in most companies is classed as a cost overhead. This means that a fixed sum is allocated to maintenance each year as a budget, which covers the cost of staff wages, spare parts and consumable items etc. The maintenance department is then judged for performance, financially or on its ability to work within its budget. Overspending is classified as 'bad', and may result in restricting the department's resources even further in future years. While underspending is classified as 'good', in that it contributes directly to company profits, even if equipment maintenance is neglected and manufacturing quality or throughput suffers as a result.

Let us just suppose that a forward-looking engineer succeeds in persuading his or her financial director, who knows nothing about condition monitoring and would rather invest the money anyway, to part with the capital needed to buy the necessary CM equipment. What happens then? Our hero, by using CM, succeeds in reducing unscheduled machine stoppages drastically, but which department gets the credit? Usually production, because they have not needed to work overtime to make up for any lost production, or have fewer rejects. Alternatively, it may be the sales department which receives the credit, because of improved product quality or reduced manufacturing cost, which has given them an advantage over the firm's competitors. It is very rare indeed that the maintenance engineer is seen as the person who has added to the organization's improved cash flow by his or her actions.

Thus, a company which does not have a system of standard value costing cannot hope to isolate the benefits of efficient plant engineering, and persuade the board of directors to invest in an effective arrangement for equipment purchasing and maintenance. This presents a bleak picture for the person who has to make out a good financial case for installing a particular condition monitoring technique. Yet it is a very familiar situation and one which my company has met time and time again. Indeed, it occurs in the majority of organizations where we have received initial enquiries regarding the installation of our software.

The very expense of a computer system, for example, to collect and analyse plant data, without which an accurate cost justification is impossible, is often in itself treated as a non-productive overhead. This is a classic 'catch 22' situation which has been stated in the past as:

> We need the computer system to calculate whether we need the computer system, even though we know that it is essential before we start.

So, in order to cost justify a particular condition monitoring project, we need first to persuade the appropriate person in the financial control hierarchy, that it should be treated as a capital investment charge in its own right, and not as an item of expenditure from the maintenance department's annual budget. Obviously, this will place us in competition with other capital investment projects for the organization's limited resources. Accordingly, the case for justifying any condition monitoring equipment must be good, and show a tangible return in a very short period of time.

19.3   COST JUSTIFICATION

To produce a good case for financial investment in condition monitoring equipment, it is therefore important to obtain reliable past performance data for the plant under review. In addition, information relating to other equipment, whose operations may be improved by better performance from the plant whose failures we hope to prevent, must also be gathered. It is also essential to establish an effective financial record of actual condition monitoring achievement. This is especially true after the installation of any original equipment, so that it is possible to build on the success of an initial project.

The performance data relating to CM must therefore be quantified financially. Which in effect, can mean going around all the departments involved, and persuading managers to put a cost figure on the various factors that come within their responsibility. Many managers, who may have criticized maintenance engineering in the past for poor production plant performance, by statements such as:

It is costing the company a fortune,

can suddenly become very reluctant to put an actual cost value on the loss, particularly when asked for precise data. However it is in their interest to try, because without financial data there can be no satisfactory cost justification for CM, and hence no will or investment to improve the maintenance situation. Ultimately, it will be their department and the company who is the loser, if poor maintenance leads to an uncompetitive position in the marketplace.

Some of the factors relevant to maintenance engineering which can have an adverse effect on the company's cash flow are as follows.

- **Lost production and the need to work overtime to make good any shortfall in output**   Some organizations will find this factor relatively easy to quantify. For example, an unscheduled stoppage of three hours could mean 500 components not made, plus another 200 damaged during machine stoppage and restart. The production line would perhaps have to work an extra half shift of overtime to make up the loss, and thereby incur all the associated manning, heating and other facility support costs involved.

Alternatively, the cost of a sub-contract outside the company to make good the lost production is usually obtainable as a precise figure. This is normally quite easy to obtain and in real expenditure terms, as opposed to the internal cost of working overtime, which may not be so precisely calculated. Other costs may also be difficult to quantify accurately, such as the sales department's need to put a value on the cost of customer dissatisfaction if a delivery is delayed, or the cost of changing the production schedule to make good the loss in production if the particular product involved has a very high priority. The cost of lost production is a random set of peaks in the cash flow diagram, which as shown in Figure 19.1, if treated independently can appear as a minor problem, but if aggregated the result can be quite startling.

However, even if we are able to accurately calculate the cost of lost production, we are still left with estimating the frequency and duration of future breakdowns, before we can come up with a cash flow statement. Accordingly, it is important to have good past records if we are to do any better than guess at a value. Should breakdowns be purely random occurrences, then past records are not going to give us the ability to predict precise savings for inclusion in a sound financial case. They may, however, give a feel for the likely cost when a breakdown happens. At best we could say for example, the likely cost of a stoppage is £8000 per hour, and a likely breakdown duration is

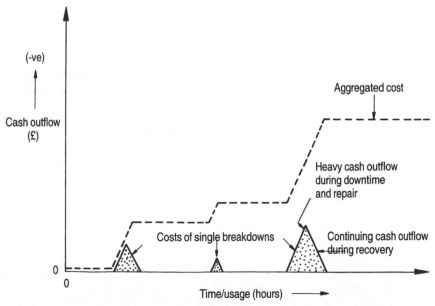

Fig. 19.1 Typical cash-flow diagram illustrating the cost of lost production.

going to be two shifts as a minimum. The question senior management then has to face is, 'Are you willing to spend £10 000 on this condition monitoring device or not?'

- **Poor quality product as plant performance deteriorates** As a machine's bearings wear out, its lubricants decay or its flow rates fluctuate, the product being manufactured may suffer damage. This can lead to an increase in the level of rejects, or to growing customer dissatisfaction regarding product quality. Financial quantification here is similar to that outlined above, but can be even less precise as the total effect of poor quality may be unknown. In a severe case, the loss of ISO-9000 certification may take place, and this can have financial implications well beyond any due simply to increased rejection rates.

- **Increased cost of fuel and other consumables as the plant condition deteriorates** A useful example of this is the increased fuel consumption as boilers approach their time for servicing. The cost associated with servicing can be quantified quite precisely from past statistics, or a service suppliers data. The damaging effects of a vibrating bearing or gearbox are, however, less easy to quantify directly, and even more difficult as one realizes that they can have further consequential effects which compound the total cost. For example, the vibration in a faulty gearbox could in turn lead to rapid wear on clutch plates, brake linings, transmission bushes or conveyor belt fabric. Thus the component replacement costs rise, but maintenance records will not necessarily relate this to the original gearbox defect. Figure 19.2 shows how

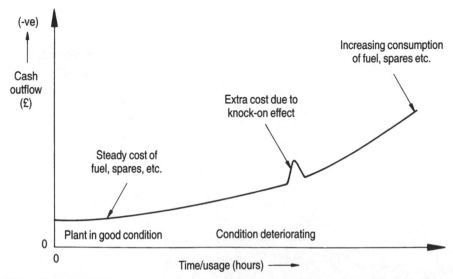

Fig. 19.2 Typical cost of deterioration in plant condition.

the cost of deterioration in plant condition rises as the equipment decays, with the occasional sudden or gradual increases as the consequential effects add to overall costs.

- **Cost of current maintenance strategy** The cost of a maintenance engineering department as a whole should be fairly clearly documented, including wages, spares, overheads and so on. However, it is usually very difficult to break this down into individual plant items and virtually impossible to allocate an accurate proportion of this total cost to a single component's maintenance. In addition, the general picture will be one of a steadily rising cost in respect of routine maintenance on the plant as the equipment deteriorates with age, and needs more careful attention to keep it running smoothly. Figure 19.3 outlines the cost of a current planned preventive maintenance strategy and shows it to be a steady outflow of cash for labour and spares, increasing as the plant ages.

If condition monitoring is to replace planned preventive maintenance, there may be a considerable saving in the spares and labour requirement for the plant which may be found to be 'over-maintained'. This is more common than one might expect, because there has always been a strong feeling in maintenance that regular prevention is much less costly than a serious breakdown in service. Unit replacement at weekends or during a stop period is not reflected in lost production figures, and the cost of stripping and refurbishing plant is often lost in the maintenance department's wage budget for the year. In other words, the cost of planned preventive maintenance on plant and equipment can be a constant drain

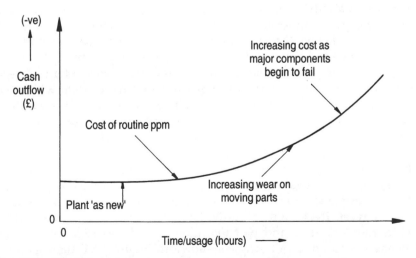

Fig. 19.3 Typical cost of a current preventive maintenance strategy.

on resources which goes undetected. Accordingly, it should really be made available for comparison with the cost of monitoring the unit's condition on a regular basis and applying corrective measures only when needed.

19.4   JUSTIFYING CM AT PRESENT

In general, the cost of any current maintenance position is largely vague and unpredictable. This is true, even if enough data is available to estimate past expenditure, and allocate this precisely to a particular plant item. Thus, if we are to make any sense of financial justification, we must somehow overcome this impasse. The reduced cost of maintenance is usually the first factor which a financial manager looks at when we present our case, even though the real but intangible savings come from reduced downtime. Ideally, past worksheets should give the aggregated maintenance hours spent on the plant. These can then be costed pro rata against total manhours. Similarly, the spares consumption recorded on the worksheets can be multiplied by unit costs. The cost of the maintenance strategy for the plant will then be the labour cost plus the spares cost plus an overhead element.

Unfortunately, the nearest we are likely to get to a value for maintenance overheads, will be to take the total maintenance department's overhead value and multiply it by the plant's maintenance labour cost, divided by the total maintenance labour cost. Even if we manage to arrive at a satisfactory figure, its justification will be queried if we cannot show it as a tangible saving, either due to reduced manning levels in the maintenance department, or through reduced spares consumption, which would also be acceptable as a real saving. The estimates will need to be aggregated, and grouped according to how they can be allocated, for example, whether they are downtime-based, total cost per hour the plant is stopped, frequency-based, recovery cost per breakdown, or general cost of regaining customer orders and confidence after failure to deliver. By using these estimates, plus the performance data which has been collected, it should then be possible to estimate the cost of machine failure and poor performance during the past few years or months. In addition, it should also be possible to allocate a probable saving if machine performance is improved by a realistic amount.

It may even be possible to draw up the traditional cash flow diagram showing expenses against savings and the final breakeven point, although its apparent precision is much less than the quality of the data would suggest. If we aggregate the graphs for the cost of the current maintenance situation, and plot that alongside the expected costs after installing condition monitoring, as shown in Figure 19.4, then the area between the two represents the potential savings. Figure 19.5 on the

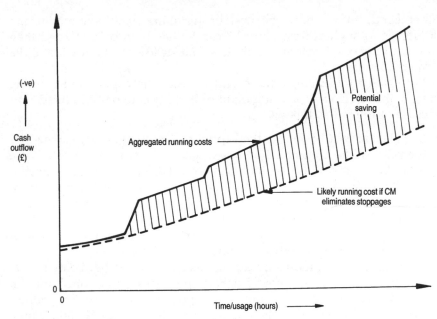

Fig. 19.4 Typical potential savings produced by use of condition monitoring.

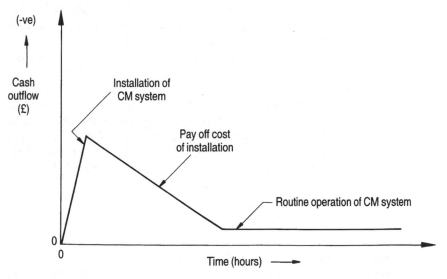

Fig. 19.5 Typical cost of condition monitoring installation and operation.

other hand, shows how the cost of installing condition monitoring equipment is high at first, until the capital has been paid off, and then the operating cost becoming fairly low, but steady during the life of the CM equipment.

Put against the savings, there will be both the capital and running costs of introducing a condition monitoring project to be considered, and these are outlined as follows.

- **Installation cost**   Some of the capital cost will be clearly defined by the price of the equipment and any specialist installation cost. There may also be preliminary alterations required, such as creating access, installing foundations, covering or protection, power supply, service access and so on. Some or all may be subject to development grants or other financial inducement, as may the cost of consultancy before, during, or after the installation. This could well include the cost of producing a financial project justification.

  The cost of lost production during installation may be avoided if the equipment is installed during normal product changes or shutdown periods. However, in a continuous process this may be another overhead to be added to the initial capital investment. Finally, it may be necessary to send staff on a training course which has not been included in the price of the equipment. The cost of staff time and the course itself may be offset by training grants in some areas and this should be investigated. It is also possible that the vendor will offer rental terms on the CM equipment, in which case the cost becomes part of the operating rather than the capital budget.

- **Operating cost**   Once the unit has been installed and commissioned, the major cost is likely to be its manpower requirement. If the existing engineering staff have sufficient skill and training, and the improved plant performance reduces their workload sufficiently, then operating the equipment and monitoring its results may be absorbed without additional cost. In our experience, this time-saving factor has often been ignored in justifying the case for improved maintenance techniques. However, in retrospect, it has proved to be one of the main benefits of installing a computer-based monitoring system.

For example, a cablemaker found that his company had increased its plant capacity by 50% during the year following the introduction of computer-aided maintenance. Yet the level of maintenance staff needed to look after the plant had remained unchanged. This amounted to a 60% improvement in overall productivity. Another example of this effect was a drinks manufacturer who used AMC's computerized scheduler to change from time-based to usage-based maintenance. This was done because demands on production fluctuated rapidly with changes in the weather. As a result, the workload on the maintenance trades fell so far that they were able to maintain an additional production line without any manning increase at all.

If these savings can be made by better scheduling, how much more improvement in labour availability there would be if maintenance could be related to a measurable plant condition, and the servicing planned to coincide with a period of low activity in the production or maintenance schedule? So, the ongoing cost of manpower needed to run the condition monitoring project must be assessed carefully, and balanced against the potential labour savings as performance improves. Other continuing costs must also be considered, such as the fuel or consumables needed by the unit. However, these are normally small and recent trends have shown that consumable costs tend to decrease as more and more companies turn to this type of equipment.

Combining the above initial costs and savings should result in an early outflow of cash investment in equipment and training, but this soon crosses the break-even point within an acceptable period. It should then level off into a steady profit, which represents a satisfying return on the initial investment, as reduced maintenance costs, plus improved equip-

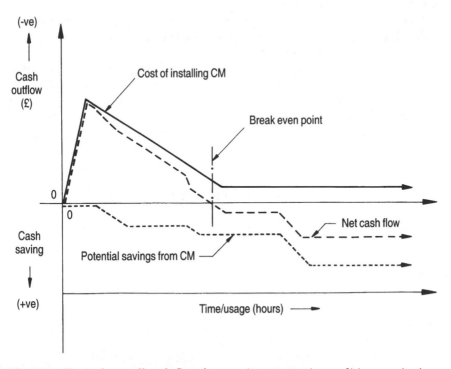

Fig. 19.6 Typical overall cash flow from an investment in condition monitoring.

ment performance, are realized as overall financial gains. Figure 19.6 indicates how the cash flow from investment in condition monitoring moves through the breakeven point into a region of steady positive financial gain.

19.5 CONCLUSIONS

In conclusion, it is possible to say that the financial justification for the installation of any item of condition monitoring equipment should be based on a firm business plan, where investment cost is offset by quantified financial benefits. However, the vagueness of the factors available for quantification, the lack of firm tangible benefits and the financial environment in which maintenance engineers operate all conspire to make the construction of such a plan very difficult indeed.

Until the engineer is given the facilities to collect and analyse performance data accurately and consistently, until the engineering and manufacturing departments are integrated under a precise standard value costing system, and until the maintenance engineering function is given the status of a profit centre, then financial justification will never become the precise science it should be.

Instead, the more normal process is one in which an engineer makes a decision to install a condition monitoring system, and then backs it up with precise-looking figures based on very imprecise data. Fortunately, once the improved system has been given the OK, it is only rarely that any thought is given to monitoring its performance against that estimated in the original business plan. This is largely because the financial values or benefits achieved are even more difficult to extract and quantify in a post-installation audit, than those in the original business plan.

19.6 REFERENCE

Eade, G. (1989) Costing maintenance, *Maintenance*, 4(1), 30–1.

CHAPTER 20

Technique selection and implementation in condition monitoring

P.R. Drake

Systems Division, School of Engineering,
University of Wales Cardiff (UWC), PO Box 688, Queens
Buildings, Newport Road, Cardiff, CF2 3TE, UK

20.1 INTRODUCTION

The selection of suitable condition monitoring (CM) methods for a particular application can be either a bottom-up, or a top-down process. The bottom-up approach would say:

These are the available methods – which one is appropriate?

To formalize this approach, the Analytic Hierarchy Process (AHP) is presented in this chapter as a case study (Drake and Harrison, 1993). While a more systematic top-down approach is also demonstrated by use of a Failure Modes and Effects Analysis (FMEA), followed by method inference (Drake *et al.*, 1995). This is based on the derived effects, and by maximizing fault coverage while minimizing the number of signals/system access points – the classic analogue systems test objective.

In addition to the above case studies, this chapter also presents a summary of the relative merits of the four commonly used 'general-purpose' methods of condition monitoring. That is the family of techniques which sit under the headings of 'Visual Inspection', 'Vibration Signal Analysis', 'Fluid and Wear Debris Analysis', and the more

Handbook of Condition Monitoring
Edited by A. Davies
Published in 1998 by Chapman & Hall, London. ISBN 0 412 61320 4.

recently introduced method of 'Acoustic Emission/Stress Wave Sensing' which is becoming increasingly popular. Table 20.1 provides a summary of the CM methods available and their applications. This table is reproduced here by kind permission of the British Institute of Non-Destructive Testing and is an extract from their yearbook.

20.2   CONDITION MONITORING METHODS

The visual inspection of machinery using the unaided eye or some form of optical assistance is arguably the simplest and most cost-effective method of condition monitoring. Indeed, the continuous use of all human senses should form an integral and formal part of any condition monitoring programme. The main advantage of visual inspection is that it can be very cheap, and free if carried out by an operator using the unaided eye. It is also direct, immediate, non-contact, remote and requires relatively low levels of skill, depending on the particular method used.

Its application may be limited to stationary equipment – particularly when using the unaided eye. However, methods are available for visual inspection of moving parts, e.g. stroboscopic techniques. Visual inspection generally requires a more direct line of access to the monitored components than is required by other methods. However, probe type systems using borescopes and fibrescopes can be used to access otherwise inaccessible components, and techniques such as infrared thermography and radiography can be used to monitor 'hidden' internal components as shown in Chapters 3, 4, 5 and 6.

Although visual inspection may require more direct access than other techniques, it has the great advantage that it does not require physical sensor contact, and can be performed in a very remote manner. For example, thermal imaging can be used for the accurate inspection of small components at long distances. This has been of particular advantage in the online monitoring of high-voltage equipment such as transformers, bus-bars and overhead power lines where for safety, as well as practical reasons, it is a great advantage to be at some distance away from the monitored system.

Vibration signal analysis is well supported by equipment that is readily available and relatively cheap. Its implementation benefits from the use of non-intrusive transducers which can be packaged as small hand-held portable devices. This, combined with the feature that, once mastered, it can be used in a wide range of applications, means that many measurements may be taken usefully at several access points, resulting in wide fault coverage for very little equipment cost. The utility of the method obviously has great organizational advantages, since the relatively low cost of training and equipment can give an excellent payback. It also reduces the need for other CM methods to be used, which in turn lowers the total skill level and training required by maintenance personnel. In addition, it decreases the complexity of the overall CM system (Drake and Buck, 1991).

These advantages are further enhanced by the method's well-proven effectiveness, the relative ease with which vibration signals can be interpreted, and the reliability of the results. Permanently mounted sensors can be used in on-line applications and in hazardous or inaccessible environments. The sensors are easily retro-fitted, maintained and repositioned due to being small and non-intrusive in their implementation. The latter characteristic also means that they are very unlikely to interfere with the performance or reliability of the monitored system. This is not necessarily the case with other condition monitoring methods requiring intrusive devices such as flow sensors.

Manufacturers produce a wide range of accelerometers. The principle selection criteria being sensitivity, mass, frequency response, dynamic range and connector position. Special purpose types are available for applications such as simultaneous measurement in three planes, high-temperature environments, very low vibration levels, and high-level shocks. Miniature types are also produced for mounting on delicate structures such as printed circuit boards and for high-frequency measurements. Usually, however, a small group of relatively cheap general purpose types will normally satisfy most needs. The piezoelectric accelerometers typically used to measure vibration produce a small output charge that requires an amplifier to be positioned close to, or built into, the sensor. This proximity requirement may exclude this type of sensor from some hazardous applications (Webborn, 1982).

In contrast, the fluid and wear debris analysis approach to condition monitoring is restricted to circulatory arrangements such as lubrication or hydraulic power systems. Where it can be applied, a wide range of items may be monitored, such as bearings, gears, cams, tappets, piston rings, cylinders and seals. If excessive wear debris are present in a fluid, then a fault or abnormal condition has already started to develop. All that can be done, if this is the case, is to take action which will avoid or minimize any further damage, and subsequently after diagnosis to correct the fault and its cause.

However, monitoring the physical condition of the fluid and checking for the presence of contamination effectively does go a step further, since it can predict faults before they occur by detecting the conditions which will cause increased wear and reduced component life-cycles. Furthermore, if the fitness of the oil is being monitored, then a condition-based oil changing policy can be implemented according to additive depletion, contaminant build-up and/or other changes in its physical state. This feature is a major financial benefit of the fluid analysis method of condition monitoring.

Table 20.1 Condition monitoring method selector

	Vibration analysis	Noise analysis	Acoustic emission	On-line debris monitoring	Debris analysis	On-line oil condition monitoring	Oil condition analysis
Bearings	X	X		X	X	X	X
Belts							
Blowers/pans	X	X					
Boilers/heat exchangers			X				
Brazing/welding equipments							
Casting/forging machines							
Compressors/pneumatic machines	X	X		X	X	X	X
Couplings	X	X					
Guillotines/cutting machines	X	X		X	X	X	X
Earthmoving/excavating plant	X	X		X	X	X	X
Electric motor generation	X	X					
Elevators/hoppers/conveyers	X	X					
Escalators	X						
Filters/separators/valves				X	X	X	X
Gearboxes	X	X		X	X	X	X
Vacuum equipment	X	X		X	X	X	X
Incinerators/furnaces/autoclaves				/	/	/	/
Internal combustion engine				X	X	X	X
Loaders/stackers				X	X	X	X
Machine tools mechanical	X	X		X	X	X	X
Machine tools hydraulic	X	X		X	X	X	X
Pressure vessels/accumulators			X				
Pumps	X	X		/	/	/	/
Structural/rigging	X						
Transformers						X	X
Turbines/aero engines	X	X		X	X	X	X
Wire/cable making							
Winding/lifting machinery			X				

In modern fluid power systems, running clearances are being made as small as possible to achieve maximum efficiency, power and to reduce the leakage caused by increased system pressures. This has resulted in a demand for increased levels of fluid cleanliness, which in turn has produced a need to monitor for particle contamination so as to ensure a low pollution value. Stringent maximum permissible contamination levels are being applied to fluid systems to achieve greater reliability.

Water in oil detection	Electric motor insulation and winding monitoring	Optical detection systems	Optical alignment systems	On-line pressure monitoring	On-line temperature monitoring	Thermal imaging	Stress/strain analysis	Erosion/corrosion monitoring	Performance monitoring	Orifice restriction monitoring
		X			X	X		X		
							X		X	
	X		X	X	X	X			X	/
				X	X	X	X	X	X	
	X	X		X					X	
					X	X			X	
X	X		X	X	X	X			X	
			X				X		X	
X	X			X			X		X	
X					X	X		X	X	
	X				X	X			X	
	X				X	X		X	X	
	X						X	X		
X				X				X	X	X
X		X	X	X	X	X	X	X	X	
X				X	X	X		X	X	
/		X		X	X	X		X	X	
X		X		X	X	X		X	X	
X	X			X				X	X	
X					X	/			X	
				X	X	/			X	
				X	X	X	X	X	X	
/				X	X	X		X	X	/
							X	X		
X	X			X	X	X		/	X	
X		X	X	X	X	X		X	X	
							X		X	
							X		X	

These levels are achieved by appropriate filtration of the fluids and this critical filtering is now subject to condition monitoring using 'post-filter' debris analysis.

The usual criteria by which a condition monitoring method is selected include cost, ease of use, degree of technical expertise required, accuracy and repeatability. However, there are two other criteria which are particularly important in the selection of a fluid analysis method. The first of these is the speed of turnround of the analysis, with the fastest methods operating in an on-line mode. Some of the more sophisticated and well-established techniques, however, require off-line analysis, possibly even off-site at specialist laboratories. In general, therefore, there is a trade-off between the analysis turnround time and the amount of information provided by fluid analysis methods.

On-site analysis will require the provision of greater on-site skills and also the purchase of appropriate evaluation equipment. While the use of computerized analysis systems may help in the implementation of very sophisticated evaluation techniques. The other critical selection criterion relates to:

'What the analysis method can provide information on'.

For example, some methods can only detect and quantify magnetic particles in a fluid, while others give a detailed analysis of a wide range of elements which may be present. Also, different methods can detect debris over different ranges of particle size. This means that the size and chemistry of the particles to be detected and analysed must be known before a specific method can be selected.

One of the advantages of wear debris monitoring over vibration monitoring is that it can provide a much clearer indication of the source of a fault. The component of a vibration signal caused by a fault can be masked by the vast array of large values emanating from a perfectly healthy machine. This is in contrast to wear debris monitoring, where only the unhealthy machine components contribute excessive amounts of particulate waste. Another advantage is that the path of the debris from source to sampling point is much clearer than that of a vibration signal, and this makes it easier to determine where the 'fault indication' is coming from. It also has the further advantage, that its measurements can be independent of the speed and load at which the monitored machine is operating.

Another key advantage of fluid analysis is that, by monitoring contamination and fluid state, it can detect the conditions which will cause severe wear, or malfunction, before such effects have taken hold. That is, it is a fault predictor as well as a failure detector. From a practical point of view, it can be argued that vibration monitoring is easier to implement in terms of fundamental signal acquisition. Together with the fact that vibration monitoring lends itself more readily to online, real-time monitoring. Also, it can be difficult or impossible to achieve a detailed fault diagnosis when using the currently available online methods of fluid and wear debris analysis. In addition, another big disadvantage with wear debris monitoring is that it can only monitor those machine components that are 'washed' by a fluid. This restriction means that it may not provide as wide a fault coverage as vibration monitoring.

Accordingly, the condition monitoring industry is increasingly taking the view that fluid and vibration analysis should be seen as complementary techniques rather than competitors. Several authors have been cited who endorse this view in the light of practical experience (Hunt, 1993), with the use of both techniques providing a much wider fault coverage and corroboratory information for critical decision-making. Having the skills and facilities necessary to apply both techniques can enable users to cater with a much wider range of monitoring situations and in some cases only one of the techniques may be applicable. For example, slow machinery may be more effectively monitored using fluid evaluation. The source of wear debris in a series of hydraulic pumps, connected to the same liquid supply, cannot be pinpointed by fluid analysis.

By comparison, stress wave sensors for acoustic monitoring are very easy to install, not only due to their non-intrusive implementation, but also because their precise positioning and orientation is usually unimportant. This is due to multiple reflections of the stress waves within a structure resulting in an even spread of the sound. This is in complete contrast to the use of accelerometers for vibration monitoring, where the positioning and orientation of the sensor can be critical to the successful detection and diagnosis of faults.

The stress-wave sensing method is based on the measurement of high-frequency stress waves using resonant sensors. These have the inherent advantage of rejecting unwanted audible quiet sounds and low-frequency mechanical vibrations. A problem which would otherwise mask the acoustic emission information of interest and result in an extremely poor signal to noise ratio. However, the incoming signal from the stress wave sensor cannot be intensified, and unfortunately the noise may not be reduced by signal averaging as the information is changing with time. Therefore, it is important that the transducer and preamplifier combination produce the minimum of background electronic noise, and care must be taken to reduce ambient acoustic sound and radio frequency interference.

Although the precise positioning and orientation of a stress-wave sensor is not important, it must be remembered that large geometry changes and interfaces within the machine assembly form a partial barrier to the propagation of structure-borne sound. Therefore, on a macro scale, careful consideration needs to be given to the positioning of the sensor. Furthermore, the range of detection of structure borne sound is restricted at high frequencies. Thus, in order to characterize a defect source fully in terms of its time scale and/or frequency content, the transducer bandwidth must cover the source bandwidth. This can be a difficult condition to fulfil because source bandwidths extend typically from direct current (DC) to 10 megaHertz (mHz).

20.3   CM METHOD SELECTION USING AHP

From the above discussion, it is clear that the condition monitoring method selection process is a critical step in the successful implementation of machine surveillance. Therefore it is essential that the process is systematic, formalized and transparent, such that it is amenable to detailed analysis for the purposes of verification, sensitivity evaluation and optimization. To achieve this, the Analytic Hierarchy Process can be used (Saaty, 1980). This is a technique which has been applied as a selection process in many different fields (Golden *et al.*, 1989), although its application within engineering has not been so widely reported. The technique is demonstrated here through its application to the selection of a condition monitoring method for a hydraulic pump.

The hydraulic pump to be monitored is a positive displacement gear pump that operates the machine tool eject mechanism, toolchanger jaw swivel, pallet change clamps/locks and vertical Z axis clamping on a Wadkin V4-6 vertical milling machine. It should be noted that the different CM methods considered here were actually tested in a laboratory, (Harrison, 1992). Also, it should be stressed that the relative merits of any particular method will depend upon the intended application. Therefore, the results produced here are not claimed to be true for all hydraulic pump monitoring applications. So the primary objective of this example is to demonstrate the application of AHP to the formalization of the condition monitoring selection process only.

Four signals were assessed in the time and frequency domains for condition monitoring of the hydraulic pump (Hoh *et al.*, 1990) these were:

- **pump outlet pressure:** the signal ripple was sensed by mounting a strain gauge on the outlet side of the pump;
- **vibration:** axial and radial vibration were sensed by a piezo-electric accelerometer; the sensor was screwed to a magnet and mounted on the outside casing of the pump;
- **motor current:** the current being drawn by the motor driving the pump was sensed by a current probe clipped to the lead carrying the supply; this avoided the need for any dismantling;
- **acoustic emission:** a stress wave sensor was bolted to the top of the pump.

The AHP technique consists of four steps as follows:

- Decide upon the criteria for selection.
- Decide upon the relative importance of these criteria.
- Decide how well each potential choice of condition monitoring method, relative to each other choice, meets each criterion.

- Combine the information derived in steps 2 and 3 to obtain a relative rating for each potential choice.

To describe the AHP technique in detail, and its application to the selection of a condition monitoring method for the hydraulic pump, the process is now presented below. Four selection criteria were considered to be relevant to this particular application.

1. **Signal usefulness** This relates to how much useful CM information, including fault diagnostics was obtained from the fundamental signals used in the CM methods when they were tested in the laboratory. It also relates to the signals' resistance to outside interference, and increases in pump operating pressure or changes in operating point.
2. **Ease of maintenance** Any CM method will require maintenance of the associated equipment. This criterion is particularly important for an online application, as considered here.
3. **Ruggedness** Many applications require CM methods to work in harsh environments. The relative importance of this criterion will obviously depend on the particular application environment. In the application considered here, it is deemed to be important because the target environment is the factory floor, where coolant and swarf contamination is possible.
4. **Sensor mounting** This criterion relates to the ease with which the necessary sensor can be mounted and accessed on the pump within its target environment. It takes into consideration the size and weight of the sensor and the physical space into which it must be placed. The motor current sensor is awkward to place due to its bulky size relative to the small space into which it had to fit. The acoustic emission sensor's size causes some problems, as it is built onto an aluminium board, and this means that it can only be fixed to certain bolts. The vibration and pressure sensors are the smallest and least bulky items, and consequently they cause the fewest size and accessibility problems.

The first step in the AHP technique is the pair-wise comparison of the selection criteria to establish their relative importance. For each pairing the user is asked:

Which is the most important criterion?

The answer to this question is a relative importance weighting on a scale between 1 (equal importance) and 9 (absolutely more important), for the more important criterion. The less important criterion is given a weight equal to the reciprocal of the latter value. The results of this pair-wise comparison are recorded in a matrix as shown in Table 20.2. Each weighting in the matrix records the importance of the corresponding row criterion with respect to the corresponding column criterion.

Table 20.2   Pair-wise rating of selection criteria

Criterion	Signal usefulness	Ease of maintenance	Ruggedness	Ease of mounting
Signal usefulness	1	3	5	6
Ease of maintenance	1/3	1	3	5
Ruggedness	1/5	1/3	1	4
Ease of mounting	1/6	1/5	1/4	1
Column sum	1.70	4.53	9.25	16.00

For example, in this particular application, ease of maintenance is considered to be more important than ease of mounting with a weight of 5. Ease of mounting is considered to be less important than ruggedness with a weight of 1/4. The weightings in Table 20.2 are then normalized by dividing each entry in a column by the sum of all the entries in that column, so that the entries in a column sum to one. Following normalization, the weights are averaged across the rows to give an average weight for each criterion. The normalized ratings are given in Table 20.3.

The next step is the pair-wise comparison of the CM methods to evaluate how well they satisfy each of the criteria. For each pairing of the CM methods within each criterion, the user is asked to decide which method best satisfies the criterion. The better method is awarded a rating on a scale between 1 (equally good) and 9 (absolutely better). The other method in the pairing is awarded a rating equal to the reciprocal of this. The results are then recorded in a matrix for each criterion.

The results for the 'signal usefulness' criterion are given in Table 20.4. Each rating in this matrix records how well the method corresponding to its row, meets the 'signal usefulness' criterion when compared to the method corresponding to its column. For example, the pump outlet pressure was found to be a far more useful CM signal than acoustic emission. Consequently, a score of 5 is entered in Table 20.4 at the intersection of the row for pump outlet pressure and the column for acoustic emission.

Table 20.3   Normalized pair-wise rating of selection criteria

Criterion	Signal usefulness	Ease of maintenance	Ruggedness	Ease of mounting	Row average
Signal usefulness	0.588	0.662	0.541	0.375	0.541
Ease of maintenance	0.196	0.221	0.324	0.313	0.263
Ruggedness	0.118	0.074	0.108	0.250	0.137
Ease of mounting	0.098	0.044	0.027	0.063	0.058
Column sum	1.000	1.000	1.000	1.000	1.000

Table 20.4 Pair-wise rating of alternative CM methods with respect to signal usefulness

CM method	Pump outlet pressure	Motor current	Vibration	Acoustic emission
Pump outlet pressure	1	4	2	5
Motor current	1/4	1	1/2	3
Vibration	1/2	2	1	3
Acoustic emission	1/5	1/3	1/3	1
Column sum	1.95	7.33	3.83	12

The ratings in these matrices are normalized, as before, by dividing them by their respective column totals, so that the column totals for the normalized ratings are equal to one. The normalized ratings are then averaged across the rows to give an average normalized rating by criterion for each CM method. Table 20.5 gives in detail the normalized ratings for the 'signal usefulness' criterion. Table 20.6 gives the average normalized ratings by criterion for all of the methods.

The final step is to combine the average normalized CM method ratings, given in Table 20.6, with the average normalized criterion weights, given in Table 20.3, to produce an overall rating for each CM method. That is, the extent to which the methods satisfy the criteria are weighted according to the relative importance of the criteria. This is done as follows:

$$a_j = \Sigma_i(w_i k_{ij}) \tag{20.1}$$

where: a_j = overall relative rating for CM method j
          w_i = averaged normalized weight for criterion i
          k_{ij} = averaged normalized rating for CM method j with
                respect to criterion i.

Table 20.5 Normalized pair-wise rating of alternatives with respect to signal usefulness

CM method	Pump outlet pressure	Motor current	Vibration	Acoustic emission	Row average
Pump outlet pressure	0.513	0.545	0.522	0.417	0.499
Motor current	0.128	0.136	0.130	0.250	0.161
Vibration	0.256	0.273	0.261	0.250	0.260
Acoustic emission	0.103	0.045	0.087	0.083	0.080
Column sum	1.000	1.000	1.000	1.000	1.000

Table 20.6 Average normalized ratings of CM methods with respect to each criterion

	Signal usefulness	Ease of maintenance	Ruggedness	Ease of mounting
Pump outlet pressure	0.499	0.137	0.074	0.170
Motor current	0.161	0.067	0.284	0.069
Vibration	0.260	0.515	0.471	0.455
Acoustic emission	0.080	0.281	0.171	0.306

Table 20.7 gives the results of this final step in the AHP technique. These results show clearly that vibration analysis and pump outlet pressure are the preferred CM methods for this particular hydraulic pump application. Having established the above tables in a computer package such as a spreadsheet, it is a simple matter to try out 'what-if' scenarios, i.e. the weights and ratings awarded by the user can be modified and the results automatically updated. This will allow a sensitivity analysis to be performed if required. In the particular application presented here, the obvious next step would be to verify the ratings awarded to pump outlet pressure and vibration analysis. Also, the weights of any criteria for which only one of the methods gets a high or low rating should be verified.

20.4 CM METHOD SELECTION DURING DESIGN

If condition monitoring is to be a 'designed-in' feature of any new machine system, then a true 'systems engineering' approach to Data Acquisition System (DAS) design must be adopted. This means that a wide or global view is taken to ensure that a totally appropriate and complete solution is provided. In the first instance, condition monitoring methods should be chosen according to their ability to detect and

Table 20.7 Overall CM method ratings

CM method	a_i
Pump outlet pressure	0.326
Motor current	0.148
Vibration	0.367
Acoustic emission	0.158

diagnose the potential faults in the specific monitored system, together with providing as wide a fault coverage as possible, in order to reduce the total number of access points required. It is very subjective, and therefore not good systems engineering practice, to compare and contrast condition monitoring methods directly, without any reference to, or analysis of, the specific system to be monitored.

The condition monitoring method selection process should start with a technical audit of the system to be monitored, to establish the presence and functionality of all its sub-systems and the components therein. Each sub-system is then subjected to failure modes and effects analysis to establish what is to be detected. From this is determined the CM methods, and the corresponding signals/transducer positions within the machine that will detect the potential failures and their effects.

Having determined the signals to be acquired and processed, the hardware and software requirements of the CM system can then be defined. To demonstrate this approach it is applied here to the hydraulic tool changing system on a Heckler and Koch BA25 vertical machining centre. This particular example is taken from the results of a European Union project (BREU-CT91-0463), funded under the BRITE initiative, and entitled 'Machine Management for Increasing Reliability Availability and Maintainability–MIRAM' (Jennings and Whittleton, 1993).

A Bosch hydraulic power unit (Figure 20.1) supplies oil at high pressure to actuators in the tool-changing system (Figure 20.2).

The unit consists of a reservoir into which is submerged a vane type pump driven by a directly coupled electric motor and pressure control valves. A contactor that is energized by the Computer Numerical Controller (CNC) controls the supply of three-phase power to the motor. Within the pump, the eccentricity of the annulus, which contains the 13 vanes and hub, can be altered by the pump outlet pressure.

The associated control valves in the hydraulic unit limit the pressure to 115 bar and the flow to 3 litres/min. The hydraulic unit also contains an accumulator which is filled with nitrogen at 100 bar. This can store up to 0.7 litres of oil when the accumulator is pressurized above 100 bar. Oil from the actuators is returned to the tank via a filter and/or bypass one-way valve. The hydraulic system is continuously pressurized, although the pressure oscillates due to leakage through the many valves and the opening and closing of the pressure relief valve.

The machine tool has three axes of interest, X and Y for the table and Z for the spindle. To change a tool, the Y and Z axes are moved to their tool change positions. The Z axis is moved upwards, to locate the vertically oriented gripper around the tool to be used in the next machining sequence. Solenoid 8-Y5 is then energized so that hydraulic pressure is applied to the tool clamp and the gripper pliers are closed.

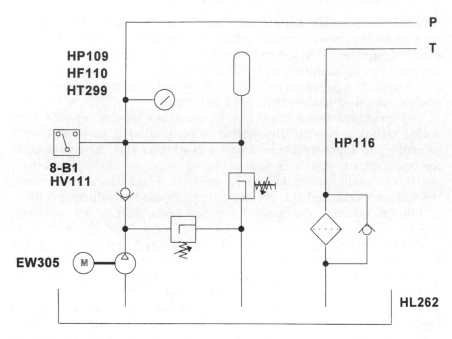

Fig. 20.1 The hydraulic power unit.

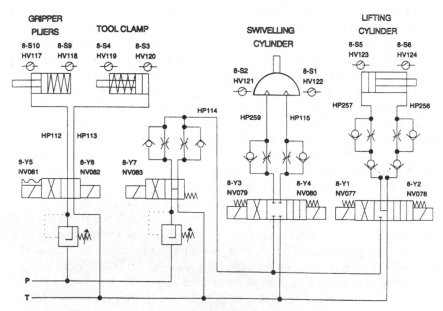

Fig. 20.2 The hydraulic tool changing system.

The tool clamp consists of a stack of 'Belleville' washers (a spring), that exert a 12 kiloNewton (kN) force. Ejection of the tool occurs after the gripper pliers have closed. Next, solenoid 9-Y1 in the pneumatic system is energised to unlock the tool from the tool magazine.

Following this, solenoid 8-Y2 is energized, allowing hydraulic pressure to be applied to the lifting cylinder. If the CNC programme defines the tool as 'large', then the controller energizes solenoid 8-Y7 to slow the changer, by bringing in a reducing circuit that lowers the hydraulic pressure. The lifting cylinder lowers the attached swivelling cylinder and gripper pliers away from the tool spindle. This action also removes the next tool to be used from the tool magazine. Next, solenoid 8-Y3 or 8-Y4 is energized allowing hydraulic pressure to be applied to the swivelling cylinder. This rotates to either the zero degree or 180 degree position, depending upon which solenoid is energized.

Following this, solenoid 8-Y1 is energized, so that hydraulic pressure is applied to the lifting cylinder to raise the tools into the spindle and magazine. The tool is now locked into the magazine by the energizing of solenoid 9-Y2. Finally, solenoid 8-Y6 is energized so that the tool clamp is released and the gripper pliers opened. Now, the Z axis can be moved from the tool-change position. This tool-changing process takes typically 5.5 seconds, but each individual activity may take up to 3 seconds before the computer numerical control reports and issues an error message.

A detailed failure modes and effects analysis for the hydraulic system outlined above is given at the end of this chapter. While Table 20.8 summarizes the potential failure modes, and the strength of their effects on six simple and measurable system performance parameters that together give complete fault coverage. In this table the strength of an effect is defined on a scale between 1 strong and 3 weak, with a blank indicating no effect. One of the measurements used is 'WD = work done to move an actuator' which is defined as follows:

$$WD = \int_0^T P(t) \cdot V(t)\, dt \qquad (20.2)$$

where: $P(t)$ = hydraulic pressure at time t;
$V(t)$ = volume of flow at time t;
T = total time to move actuator.

This is approximated in a sampled data system by:

$$WD = \sum_{n=0}^{T} P_n \cdot V_n \cdot \delta t \qquad (20.3)$$

where: P_n = hydraulic pressure at sample time n;
V_n = volume of flow at sample time n;
δt = sampling interval;
T = total time to move actuator.

Table 20.8 Failure modes and strength of effects for hydraulic system

Component/Failure	Effect					
	T	P_{mn}	P_{max}	WD	FL	MP
Entire system						
● leakage					1	
Reservoir						
● low oil level					1	
● oil viscosity	3	2		2		3
Motor and pump						
● supply voltage or frequency		3		2		1
● restricted flow		2	3	2		2
Pressure relief valve						
● opening pressure			1			2
One-way valve						
● restricted flow	1	1	2	1		2
Accumulator						
● pressure not stored	1	1	2	2		3
Pressure relief valve with manual unload						
● opening pressure			1			2
Filter and one-way valve						
● filter blocked		1		1		2
● no restriction		1		1		2
Piping						
● restricted flow	1	2		1		2
Solenoid valves						
● low voltage	1	2		1		
● slow operation	1	1		1		
● restricted flow	1	2		1		3
Flow restrictors and one-way valves						
● restricted flow	1			2		3
● no reduction	1			2		3
Pilot operated one-way valves						
● restricted flow	1			2		3
● slow operation	1			2		
Gripper pliers/tool clamp						
● string stiffness	1	3		2		
● friction	1	3		2		
Swivelling cylinder						
● inertia	1	3		2		
● friction	1	3		2		
Lifting cylinder						
● load	1	3		2		
● friction	1	3		2		

T–time for an actuator to move;
P_{mn}–mean hydraulic pressure during movement of actuator;
P_{max}–maximum hydraulic pressure;
WD–work done to move an actuator;
FL–level of fluid in reservoir;
MP–apparent motor power.

Additional sensors are not required to measure the time for an actuator T to move since this can be done using existing control signals. In particular, the energizing of a solenoid marks the start of a movement and the activation of a switch marks the end. The most appropriate position for the pressure P_{mn} and P_{max} transducer is at the output of the hydraulic unit, signal HP109 in Figure 20.1. At this position the pressure in different branches of the circuit can be measured according to the solenoid settings. Furthermore, this is the best position to measure pump pressure ripple, which is known to be a very good signal for health monitoring as shown in Chapter 10 and in Watton (1992).

It should be noted that at positions away from the pump the ripple is distorted. For example, attenuation of the ripple is caused by rubber pipes and the hydraulic fluid. Flow V from the pump and accumulator is represented by signal HF110 in Figure 20.1, which is measured by a flow transducer. While the fluid level FL in the reservoir, is measured by a float-type transducer. This can be affected by liquid surface oscillation due to ripples generated as fluid is returned to the reservoir.

To eliminate this effect from the measurement, the fluid level is averaged over a short period of time. Finally, the motor power MP is measured by a power meter connected to the DIN rail inside the machine electrical cabinet.

To complete the analysis, the following is the FMEA for the hydraulic system by component:

- **Reservoir** An oil level below the pump inlet will allow air to be drawn into the system. This will lower the system pressure and increase the time for the actuators to operate. An increase in oil viscosity will increase resistance to flow and thus increase the cycle times. Likewise, a decrease in viscosity will reduce the cycle times. Leakage from the reservoir will reduce the oil volume, which could allow air to be drawn into the pump.
- **Motor and pump** Reduced supply voltage or frequency, or a damaged winding, may decrease the rotational speed and lower the pump outlet pressure. This will increase the time for the actuators to move. A partially blocked suction pipe will lower the pump outlet pressure due to restricting the flow. Leakage from the pump will return the oil to the reservoir and reduce the output pressure.
- **Pressure Relief Valve** If the set pressure of 115 bar is altered, then the tool clamp holding force will be changed in the same direction. A reduced set pressure will lower the volume of oil in the accumulator. Failure of the valve to open may result in damage to components. Failure of the valve to close will prevent the system's being pressurized. Leakage from the pressure relief valve will increase the cycle times and reduce the pressure in the system.

- **One-way valve** A one-way valve that only opens slightly, or is partially blocked, will restrict the flow and thereby cause actuator response times to increase. This in turn will eventually cause the CNC to 'time-out' as the condition becomes more severe. It will certainly 'time-out' if the valve becomes completely stuck closed. Leakage from the valve will lower the pressure applied to the actuators and increase the flow from the pump.
- **Accumulator** Leakage of nitrogen from the accumulator will prevent it storing hydraulic pressure. Leakage of oil will lower the tank level and reduce the service interval.
- **Pressure relief valve with manual unload** If the set pressure is altered, then the tool clamp holding force will be changed in the same direction. A reduced set pressure will reduce the volume of oil in the accumulator. Leakage of oil will reduce the pressure in the system and increase the tool change cycle time.
- **Filter and one-way valve** As the filter becomes blocked, the flow will be increasingly restricted until the one-way valve opens. If the one-way valve is stuck closed, then the pressure drop across the filter will continue to increase as debris are collected. If the one-way valve is stuck open, then there will not be a pressure drop across the filter.
- **Piping** A partial blockage of the piping will restrict the flow and eventually cause a 'time-out' as it becomes more severe. Leakage from the piping will reduce the pressure in the system and increase the flow from the regulator.
- **Pressure limiting valves** A change in the set pressure to the valve in the gripper pliers and tool clamp circuit will affect the holding force. A change in the set pressure to the reducing circuit will affect the time for the swivelling and lifting cylinders to operate. A rise in the pressure drop across these valves will increase the time for the actuators to move.
- **Solenoid valves** If the supply voltage changes, then the time for the solenoid valve to operate will change. As the voltage reduces it will eventually result in a CNC 'time-out'. A resistance or inductance change in the coil will alter the current. This will change the operating time for the solenoid. If the current increases enough, a fuse will blow and the CNC will produce an error message. An increase in friction between the shuttle and the sleeve will increase the operating time of the solenoid valve, and eventually cause a 'time-out' as it becomes more severe. A damaged spring in the solenoid valve will restrict the movement of the shuttle, thus increasing the pressure loss across the valve. This will result eventually in a 'time-out'. For centrally biased valves, the shuttle may not return to the correct position, possibly resulting in a leakage across the valve. If the detented solenoid is damaged, then the shuttle could unexpectedly move, resulting in

damage to the tool, the workpiece or the machine. Leakage from the valve will reduce the pressure applied to the actuators.

- **Flow restrictors and one-way valves** A partial blockage of the flow restrictor will increase the time for an actuator to move. A partial blockage of a one-way valve will increase the operating time for an actuator. A one-way valve that is jammed open will result in a reduction in the operating time for an actuator. Leakage from a flow restrictor or one-way valve will reduce the pressure applied to the actuators.
- **Pilot operated one-way valves** A blockage of a pilot line to a one-way valve will prevent the valve operating, resulting in a 'time-out' and the CNC producing an error message. Partial blockage of a pilot line to a one-way valve will make the valve open more slowly, and thus increase the operating time for the lifting cylinder. If the pilot operated one-way valve is jammed open, then leakage within the lifting circuit will cause the cylinder to move downwards.
- **Gripper pliers** A change in the spring stiffness, or the friction between the sliding surfaces, will affect the time for the actuator to move.
- **Tool clamp** Failure of a 'Belleville' washer will lower the force required to eject the tool-holder. If a claw fails, then the tool will not be held in the correct position. Leakage from a seal will allow the oil to escape, either through the draw bar or down the inside of the spindle. Both cases will result in oil being blown out during a tool change due to activation of the blow-out valve.
- **Swivelling cylinder** Failure of the swivelling cylinder to rotate, or partial rotation into an incorrect position, will result in the CNC producing an error message. A change in friction in the bearings, or between the sliding surfaces within the unit, will change the time to rotate. Leakage from the unit will reduce the volume of oil in the reservoir.
- **Lifting cylinder** Failure to move, or partial movement, will result in an error message from the CNC. A change in friction will change the time for the actuator to move. Leakage from the actuator will reduce the volume of oil in the reservoir.
- **Switches** The failure of a switch will result in the CNC producing an error message.

20.5 CONCLUSIONS

The above discussion has indicated the advantages and disadvantages of the main condition monitoring methods and also some of the dangers and subtle factors involved in their direct comparison for selection. It has also shown how method selection can be achieved, by use of the analytic hierarchy process, and by the use of failure mode and effects analysis. Both these techniques have been presented via illustrative examples to show the veracity of their application to the condition monitoring area.

20.6 REFERENCES

Drake, P.R. and Buck, A.A. (1991) Quality at the controls, *Manufacturing Engineer*, **70**(4), 49–51.

Drake, P.R. and Harrison, P. (1993) Condition monitoring method selection using the analytic hierarchy process, *Proceedings of the 5th International Congress on Condition Monitoring and Diagnostic Engineering Management*, UWE, Bristol, UK, 1–6.

Drake, P.R., Jennings, A.D., Grosvenor, R.I. and Whittleton, D. (1995) A data acquisition system for machine tool condition monitoring, *Quality and Reliability International*, **11**(1), 15–26.

Golden, B.L., Wasil, E.A. and Harker, P.T. (eds) (1989) *The Analytic Hierarchy Process – Applications and Studies*, Springer-Verlag, Berlin, Germany.

Harrison, P. (1992) Condition Monitoring of Pumps, MSc Dissertation, UWCC, Cardiff, UK.

Hoh, S.M. Williams, J.H. and Drake, P.R. (1990) Condition monitoring and fault diagnostics of machine tool axis drives via axis motor current, *Proceedings of the International Conference on Manufacturing Systems and Environment*, Tokyo, Japan, 93–7.

Hunt, T.M. (1993) Vibration and wear debris analysis – how they correlate and complement each other, *Proceedings of the 5th International Congress on Condition Monitoring and Diagnostic Engineering Management*, UWE, Bristol, UK, 109–14.

Jennings, A.D. and Whittleton, D. (1993) *Identification of BA25 Machine Tool Signals*, MIRAM project report MIRAM-UWC-2.2-EXT-ADJ&DW-3-B, UWCC, Cardiff, UK.

Saaty, T.L. (1980) *The Analytic Hierarchy Process*, McGraw-Hill, New York, USA.

Watton, J. (1992) *Conditon Monitoring and Fault Diagnosis in Fluid Power Systems*, Ellis Horwood, London, UK.

Webborn, T.J.C. (1982) Vibration measurement and incipient failure detection in machines, *Noise and Vibration Control Worldwide*, **13**(6), 252–3.

CHAPTER 21

Pitfalls, benefits and collective wisdom

R. Jones

MCP Management Consultants, The Business and Innovation Centre, Aston Science Park, Love Lane, Birmingham, B7 4BJ, UK

21.1 INTRODUCTION

In 1988 a national survey of maintenance was commissioned by the Department of Trade and Industry (DTI), which showed that UK manufacturing companies could save £1.3 billion a year through more effective maintenance management and practice. The survey participants highlighted numerous areas of concern as shown in Table 21.1. These issues are all relevant in the 1990s and the DTI through their 'Managing in the 90s' programme are providing both awareness of, and insight into the challenges facing companies in the twenty-first century (Anon, 1991).

In 1987, MCP Management Consultants developed their benchmarking and continuous improvement process called Asset Management Information Service (AMIS). The DTI supported this AMIS initiative and by the end of 1995, over 1300 AMIS maintenance benchmarking audits

Table 21.1 Factors which caused dissatisfaction with maintenance department

● Planned maintenance	50%
● Slow response	23%
● Lack of spares	26%
● Lack of staff ability	27%
● Lack of management ability	21%
● Lack of staff	34%
● Plant complexity	27%
● Excessive costs	24%
● Lack of information	16%

Handbook of Condition Monitoring
Edited by A. Davies
Published in 1998 by Chapman & Hall, London. ISBN 0 412 61320 4.

have been carried out worldwide. The results from the AMIS database, and the consequential improvements carried out by numerous participants, are the basis for the comments contained in this chapter.

The AMIS process is to provide senior management with the facility to carry out a self-appraisal of their maintenance effectiveness. This audit will establish the current position of the company before embarking on a programme of change. It allows all aspects of manufacturing best practice to be considered before the introduction of a new philosophy, or some three-letter acronym 'flavour of the month' initiative. The optimal solution and prioritized action programme can then be prepared based on the Business Goals and Critical Success Factors (CSF).

Analysis of the AMIS database some seven years after the initial DTI survey indicates that many of the original findings are still relevant. Some of the statistics arising from the AMIS database analysis are shown in Table 21.2. The issues associated with each of these categories are best summarized as:

- **cost control**   little or no idea of how, why or where maintenance costs are incurred;
- **workload planning**   lack of an effective system to record work, or to schedule the total workload, hence no idea of the resources required;
- **maintenance effectiveness**   lack of data to enable any objective measure of performance to be made and few methods available to determine actual labour performance/output.

In addition, there are a number of fundamental issues which have emerged during the audit process, rather than through the individual audit categories. These are:

- **Maintenance strategy and objectives**   In general, manufacturing and service companies devote a considerable amount of time developing a strategy, or plan, to achieve their business objectives. However, few organizations apply the same process to their maintenance activities.

Table 21.2   AMIS database analysis

• No. of audits	1300
• Average score	43%
• Number of Class A companies (score >75%)	28
• $\dfrac{\text{Maintenance cost}}{\text{Sales turnover}}$	3.7% Average
• Percentage companies with computerized systems	80%
• Percentage companies satisfied with computer system	30%
• Percentage companies with maintenance strategy	10%
• Percentage companies with initiative fatigue	85%
• Average maintenance work profile	>60% reactive
• Condition monitoring as percentage of planned work	<3%
• Spares stock turn by value	>3 years

Less than 10% of the participants have a clearly defined maintenance strategy or, in fact, any maintenance objectives at all. Where objectives are quoted, they are usually meaningless statements concerned with 'keeping the plant going for the minimum amount of expenditure'. When the strategic planning process has been completed and changes affected, no one person is appointed to continuously review its effectiveness or to monitor the achievement of the stated objectives and take the appropriate actions.

- **Accountability for operation and maintenance activities**   Some companies decide to transfer maintenance craftsmen to a specific production area, but they fail to define exactly the production management's accountability in terms of maintenance. All too often production staff simply let the craftsmen 'get on with the job' which tends to be reacting to breakdowns rather than some form of proaction.

- **Maintenance budgeting**   Too many companies are still relying on 'an historical approach', which means taking last year's figures and adding something for inflation and then waiting for the accounts department to reduce the figures!

- **Computerized maintenance management system**   Some AMIS participants see the purchase of a computerized maintenance management software package as the solution to all their problems. Other companies believe that the way to increase their performance is to buy a computer system. This action taken in isolation rarely changes anything, on the contrary, it can preclude the use of such a system in the future because of the poor implementation experienced initially. Over 60% of companies with computerized maintenance management systems are not satisfied with their choice of package. Accordingly it should be noted that a good maintenance package is only a tool which should be used as an integral part of an effective and well-thought-out management system.

- **Reliability Centred Maintenance (RCM)**   Many companies have adopted RCM in the belief that it will change the way maintenance is managed, and that it is devloping into the all encompassing answer for every maintenance manager. RCM can be highly effective for establishing the most appropriate maintenance techniques to be applied in a particular company's circumstances. However, it cannot provide the management controls required for the maintenance function and is not in itself a management system.

- **Team briefing**   Audit results show that team briefings are deteriorating into five-minute reporting sessions on how the company is performing, rather than their intended role of a participative discussion. Effective team briefings should be multi-level as well as multi-disciplined and, managed effectively, they can be an extremely valuable means of identifying and discussing how maintenance problems and issues can be resolved. Accordingly, maintenance managers

who devote little attention to such communication methods are missing a valuble opportunity to improve their departments' performance.
- **Total Quality Management (TQM)**   Again many companies have embarked on the road to excellence using TQM as the vehicle. However, statistics show that up to 70% of these companies have not achieved their stated objectives after a three-year period.
- **Total Productive Maintenance (TPM)**   This approach is producing excellent results where companies have understood the major culture changes required for its implementation. TPM can be regarded as a subset of TQM but concentrates on 'doing' things with the man/machine system.

Having identified these problems, some reflections on the issues raised are contained in the sections which follow.

21.2   MAINTENANCE POLICY

Most manufacturing companies consider it important to develop a manufacturing policy, with well-defined business objectives. The policy should define how the business objectives will be met and it is usually widely communicated within the company. By contrast, few organizations put the same emphasis on developing a maintenance policy. The maintenance function is rarely considered an activity worthy of a separate policy or of clearly defined objectives.

Accordingly, maintenance is considered the 'Cinderella' of the organization, and as such, often assumed to be the exclusive concern of maintenance engineers and nothing to do with the business or manufacturing policy. As a result, it is most unlikely that any existing maintenance policy forms an integral part of or supports the overall business goals. Most maintenance organizations, therefore, have a disturbing lack of strategic direction in how to maintain and care for their assets.

A manufacturing service industry should be an interaction of people, systems and assets. Thus, the maintenance policy applied to those assets is clearly an essential element of achieving the overall business objective or plan. The block diagram outlined in Figure 21.1 clearly shows the correct relationship between a maintenance policy and the overall business objective. It should be noted that the maintenance policy is required to cover all the aspects of engineering required, including the procurement of new equipment, capital projects and asset care.

The policy, when formulated, should contain a brief statement of the overall objective of the maintenance function, such as the mission statement below, or something similar:

> To ensure the required availability and performance of all assets at an economic cost throughout their life cycle.

POLICY/STRATEGY HIERARCHY

Company business objective

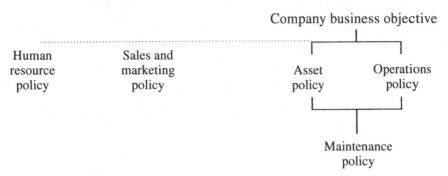

Fig. 21.1 Structural relationship.

So, developing the overall maintenance policy must start with an understanding of the business policy, or plan, and the overall operating environment. It cannot be developd in isolation, but must be developed jointly with the operations policy, since the two are closely related and interact with each other. In some organizations this is reflected in a combined manufacturing and operations policy that includes maintenance.

The maintenance policy must also include technical and managerial strategies that specify how the maintenance approach is selected, how labour and material resources will be planned and controlled, along with what information systems are required, and how maintenance performance will be monitored and improved. Figure 21.2 outlines the interaction of these aspects of maintenance policy in a hierarchical block diagram. To help in the formulation of a maintenance policy it is useful to refer to the British Standards Glossary of Maintenance Management Terms – BS3811 (1992) which defines maintenance in the following manner:

> The combination of all technical and associated administrative actions intended to retain an item in, or restore it to, a state in which it can perform its required function.

Accordingly, to develop a comprehensive maintenance policy the following points must be considered:

- **Establish the Critical Success Factors (CSF)** Confirm the actions or goals which must be achieved. These are normally the factors which

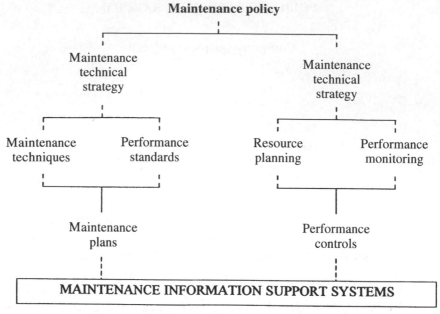

Fig. 21.2 Maintenance policy hierarchy.

have the greatest impact on the business, such as minimizing the total cost of maintenance, maximizing the availability of key plant or achieving a higher level of safety.

- **Quantify the objectives** Each identified critical success factor must be quantified together with an appropriate timescale for achievement. The necessary requirements should be agreed jointly between the maintenance and production departments.
- **Requirements to achieve the objectives** The process to achieve the objectives, the strategy, should consider the following aspects. Available maintenance techniques, the skills and people necessary, what material resources are needed, the organizational structural involved, the planning, cost and performance control, and any supporting resources. The later includes spares, technical data, management information and control etc.

The maintenance policy can now be developed from known and agreed objectives. It should define the process for identifying and achieving sound operational goals, selecting the maintenance approaches, monitoring performance and providing effective management control. In the following sections of this chapter, we will now look at the development of the technical and management strategies involved.

21.3 MAINTENANCE TECHNICAL STRATEGY

This strategy is fairly obviously concerned with the technical aspects of maintenance, and specifies how scientific knowledge and experience are used to:

- identify the best proactive maintenance, repair, service and replacement of all maintainable items;
- implement the chosen maintenance approach.

The key consideration of a maintenance technical strategy is to minimize emergency or breakdown repair actions, in order to maximize equipment availability through its inherent reliability and maintainability, or:

To retain an item in, or restore it to, a state in which it can perform its required function.

A maintenance department has a number of different approaches available to maintain the equipment in their care. The type of maintenance used will depend on a series of requirements and decisions which will have to be made as part of the formulation of the maintenance strategy. That strategy will usually be a combination of maintenance activities as shown in Figure 21.3, wherein the definitions of the various types of maintenance are as follows:

- **unplanned corrective maintenance** work carried out after a failure has occurred and intended to restore an item to its original state;
- **emergency breakdown maintenance** work carried out with no pre-planning or forewarning that a failure would occur;

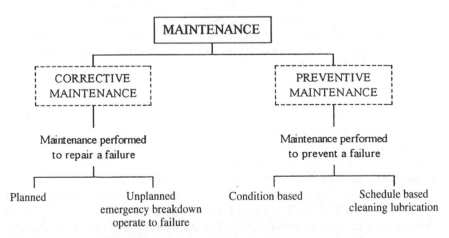

Fig. 21.3 Types of maintenance.

- **operate to failure maintenance**   work carried out after a failure has been allowed to occur.
- **planned corrective maintenance**   planned work to replace a temporary equipment repair and restore the plant to its original condition;
- **preventive maintenance**   work carried out at predetermined intervals or corresponding to prescribed criteria, intended to reduce the probability of failure or loss of performance of an item.
- **condition based maintenance**   preventive maintenance work based on periodic or continuous monitoring of the deteriorating condition of an item;
- **schedule based maintenance**   preventive maintenance work carried out to a predetermined interval of time, number of operations, or mileage etc.

The aim of the strategy should obviously be to reduce the level of unplanned work to the absolute minimum. In addition to the above procedures there is a further option available:

- **design out maintenance**   work undertaken to remove or improve an item or component that is a recurring problem in plant or equipment.

The maintenance department may also be involved in other engineering work such as modular changes, modification to plant and buildings, installation of new plant and equipment. In addition, it may provide a service to the production department involving cleaning, setting up for manufacture, tool changing, machinery shutdown and turnaround.

Maintenance principles are the same whether the business provides a service or manufactures a product. It is just as important to provide the required working environment, such as lighting and heating for a service organization, as it is to satisfy the maintenance requirements of a manufacturing process. A large number of companies, however, fail to apply even the most basic maintenance principles and this results in a totally reactive situation to equipment or plant failures. An unmanaged or reactive approach is by its very nature uncontrolled and therefore also expensive to operate. A well-managed maintenance activity on the other hand is proactive, planned, controlled and far less costly.

As a combination of different skills, technical knowledge and experience is required to identify maintenance needs, adopting the most appropriate solution is an important consideration. The required approach must consider the maintenance characteristics, production requirements, economic and safety factors of each item of plant and equipment, and do so within the framework and understanding of sound management procedures. It is necessary to analyse equipment failures in terms of their mode of failure and time to failure, so as to determine the most effective technical approach. The modes of failure must also relate to the existing preventive maintenance schedules.

The technical strategy should thus be based on a combination of the following approaches:

- Operate To Failure (OTF)
- Fixed Time Maintenance (FTM)
- Condition Based Maintenance (CBM)
- Opportunity Maintenance (OM)
- Modular Changes (MC)
- Shutdowns (SD)
- Design Out or Improvement Maintenance (DOIM)

A close investigation into the characteristics of each machine must be carried out, to identify the behaviour or performance of each plant unit, the production pattern and the type of equipment involved. The behaviour or performance of the machine can be measured by its failure rate or reliability. A typical parameter being Mean Time Between Failure (MTBF). In addition, its maintainability or Mean Time To Repair (MTTR) can be recorded plus the machine's availability value.

The manufacturing pattern also needs to be analysed into its sequence of production and non-production intervals. That is, are there any available time windows for maintenance within the production pattern? Other questions to be asked are:

- What are the number of standby units or is there any redundant equipment available to allow the machine to be taken out of service without interrupting production?
- What is the best balance between planned preventive maintenance and shutdown work?

21.4   LESSONS FROM BENCHMARKING RESULTS

In the late 1980s many engineers bought the latest tools or gadgets without considering how they would fit in to their maintenance activities. As a consequence, many companies have since discarded expensive condition monitoring equipment, and even suggested that CM is not a viable or cost-effective policy to adopt. From our experience with over 1300 companies worldwide, we believe that a strategic approach is required, to ensure maximum success from a well-managed maintenance function. A typical model for good strategic asset management is shown in Figure 21.4, and this illustrates the interaction necessary at both the corporate and operational level.

Figure 21.5 shows the linkage of condition monitoring to equipment performance within the firm's equipment strategy, and also to computer-based information which lies within the information strategy adopted. In addition, the role of systems, condition monitoring, Computer Aided

STRATEGIC ASSET MANAGEMENT

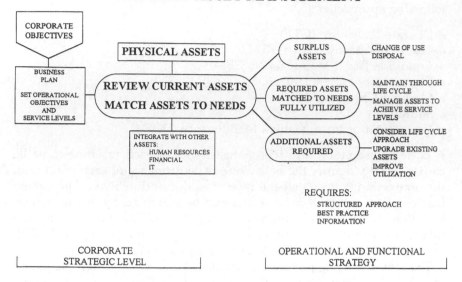

Fig. 21.4 Strategic asset management model.

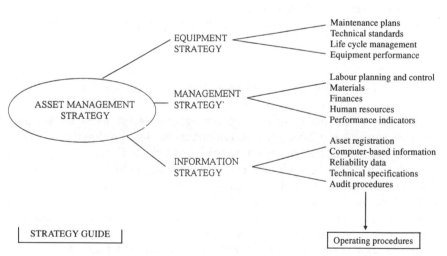

Fig. 21.5 Relationship of condition monitoring to equipment strategy and computer-based information.

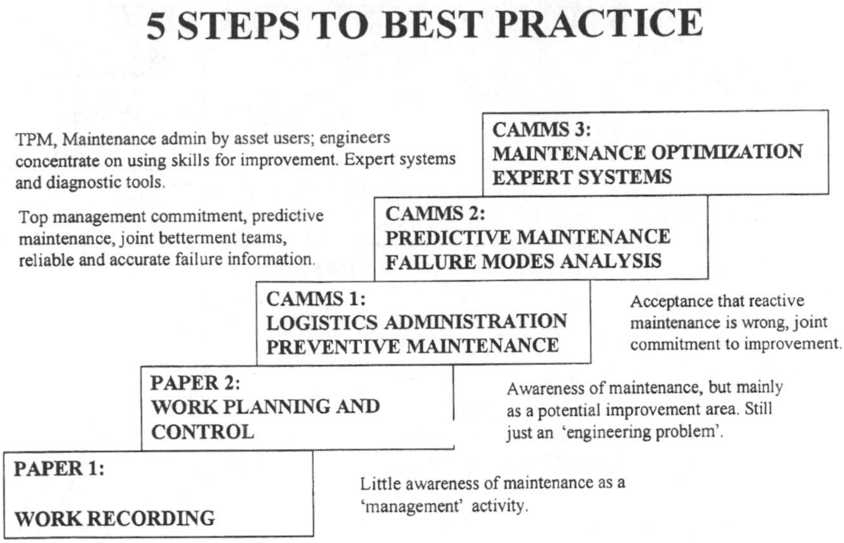

5 STEPS TO BEST PRACTICE

TPM, Maintenance admin by asset users; engineers concentrate on using skills for improvement. Expert systems and diagnostic tools.

CAMMS 3:
MAINTENANCE OPTIMIZATION
EXPERT SYSTEMS

Top management commitment, predictive maintenance, joint betterment teams, reliable and accurate failure information.

CAMMS 2:
PREDICTIVE MAINTENANCE
FAILURE MODES ANALYSIS

CAMMS 1:
LOGISTICS ADMINISTRATION
PREVENTIVE MAINTENANCE

Acceptance that reactive maintenance is wrong, joint commitment to improvement.

PAPER 2:
WORK PLANNING AND
CONTROL

Awareness of maintenance, but mainly as a potential improvement area. Still just an 'engineering problem'.

PAPER 1:

WORK RECORDING

Little awareness of maintenance as a 'management' activity.

Fig. 21.6   The route to best practice.

Maintenance Management (CAMM) etc. all need to be clarified within a clear team philosophy. For example, the use of Operations Driven Reliability or Total Productive Maintenance versus the traditional dedicated maintenance approach.

Along the route to best practice, the demand for tools and techniques will change as shown in Figure 21.6. This 'step-by-step' approach may be translated into the system requirements when necessary, to encompass the various techniques available and philosophies which can be adopted. Figure 21.7 shows the steps to take from innocence to excellence; from, in effect, reactive to proactive maintenance; or from firefighting to a class 'A' excellence benchmark.

When managing the changes from reactive to proactive maintenance, it is important to arrange to change all the key factors, not just to buy technology. Figure 21.8 highlights this need to manage the changes in organization, that is use of natural autonomous teams, people or operator care, job design and technology. The use of condition monitoring has now become more sophisticated and selective. Ten years ago for example, senior management would ask whether to use TPM, RCM or CM. In today's environment, management understand the need for an integrated approach, tying in strategies, philosophies, techniques, systems and tools, while ensuring that the relevant skills are also available.

Many companies are now carrying out RCM and Failure Mode Effects and Criticality Analysis (FMECA) reviews on any major failures. Local natural teams, including craftsmen and operators, would then select the

5 STEPS TO BEST PRACTICE

FEATURES OF MAINTENANCE APPROACH					EXCELLENCE
Proactive - 'TPM' - concentration on technical excellence only			CAMMS 2	CAMMS 3	CAMMS 3
Predictive - CBM - FMECA - commitment - betterment			CAMMS 2	CAMMS 3	CAMMS 3
Preventive - planning + scheduling - good records	CAMMS 1	CAMMS 1	CAMMS 1	CAMMS 2	
Mainly reactive - some plans little data	Paper 2 Work Control	CAMMS 1	CAMMS 1		
Totally reactive Few records	Paper 1 'Task' Recording	Paper 2 Work Control			
INNOCENCE	30 ≤ 30	45 > 30-	60 > 45	75 > 60-	> 75 ◄ MAINTENANCE EXCELLENCE BENCHMARK

Fig. 21.7 Five steps from innocent to excellence.

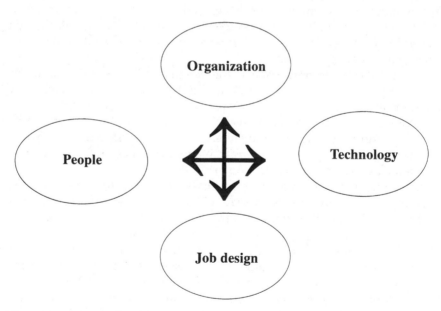

Fig. 21.8 Organizational change management.

most appropriate approach to maintaining an item. Figure 21.9 shows a typical algorithm for this type of decision making. The use of condition monitoring will be considered if a deterioration characteristic is evident. This characteristic will determine the best tools to provide the statistics required by monitoring, e.g. process variables, vibration, thermography, laser alignment or oil analysis etc.

During this analysis the team members will select the most appropriate tool and who best to collect the data, the operator, craftsman or specialist. The question of training and provision of tools/technology can then be addressed by the company. Condition monitoring will play a more important role in future maintenance activities, as process machinery and equipment become more automated and sophisticated.

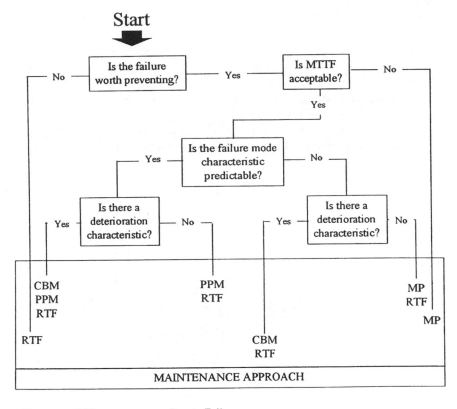

Key	RTF	=	Run to Failure
	MP	=	Improvement Maintenance
	PPM	=	Planned Preventive Maintenance
	CBM	=	Predictive or Condition Based Maintenance
	MTTF	=	Mean Time to Failure

Fig. 21.9 Decision-making algorithm.

Predictive techniques and transducer technology are improving daily, and accordingly the maintenance strategy must cater for this change and be capable of demonstrating a continuous improvement. A typical Computer Aided Maintenance Management Scheme (CAMMS) model is shown in Figure 21.10, and this outlines the interaction between the issues discussed and the use of a computer system to control maintenance.

The more sophisticated condition monitoring systems on the market today claim that huge cost benefits can be achieved through their use. However, our database shows that over 50% of all CM systems bought in the UK during the last five years are now abandoned. The benefits claimed can only be demonstrated, and realized, where clear strategies are adopted, plus a formal analysis of maintenance activity and the various approaches involved. In addition, the identification of the philosophy behind the tools selected to support a companies TPM concept is required.

As are clear objectives for carrying out feasibility studies into the purchase of expensive, high-tech condition monitoring equipment. It should be noted that condition monitoring and the application of Condition Based Maintenance does have an important contribution to make to a cost-effective maintenance strategy. This is shown by the list of successful case studies provided in the next section, which highlight the benefits and the breadth of CM success throughout all industrial sectors as well as in building services.

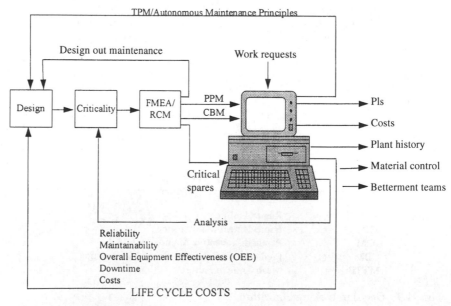

Fig. 21.10 A typical CAMMS model.

21.5 CASE-STUDY EXAMPLES

The following information relating to actual case-study examples of condition monitoring implementation are presented by courtesy of Entek Ltd.

- A paper plant in the UK successfully implemented a vibration-based condition monitoring system on major paper-making machinery. One year's activity produced 72 instances of corrective maintenance action based on CM recommendations. Sixty-eight of these were carried out at a convenient maintenance time window, and four were deferred for further investigation. This additional maintenance workload increased the line availability from 92% to 97%, although the representative cash savings resulting from this improvement were not released!
- A specialist pump manufacturer implemented production test monitoring for performance and acoustic acceptability. Subsequent defect diagnosis resulted in a change to build procedures which decreased the rejection rate from 29% to 6%. All rejected units were subsequently disassembled and reworked by the same team of production technicians, and an available increase of some 20% was achieved.
- A cable manufacturer has a highly automated yet simple process plant which introduced CM techniques to reduce unplanned failure and extend maintenance service intervals. Within three years most routine maintenance functions, other than first line maintenance, were abolished. Maintenance costs in the plant were decreased by 37% while no tangible reduction in production efficiency was detected.
- In an oil production facility, highly sophisticated heavy rotating machinery demanded routine shutdowns with an associated loss of field production. The introduction of process parameter trending coupled with vibration and oil analysis techniques, produced savings of 92 lost production days. This is for a class of machines which could generally be responsible for $120 000 per day of production. The saving was achieved by the effective streamlining of service intervals and content.
- Standard production machinery failures in a drinks manufacturing plant were responsible for some 21 identifiable line stoppages in a one-year period. Following the implementation of routine CM, this fell to 5 for the subsequent year. During that year, a potentially disastrous failure was averted on the filler/sealer machinery, which is considered to be of major criticality ranking in the plant. Failure on this machine would have resulted in a loss of 6000 cans per hour, for a period of up to two months while awaiting parts. Early notice of failure resulted in prior arrangement for spare parts and a planned downtime of some four days only.

As a consequence of the evidence contained in the above case-study material, it is strongly felt that condition monitoring and predictive maintenance software and hardware has to become more adaptable. It should also be flexible enough to accept varied inputs, simply and securely, and thus provide maintenance engineers with the reliable tool that their applications require.

21.6 CONCLUSIONS

This chapter has attempted to identify the pitfalls, benefits and collective wisdom surrounding the use of condition monitoring in industry. The overall impression gained from the data available is that CM is a valuable and useful tool to use in certain maintenance situations. However, consideration must be given to those issues raised in the chapter, relating to its appropriateness for a given situation, versus the alternative maintenance techniques available. Identification of a clear maintenance strategy within a company is therefore highly important, as is the place of all the tools used within the approach adopted, and this includes condition monitoring.

21.7 REFERENCE

Anon (1991) *Managing Maintenance into the 1990's*, Department of Trade and Industry, London, UK.

PART SEVEN

Research and Development in Condition Monitoring

CHAPTER 22

Knowledge-based systems for condition monitoring

R. Milne

Managing Director, Intelligent Applications Ltd, 1 Michaelson Square, Livingston, West Lothian, Scotland, EH54 7DP, UK

22.1 INTRODUCTION

One of the primary aspects of maintenance is to determine what needs to be repaired and when. This decision is based on a combination of skill and experience. Both of these are used to interpret the available data, and even to determine what information to make available. Typically, a skilled maintenance manager will use his experience on those failures which occur at particular times, as well as on what faults most often show themselves. A critical task in this area is the actual diagnosis of any particular problem. This can be especially difficult in an environment of complex machines, process measurements or more detailed condition monitoring data. It can take many years of training to determine how best to use the information available. This is because so much of maintenance relies on skill and experience, which can present a number of problems. For example, in many companies the most skilled and experienced people are also the more senior staff, and they may often be promoted into other non-maintenance line positions or they can retire.

In some situations this can lead to a severe problem, in that the only person with a complete understanding of the machine is no longer available. There is also the considerable requirement to teach the particular skills and diagnostic techniques to younger and more junior people. This is particularly difficult when a machine has various quirks of behaviour and some problems can be fairly rare. In many firms it may be difficult to get graduates or new staff with the appropriate skills. This can be further exacerbated by the constant introduction of new technology, where the skills of the existing staff may suddenly become inad-

Handbook of Condition Monitoring
Edited by A. Davies
Published in 1998 by Chapman & Hall, London. ISBN 0 412 61320 4.

equate. There is a constant pressure within industry to do more work with less people. This either means reducing staffing levels, or increasing the amount of work done without changing personnel numbers. As a result, the time available for work becomes more and more precious. There are also considerable demands on each experienced person's time, and this can make it difficult to devote adequate time to one area or to a difficult problem.

Finally, people have their good days and bad days, and they do take holidays. One of the worst nightmares a company may have is for its most important machine to break down, just after the only engineer who completely understands it goes on holiday. All of these factors imply that something is needed to capture the skill and expertise of the senior people, and make it available around the clock, 24 hours a day and through the hands of less skilled staff. Knowledge-based or expert systems (KBS) are a technology designed to precisely achieve this aim. They provide the tools and techniques to preserve the skill and experience of the more senior people, and capture them in a computer program. This computer program will behave consistently, and its skills would be available 24 hours a day. In general, these computer programs are much quicker in execution than the expert. As a result, the time needed for testing is reduced and people are released for other more important activities. Figure 22.1 outlines the major problems knowledge-based systems are designed to address.

This chapter describes in more detail knowledge-based systems, what they are, and what they are not, together with the background to knowledge-based systems and some of the ideas involved in current applications. Many of the factors described in this section are related to the introduction of computers to assist with these tasks. Thus, in all situations, it is best first to identify what goals are being sought, and how best applications could be developed to address these. In more than one situation, knowledge-based systems have provided the inspiration for a previously undiscovered traditional solution. The real emphasis should therefore be on the application to develop and not the technology. Having said that, however, this chapter focuses on the technology of knowledge-based systems, which, during their initial introduction, caused many companies some concern over the possibility that such systems would replace people. Given that the expert system is capturing

- Lack of staff
- Lack of correct skill levels
- Lack of time to perform tasks
- Inconsistent performance of tasks

Fig. 22.1 The problems which are addressed by knowledge-based systems.

the knowledge and experience of a skilled person, this was understandable. The natural question which occurs in a manager's mind, is whether the experienced person is still needed after the expert system is developed.

In over 10 years of knowledge-based systems being used, however, there have not been any situations where the experienced person lost his or her position as a result of the introduction of an expert system. This is for a number of diverse reasons, one of which is that on many occasions a company needs to capture and retain the expertise of a skilled person, possibly because that expert is about to retire or change jobs. One of the most well-known examples of this is a system General Motors developed known as 'Charlie'. It captures the experience of a senior test engineer, called Charlie, who, with regard to how to diagnose faults in automotive engines, was one of the few people able to assess accurately and quickly the state of an engine and determine what was wrong. General Motors realized that when he retired there would be a huge gap in the capabilities of the company. Accordingly, they rapidly developed a fault diagnostic expert system in order to capture his expertise. General Motors could not allow Charlie to retire in the knowledge that his experience would be lost to the company. With the development of the expert system, Charlie could now retire with the knowledge that he had made a lasting contribution to General Motors.

Another famous early expert system was developed by Campbell's Soup to diagnose problems in their soup cookers. Once again, the senior expert was about to retire and this was the only method by which they could retain his expertise. He was very co-operative and helpful, as he knew he was making a worthwhile contribution to the company. Another area in which knowledge-based systems are useful is when they perform a routine task that does, however, require a degree of skill and training. In general, experts are able to handle a wide range of tests with a high level of skill. Quite often, however, they also need to perform relatively simple mundane tasks. These tasks can, however, require some skill, perhaps in conjunction with a more complex analysis. The 'Amethyst' system is an example of this, where it uses the knowledge of an expert to perform a relatively routine and mundane task. This has the tremendous advantage of freeing up the expert for more important and complex tasks. Amethyst performs the same analysis as the expert, but in far less time. In general, it has been found that this type of staff have far too much to do and too little time available to do it.

As a result, expert systems make their task more manageable and their lifestyle less hectic. Figure 22.2 illustrates where expert systems have been used to date in manufacturing industry. A further area where knowledge-based systems are helpful is where they allow the expertise to be spread to other parts of the company. The on-line vibration diagnostic expert system, developed for Exxon Chemical, for example,

Capacity planned	Process and facility simulation
Design to manufacture	Process control and optimization
Electrical and electronic assembly	Process diagnosis
Electrical and electronic testing	Process planned
Equipment repair and maintenance	Product configuration
Faculty and equipment layout	Production planning
Financial analysis and planning	Statistical process control
Human resource management	Statistical quality control
Materials handling	Strategic planning
Mechanical assembly	Tool selection
Mechanical design	Work order writing
Mechanical testing	Work scheduling

Fig. 22.2 Expert system applications throughout manufacturing.

was pushed very strongly by the senior machinery engineer. He needed a way to make vibration diagnostics available to other plant personnel. This would free up his time for more essential tasks. It also helped the company in their programme of broadening out the operations of the company and in the allocation of work. The system saves the expert considerable time and helps the more junior people to be able to perform the machine monitoring tasks on their own. In many situations, an expert system can perform a new task which is not practical on a routine basis. For example, a KBS was developed for British Steel Scunthorpe to monitor a waste gas recovery system which was originally controlled by a Programmable Logic Controller (PLC).

With the old way of working, if anything went wrong with the plant then the logic controller returned the plant to a fail safe condition. So, by the time the engineering manager discovered the fault and went to determine the cause, the system was in a completely different state. An expert system was then installed which runs continuously, monitoring the state of the plant. The expert system is instantly able to identify the cause of a problem before the PLC returns it to a failsafe condition.

This allows a fault diagnosis to be made which was not possible before. For a person to be able to perform this task, they would have to wait by the machine day after day for the fault to occur. Finally, one other area where knowledge-based systems help with tasks, rather than hinder, is in the introduction of new equipment or new technology. There may be an expert somewhere who knows about a particularly complex machine, say a flexible machine tool. However, if a company acquires that tool, they will not have this expertise within their own company.

As a result an expert system, to help them troubleshoot it, will allow them to acquire this new capability without also having to bring in an additional expert. This is a common situation as new high-technology equipment is introduced. Even in the area of vibration-based condition monitoring, this situation may be observed. When a company begins vibration-based condition monitoring, they typically have no one trained in the interpretation and diagnosis of the vibration signatures. The Amethyst expert system provides the expertise they are missing, and allows them to introduce the new technology without difficulty. There are situations where knowledge-based systems have led to a change in the number of staff for given skill requirements. However, in these situations the expert system was not the cause but only the effect. Many companies are trying to obtain more output with the same number of staff, or reduce/use lower skilled staff in certain areas. This is a very common problem for companies growing rapidly. Knowledge-based systems become the only means by which they can achieve their objective of higher productivity with lower skill levels. Also, for many firms, there are just enough trained people, so knowledge-based systems are vital in order to achieve the company's goals.

22.2 THE ROLE OF KBS IN MAINTENANCE

As discussed in the introduction, the maintenance manager wants to solve a problem, but he also needs diagnostics. Many systems are able to detect that there is a problem, but not what the precise problem is and what to do about it. Many maintenance systems run through the use of a maintenance management package which is responsible for the issuing of work orders. Currently, all they provide is data to the effect that something is wrong with the machine and it has stopped. This is hardly the information to put on a work order to effect a repair. What is really desired is a more detailed analysis of the particular problem. For example, in rotating machinery the condition monitoring system may detect that the level of vibration on a motor-driven fan is too high. With most current maintenance systems, it would then be necessary to issue a work order so that the cause may be established.

The 'Amethyst' expert system, on the other hand, is able to look at the vibration pattern and perform a more detailed diagnosis automatically. It can then determine whether the problem is due to bad bearings, misalignment or blade problems, to name a few possible causes. The output of the expert system can also be inserted onto the work order, so that the maintenance personnel know precisely the cause before action is taken, and this obviously leads to considerable efficiency of working. Figure 22.3 provides an illustration of an Amethyst diagnostic report. As another example, if a machine tool suddenly stops it is important that the operator quickly determines whether to call the electrical or the mechanical engineer, or whether he or she can take some simple action to get the machine working again. An on-the-spot expert system provides the skill needed in a convenient program to lead him or her

```
Route STEEL              Total number of faults: 3        Alarm: IN/S
                                                          Alarm: g/SE
Machine SCREW COMPRESSR

Pos 5            Dir H      IN/S      Bad bearings with MAJOR faults
Mach Type pump-horz cent                 BSF and/or BTF Sidebands Exist
                                     Hydraulic or Aerodynamic
Rotating Speed RPM  1525             Misalignment
Overall Ampl        0.379 IN/S
Alarm Limit         0.314 IN/S
```

Fig. 22.3 An example diagnostic report from Amethyst.

through the steps needed to make these decisions. As production machines become more complex, the task of diagnosing and analysing problems is also becoming more difficult. In fact, the problems are getting so complicated that most industrial companies are having severe difficulties coping with this problem. Thus, it is vital that expert system techniques are used to capture machine diagnostics, and thereby make them available on a regular basis.

The first knowledge-based systems were developed almost twenty years ago, with Mycin being one of the initial large and successful application programs. This system captured the expertise of doctors in the area of internal infectious diseases. The Mycin software was used very successfully in early trials and in one study the program did better than the ordinary doctors at identifying the cause of internal diseases. In fact the good news was that Mycin was better than the doctors, but the bad news was that Mycin was correct only 64% of the time! Another early expert system was known as 'Dendral'. This was also developed in the medical area, and helped to interpret mass spectrographs, thereby determining the structure of the chemical compound that would have caused it. It contained the knowledge of experienced chemists, with regard to the type of chemical structures necessary to produce certain combinations of peaks on the mass spectrograph and was quite successful.

Another early system, which paid for itself many times over the first time it was produced was Prospector. This program was used to evaluate the probability of finding molybdenum deposits when given geological data. Its rules and knowledge described the type of geological features needed for a high probability of a large mineral deposit. The first time it was used, it correctly identified the location of a large deposit, and this subsequently required over a $1 000 000 of work to confirm that the mineral body was as indicated. Fortunately, it was! Since that early beginning, over 1000 different expert system applications have now been developed, many of these in the USA. Modern-day projects range in size from large multi-year developments, to very small personal-computer (PC) based applications requiring only a matter of weeks to develop.

The area of knowledge-based systems has a number of aspects, and as a result there are a variety of meanings and a wide range of interpretations/understanding as to what is involved. Unfortunately, it is difficult to define knowledge-based systems simply and precisely. There is a common joke that if you have five experts on KBS in a room, you will receive five different definitions. In actual fact, you will probably receive 10 or 20, with that in Figure 22.4 being widely accepted. The original foundation for the term 'knowledge-based systems' was from the idea that it was a computer system that captured the expertise of an expert, knowledge-based systems being an outgrowth of the field of Artificial Intelligence (AI). This seeks to understand ways in which computers can be made to perform tasks that make humans seem intelligent. Over the years however, the goal of looking for intelligent computers in the form of human intelligence has been abandoned. The emphasis instead has shifted to applying these techniques to provide useful solutions.

Artificial Intelligence itself covers a whole range of topics involved with computers performing advanced and complicated tasks. AI includes areas such as:

- vision;
- robotics;
- problem-solving;
- natural language understanding;
- fundamental problem solving and search techniques;
- knowledge representation;
- machine learning.

Knowledge-based systems really cover one application area, that of applying knowledge-based programs to solve particular problems.

In general, knowledge-based systems are not expert, they are just better than traditional computer programs in specific application areas. The term 'expert' has misled many people, and it is important to note that the idea is to have a computer program which captures the skill and expertise needed for a particular problem. The whole basis of knowledge-based systems is to use more knowledge to solve a problem. Traditionally, any computer program in any language can achieve the same purpose, but this ideal is often too difficult or expensive to achieve. To put in the knowledge of an experienced machinery engineer into a vibration diagnostics program can be very difficult, in for example, a low-level computer programming language. Through the techniques of

Knowledge based systems provide the tools to automate
a skilled engineer's knowledge

Fig. 22.4 Knowledge-based systems provide the tools to automate a skilled engineer's knowledge.

knowledge-based systems, it is much easier to represent and manipulate this information. Fundamentally, we can thus say that knowledge-based systems provide tools to implement the information needed to solve a particular problem.

In many cases, the same problem could be solved with conventional technology, but as stated above, this may be too difficult to be cost-effective. Because of the emphasis on knowledge, most KBS today are referred to as Intelligent Knowledge Based Systems (IKBS). For consistency, this chapter will refer to the older term of simply knowledge-based systems. Another misconception about knowledge-based systems is that they learn or improve. There are a number of techniques for automatic machine learning; the most common is what is known as inductive learning. In this case, a computer program is given a set of well-defined examples. It then automatically constructs its own operating rules from this set of examples. In some application domains, this technique is very powerful. However, it can be extremely limited for very big problems, or areas where there are a large number of variables with a vast number of possible values for each. As a result of these difficulties, knowledge-based systems to date do not exhibit learning and consequently do not change their behaviour dramatically over time.

Throughout the world today a large number of successful expert system applications have been developed in the maintenance area. The following are a few examples produced in the UK, and which are typical of some of these working applications. In British Steel, for example, a diagnostic expert system has been implemented, which is interfaced directly to a programmable logic controller. The PLC is controlling valve movements on a waste gas recovery system. Prior to the expert system, if there was a failure, it was very difficult to work out what had gone wrong, because, as previously mentioned, the plant would revert to a fail-safe condition. The expert system now used is able to identify faults accurately within seconds of a problem occurring and do so fully automatically.

At another British Steel plant, a very large real-time expert system has been developed. This system is interfaced directly to the steel-making plant data acquisition system. As the steel-making process progresses, the expert system checks for over 400 possible faults every 15 seconds. The output of the system is passed to the maintenance manager who then knows which problems are developing, and which of these need rectification as soon as possible. A major aspect of this system is its ability to capture long-term trends and to predict failures before they occur. There are a number of other similar real-time systems which have been developed in the UK. Within the machine-tool area for example, a number of programs have now been developed on a trial basis for the diagnosis of machine-tool stoppage. Some of these systems are interfaced directly to the machine control unit, and so are rapidly able to diagnose problems based on the controller information. Other systems interact with the operator and give him or her guidance on the steps to take in order to isolate a problem.

In the area of condition monitoring of rotating machinery, the Amethyst system has been one of the most successful expert systems. the Amethyst system is interfaced to the IRD Mechanalysis vibration database. Once the vibration signatures have been collected on items of plant such as pumps, compressors, fans and motors, Amethyst automatically performs a diagnosis of any problems. Using only a simple description of the machine, it applies the knowledge and expertise of a senior machinery engineer. Amethyst is thus able to perform in a few minutes what would require a trained and experienced engineer several hours. Exxon Chemical in Scotland has a diagnostic expert system interfaced directly to their machinery protection system. As the machinery protection system scans the incoming vibration signals, the expert system performs a diagnosis and makes the results available to the senior machinery engineer. The system is able to run continually 24 hours a day checking for any developing problems. Wiggins Teape have also implemented a system that advises on quality problems at the end of the paper-making process. Although the main focus of the system is on quality monitoring, it is also used to help identify any maintenance-related problems that will affect quality.

22.3 KBS–BASIC TECHNIQUES, ADVANCED CONCEPTS AND INTERFACING

The main idea behind knowledge-based systems is to provide a more powerful means of representing the information and then manipulating this at a higher level. Figure 22.5 shows the basic elements which make up an expert system. A basic KBS implementation captures the knowledge of the experienced engineer in the form of rules. These rules are intended to be true facts, that is facts which are true in isolation and not dependent on other surrounding computer code. Very often these facts

* The **inference engine** which matches the current situation and expert knowledge to draw inferences by forward chaining or backward chaining.

* The knowledge base which contains:
    - all expert domain knowledge included in the system
    - the information about the current situation

* The user interface which accepts inputs from the user and generates the displays and reports in support of dialogue and results.

Fig. 22.5 Expert system elements.

are called rules, production rules or if–then rules, and the total system is called a rule-based system. This collection of information is most commonly referred to as the rule base or the knowledge base. A basic rule in such a system is of the form:

If a number of items are true, then a conclusion must be true.

For example:

If the output flow rate is very low, and the pump is on, then the output filter is clogged.

Or as a more complex example:

If the chemical reactor temperature is too high, and the output flow-rate is too low, and the output flow has stopped and the pump is on, then the compound has congealed, so stop the agitator immediately, and notify the operator.

Figure 22.6 illustrates another simple rule as it may appear in an expert system program listing.

The Mycin expert system application referred to earlier, consists of many hundreds of rules of this style. The use of the if–then rule can be contrasted with other programming techniques that use a variety of computer commands and instructions. One of the key ideas of knowledge-based systems is exactly that, but only if the rule style is used. Theoretically these rules can be written down in any order and the expert system software will automatically bring them together to reach conclusions. This is one of the potential powers of an expert system where the true facts can be specified and the program will work out how to bring them together to reach a result. In actual fact, it never works this simply, and care must be taken to determine which rules are needed and how they link together.

Another key aspect of knowledge-based systems is that the knowledge is represented symbolically rather than numerically. In the example above, we use the statement 'if the output pump is on'. We could have also used the statement 'if B3 = 1'. Clearly the English phrases make it much easier to read and understand. In general, knowledge-based systems use the higher level of representation with symbols rather than numbers. A complete expert system application will consist of two items,

```
IF the overall vibration is too high
AND a MULTIPLE of RPM is too high
AND the overall axial vibration is more than 50% of the horizontal
THEN misalignment is likely
```

Fig. 22.6 A typical simple rule.

the rule base containing the production rules which describe the knowledge and an 'inference engine'. The inference engine is that part of the computer program which manipulates the rule base. It understands the process by which the rules are used to reach a possible conclusion. Figure 22.7 outlines a typical expert system structure.

Inference engines are standard and independent of a particular application. It is only the rule base that changes. One could imagine taking a complete application and removing the rules and the knowledge base. What one is then left with is an inference engine and the mechanism for editing and building up the rule base. This type of computer software is known as an expert system shell. It is the shell of a completed application. Shells are used as the standard tools to develop expert system applications. They provide the mechanism for entering, testing and debugging rule bases, as well as the inference engine to manipulate them. There are a wide variety of expert system shells available, from large mainframe systems using advanced techniques to very simple low-cost PC-based systems. Most modern shells also contain extensive tools for browsing and examining the knowledge base and graphically displaying what is happening. Modern tools also have standard features you would need for a PC software package, such as interfaces to databases and screen paint packages.

When it is necessary to solve a problem there are two basic approaches; one can either start with the data and see what conclusions can be reached, or one can start with an objective and see whether the data is available to support that goal or conclusion. These two approaches are known respectively as forward and backward chaining. For example, if you know a particular animal is large, grey, warm-blooded and has a trunk, you can deduce that it is an elephant. In many situations, forward chaining seems very natural. One starts with the data available and sees what conclusions are reached. In other situations, it is more straightforward to organize a computer program by having a number of possible

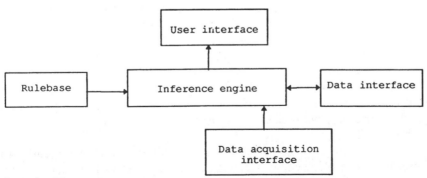

Fig. 22.7 Expert system structure.

conclusions and seeing if each is true. If we were to play a game where you needed to guess which animal I was thinking of, you might use the strategy of guessing that I am thinking of an elephant and asking questions to determine whether this is true. This is the main idea of a backward-chaining system. It starts with a list of possible conclusions or goals and tests whether they are true or not.

In general, if there are a small number of goals and a large amount of data available, backward chaining is the most efficient strategy. If there is a small amount of input data and a large number of goals, then forward chaining is the most efficient strategy. Most expert system shells provide for both forward and backward chaining. In reality, which one is chosen is not very important. With experience, the same results can be achieved with either approach. Also, if one wants to check exhaustively for all possible conclusions rather than stop at the first conclusions, it does not matter which approach is used. Simple rule-based KBS use an item–value relationship such as output temperature is high. Many items, however, have a set of relationships applicable to them. For example, a pump will have a state of either on or off. It will have a motor current value of low, normal or high. It may also have other characteristics such as its name, location and the number of blades on its impeller. It seems very natural to collect all these facts together as the description of an object. This fact is particularly true when one remembers that knowledge-based systems are about using information to solve a problem.

Grouping objects as items can make it much more efficient to represent and write down the knowledge needed. As a result, object-oriented systems are a very powerful way of capturing complex knowledge. For example, most valves are very similar, so one could imagine a valve object that contains the essential although general facts of a valve. In most systems this general description of a valve is called a class. There are then many instances of the valve such as, valve 1, valve 2 and valve 3, these would all have many of the common properties of the generic class valve, but some of them would be different in detail. Object-oriented techniques provide for the notion of inheritance, that is providing a set of standard properties for a class of objects that each instance automatically inherit. When I say, 'The machine is a pump', you know many things about the machine. This is because of knowledge inheritance. The instances then may have other factors which are unique to them such as location. This ability of inheritance is one of the reasons why using objects is a very efficient way of writing down knowledge. One describes a general class and automatically each instance has a copy of those same facts.

Another major strength of objects is the ability to send messages to them. For example, one would send a message requesting the current temperature to a wide variety of objects. A chemical reaction vessel would provide a different temperature to an outside temperature measurement. But from the viewpoint of the rule base, each object could be sent a temperature message the same way. This also makes manipulating the knowledge much easier, as the idea of standard messages helps to remove some of the detail from the program. The discussion so far has focused on one relatively simple rule base, addressing one problem. In many situations it is convenient to have a number of rule bases working together, each with a different specialization or area of expertise. In this situation, the rule bases communicate through a strategy known as a blackboard system. This derives from the idea of police in a situation room, writing facts up on a blackboard for all the detectives to look at. Each detective who has a fact to contribute to the problem can add his fact to the blackboard. The other detectives can then examine the total blackboard and gradually the problem is solved piece by piece. Blackboard systems are a very powerful mechanism for solving complex problems. They are, however, much more complicated to use. Most modern tools provide mechanisms to support blackboard problem solving.

Early knowledge-based and expert-system support products were designed to be used in a consulting style. That is, when they needed information they requested it from the user of the computer system. While the expert system was running, it would ask the user a number of questions. Through this process of twenty or so questions, it was able to reach the conclusions needed. For many applications, this offline style of expert system is suitable. It is particularly useful when some of the input is very subjective or hard to measure automatically. For example, a diagnosis might involve checking for evidence of oil leaking on to the floor, or listening to unusual sounds, or visually checking in a particular place for evidence of further problems. In each case, to input the data electronically would be extremely difficult, but for a person it is a straightforward task. Knowledge-based systems of this style have the benefit of replicating expertise but still require a person to assist with the diagnosis. If the expert system is used after a problem has occurred, it is quite possible that the system is in a difficult state, and some of the diagnostics may well be difficult because of changes in plant condition. For example, from a safety shutdown.

Currently there are also a large variety of expert system products that provide good interfaces to existing data acquisition systems, databases and computer systems. These tools are particularly useful when it is desired to receive data automatically from the plant and perform a diagnosis. This technique further enables the release of skilled people, and it is also very important when large quantities of data are involved, as for example in a database. Finally, by connecting the expert system online to the real plant, it is possible to have the diagnosis performed automatically at the moment of failure, rather than later when an engineer arrives. Modern expert system tools can now be interfaced to virtually any existing computer or electronic data acquisition system. In fact, in general, the same tool can be used as a yes/no question type system or directly interfaced online to the plant.

One of the key mechanisms for integrating knowledge-based systems into other systems is through the use of computer programming languages. Most knowledge-based systems can now interface directly to software written in Fortran, Pascal, Basic or 'C'. For a company which has implemented interface software already in one of these languages, it is usually a straightforward, but a time-consuming task to interface the data acquisition system to the expert system. These interfaces do require a knowledge of computer programming, and as such is a job for a system integrator rather than the expert. Most expert system shells also provide extensive interfaces to standard databases. On the PC, interfaces are available to DBase format databases. For larger systems, most products have an interface to Oracle and SQL databases, and most VAX products interface to RDB. A related aspect of integration is similarity of hardware and software platforms. Expert system products now run on all the standard computer platforms. It is no longer necessary to buy special computers just to run the expert system. IBM PC and compatible computers are particularly well supported, with a large variety of expert system tools.

There are a number of very powerful and well proven products available for VAX systems including both VMS and UNIX. There are also a number of powerful workstation products for UNIX systems. For large corporate operations, there are now expert system products available on virtually all mainframe systems. These products integrate and interface well with mainframe database systems. It should be noted that although integration and interfacing to the real world is now very practical, it can also prove to be difficult. It should also be recognized that the interfacing problem is very much a problem of standard computer interfacing. The primary difficulties are outside the expert system environment itself. Once good data acquisition has been developed, the integration is relatively straightforward. To date however, relatively few companies have tackled the integration problem. This is partly due to the small number of companies with skills both in knowledge-based systems and in data acquisition. It should be recognized, however, that interfacing can be time-consuming, expensive, and require considerable testing. Many times these costs are viewed as expert system costs. Once again, although they are part of the total integrated expert system, they are standard hardware and software problems with straightforward solutions.

22.4 APPLYING KNOWLEDGE-BASED SYSTEMS

Given that we now understand what knowledge-based systems are and how they might benefit our position, what do we do next? There are three basic approaches. These approaches very much depend on the investment that we are prepared to make in time and skill, together with what the long-term strategy of the company is. The simplest approach is to buy an off-the-shelf application solution. Amethyst is an example of such a system. It is focused on one small narrowly defined area but performs that task in a very capable way. This approach has the benefit of not requiring any skill or training within the company relating to the area of expert systems. In general, it also requires a minimal investment. One is able just to buy and start using the expert system. By using an expert system appropriate to the company's needs, it is possible to get a good understanding of how useful it is, and how many of the other issues we have discussed actually worked out. Figure 22.8 shows the initial options in respect of expert systems.

There are a number of expert system solutions available, but many companies are sceptical or worried about obtaining and using these packages. Hopefully the background given in this chapter illustrates that there is nothing to be worried about, and these programs are actually a very sensible investment. The biggest constraint is the availability of a package of this nature. The next stage and the most recommended approach, is to work with an existing expert system solution supplier. There are a number of companies who have expertise in developing expert system applications for specific areas. In general, these companies are more than happy to visit and discuss the operation of your company, your primary problem areas and how in general knowledge-based systems might be able to fit in. These supplier companies will also work with you to outline a provisional requirement that would be suitable for an expert system and provide a costing.

Choosing knowledge-based systems and deciding the appropriate application is a complex task that is very expert in itself. Unfortunately, the knowledge-based systems experts have not been able to write a KBS system to perform this task as yet. The approach of working with a supplier also has the advantage that you can control and minimize the amount of training needed by your own company, particularly in the

- Obtain an off-the-shelf solution
- Have a supplier customize an existing system
- Have supplier develop a custom solution
- Develop your own complete system

Fig. 22.8 First options.

expertise of getting the first application off the ground. A large number of big firms have used the approach of contracting for the development of a first expert system. They use the development and installation of this with an experienced professional company, as a means of learning how best to develop such systems and understand them. After the first system they are then in a position to develop their own expert systems. Usually the first steps are in improving and evolving a delivered, professionally developed system. The largest disadvantage of this approach is that it can be relatively costly in external spending.

Expert system developers are usually very skilled, highly trained people. As a result, they are relatively expensive to contract. On the other hand, this approach gives you the highest probability that your investment will be worthwhile and minimizes the risk. Trying to do the task with minimal investment could lead to a lack of skill, and that would endanger the whole investment. The final approach is to develop the expert system by creating an expert system group within the company. There are a number of guidline booklets such as Anon (1994) and Wiig (1995). These books outline in detail the first steps you should take in identifying a possible application, in assembling a team, and the proper training that they will need. The following points summarizes the key elements:

- Knowledge-based systems are a form of computer software.
- Good computer programmers will have no trouble picking up the basic expert system programming techniques.
- Others who are reasonably competent computer users will also be successful in learning knowledge-based systems.

People who are not computer-literate, in general, have a very difficult time. This is not so much from the difficulty of using the software packages, but from the importance of being able to structure and organize a software solution. There are a large number of graduates available who have basic expert system training, and hiring one of them is sometimes the best way forward. A company will also need an expert system shell. These can be relatively inexpensive, costing just £1000 on a PC. More sophisticated applications with experienced staff can be developed using large PC or workstation software costing several thousand pounds.

A very common mistake is to assume that only the cost of the expert system shell is needed to get started. The company has to be prepared to dedicate significant personnel resources. These must come from two groups of people. The software development team, which has to be able to dedicate considerable time in order to build the KBS system, plus the expert, who must also dedicate his time in contributing his specialist knowledge. A typical small system, performing one or two tasks, will require 1–3 months to develop. A medium-sized expert system, tackling a significant but well-contained problem, may require 6–8 months to develop. Large systems well integrated with the existing plant often require in excess of one year. It will be recognized that the price of the expert system tool is almost irrelevant compared to this large labour cost. Figure 22.9 outlines the typical costs associated with an expert system project.

The first application should be selected by a combination of minimizing risk and ensuring payback. It should address an important business problem, and one that is well-contained, and well-defined. It should be possible to write on one sheet of paper the basic data inputs, the output required, and the basic purpose of the system. The problem should use well-defined and easily available knowledge. Problems that require large amounts of common sense or human background are extremely difficult if not impossible. This is one of the reasons why developing a KBS system to select expert system applications is difficult. The Amethyst problem is, however, an excellent example of a good application. The data input is clearly defined from the vibration database, and the expertise necessary requires no common sense or world knowledge. In addition the output is a well-defined report.

One test as to whether the expert system contains no common sense is the telephone test. If experts can diagnose the problem over the telephone, then you can be certain that they are not using any knowledge other than that which they are being asked about. It also means they are able to issue a request for the data they need in a clear and understandable way. With regard to the size of the problem, if the expert is able to solve a problem almost instantly, then there will be little time saved from the implementation of a KBS system. At the other extreme, if the expert takes 1–2 hours to solve the problem, then it is far too complex. Many of the best problems are ones which require an expert 2–10 minutes to solve. It is worth remembering that some of the benefit is by providing for a relatively straightforward activity in computer software 24 hours a day. It is also very important to select a problem that will have a long-term application for the company.

Many times companies have addressed particular areas only to find that on the next reorganization, or the next introduction of new technology, the problem goes away. Diagnosing the oldest section of machinery can become a useless exercise when it is replaced by more modern

```
         Expert system shell    5%
   Rulebase development time    30%
          Integration time     30%
              Testing time     25%
  Documentation and training    10%
```

Fig. 22.9 Costs of an expert system project.

equipment. On the other hand, developing the expert system to assist with the introduction of new equipment is much more useful. There have been a number of studies in the UK to look at the amount of investment required for knowledge-based systems. Many of these successful knowledge-based systems in manufacturing have required a total investment of about £20 000 to £25 000. These systems generally require 2–3 person-months of work to develop, test and install. It is unrealistic to expect knowledge-based systems to be developed for less than a few thousand pounds unless they are very small, simple applications developed by experienced staff. Large investments of over £50 000 should be avoided on a first application, unless this is contracted out to an experienced development team in an area that is already well proven.

There are a number of organizations within the UK able to provide help and assistance in learning more about knowledge-based systems and how to apply them to your own business. The primary professional body within the UK is the British Computer Society Specialist Group on Expert Systems. This group have an information pack available to new members which provides further background information on knowledge-based systems. In addition, they hold regular regional meetings to bring together companies interested in expert system activities, and they also hold the annual UK conference on expert systems and their applications. Normally, this conference is held in December in Cambridge. Quite recently, the DTI has started a programme specifically targeted at assisting companies with using expert systems in manufacturing and called Manufacturing Intelligence. They have produced the booklet detailed in the references below which is designed to assist managers or companies in understanding what expert systems are, what opportunities may exist in their business and how to get further information. The booklet is available on request.

Within the UK the suppliers of expert system software and system development capabilities have formed a professional organization known as the AI Suppliers Association. Membership in this organization implies a high level of skill and professionalism. There is a directory of all the current members available from the AI Suppliers Association. The existing suppliers of knowledge-based systems applications are probably the best single source of help and assistance. In each country there are a number of suppliers, each with experience in a particular area. By examining the list of projects completed to date by each company, it is possible to ascertain which one has the most experience in your particular area. There are a number of tool suppliers which have very general expertise in a number of application areas. The tool suppliers will have the broadest knowledge and the access to the widest number of completed projects. However it is very important to distinguish which projects have been completed using their tool, versus the projects in which the people you are dealing with have actual experience. The application suppliers will have much more in-depth knowledge, but in a more focused area. The number of application specific suppliers is steadily growing, albeit at a slow rate. The author at Intelligent Applications can provide further contacts to all of these organizations.

22.5 CONCLUSIONS

In this chapter an overview of knowledge-based systems has been presented. We discussed why there is a need to address critical skill availability issues, the basic idea behind them as well as some of the more advanced ideas. Knowledge-based systems are now becoming a standard part of the maintenance world. A few years ago there were few expert systems applications in maintenance, and it was a great risk for a manager to make the investment. At present, however, the technology is well proven and the risks are small. The number of applications is still small but steadily growing. As the technology becomes more application specific and better integrated with existing working practices, the rate of new developments will grow rapidly. Figure 22.10 outlines an approach to justifying the cost of expert system development.

Costs
- Identify all cost factors.
- Identify the amounts of money that will be expended for each cost factor.
- Identify when (by month, quarter of year) the sums will be expended.
- Verify with other sources that the numbers are reasonable.

Benefits
- Identify expected benefits (direct, intermediate and implicit).
- Work with managers and representations from departments concerned to identify how the indirect benefits may translate into direct benefits.
- Generate ranges of values for benefits (low, reasonable, high) by month, quarter or year, and indicate specific scenarios that may apply.
- Verify with managers that estimates are reasonable.

Cost-benefit evaluation
- Generate cost benefit measures for most likely case using the firm's normal measurements, for example, net present value (NPV), internal rate of return (IRR or ROI), pay-back period, etc.

Fig. 22.10 Approach to cost justification.

Over the next few years we will see an explosion of expert system applications, as the more forward-looking companies understand the commercial benefit they will receive, and as more application oriented packages become available. Knowledge-based systems are now established and the stage of spreading it throughout industry has begun. We have discussed how knowledge-based systems can be well integrated with existing systems, and how they will also play a critical role in maintenance. KBS are now a well-proven technology, as illustrated by Intelligent Applications receiving the Queen's Award for Technology in the development of knowledge-based systems for maintenance. The future is clearly coming and it is better to get involved now.

22.6   REFERENCES

Anon (1994) *Manufacturing Intelligence: A Decision Maker's Briefing*, Department of Trade and Industry (DTI), London, UK.
Wiig, K. (1995) *Expert Systems: A Manager's Guide*, Management Development Series 28, International Labour Office (ILO), Geneva, Switzerland.

CHAPTER 23

Future developments in condition monitoring techniques and systems

A. Davies, P.W. Prickett and R.I. Grosvenor

Systems Division, School of Engineering,

University of Wales Cardiff (UWC), PO Box 688,

Queens Buildings, Newport Road, Cardiff, CF2 3TE, UK

23.1   INTRODUCTION

In reviewing the likely future developments within the field of condition monitoring (CM), one cannot but be immediately struck by:

Firstly the wide diversity of work being undertaken in this area, and secondly, the intensity of the activity that is ongoing.

It is remarkable, and significant, that just these two words, 'condition monitoring', have the ability to encompass so large a range of human endeavour, not only in maintenance, but in all aspects of engineering. The reason why all this effort is taking place is not difficult to fathom, and can be explained by:

The movement away, in recent years, of industrial processing and manufacturing systems, from manual operation to full automation.

This has come about as a consequence of changes in the economics of industrial processing, together with advances in electronics, computer systems and software engineering. As we have seen in the foregoing chapters, the field of condition monitoring has embraced all these changes and developed rapidly as a consequence of them, by the

Handbook of Condition Monitoring
Edited by A. Davies
Published in 1998 by Chapman & Hall, London. ISBN 0 412 61320 4.

practical implementation of established and new techniques in cost-effective commercial systems. Naturally, each sector within the area of condition monitoring has its own research and development programme, many of which could probably occupy a chapter in their own right. Consequently, this chapter will limit itself to illustrating the advances in real-time, online, monitoring systems and the enabling aspects of current information technology (IT).

To work successfully, an automated processing or manufacturing plant must use information technology to control each item of equipment, and thereby ensure that the whole system performs as a single entity when in operation.

Simply defined, information technology is:

> The way in which we collect, store, process and use information (Zorkoczy, 1982).

Thus in the context of condition monitoring, the information technology used can be described as providing the ability to:

> Accurately sense information, encode it, transmit it, receive it, store it, interpret it, decide actions upon it, and issue appropriate system commands as a consequence of it.

Ideally, all the above activities would be done automatically, on the basis of Real Time Actual Condition Knowledge (RTACK) and without human intervention. This faculty, which implies the use of real-time computer processing, rapid data interpretation and automatic logical decision-making, allows the information technology aspect of condition monitoring to supply the 'intelligence' for the operation of unmanned industrial systems.

In the twenty-first century, it is probable that one of the main characteristics of a successful industrial company will be:

> The extensive use of highly automated and flexible production equipment (Tuenschel, 1987).

This implies that the company concerned will have achieved a high level of competence in applying the techniques of information technology to industrial operations. In such a concern economic production will only be feasible if:

- the automatic production system employed by the firm is flexible, highly reliable, easily maintainable and thus capable of achieving an excellent level of productive availability;
- the quality of the manufactured output approaches perfection and as a consequence defect levels tend towards zero;
- the company is regarded as a 'holistic' system and accordingly, structured to provide maximum logistic support to the industrial operations conducted by the firm.

In this scenario maintenance has to evolve into an efficient organization, which utilizes the capabilities of information technology, to service extremely complicated high-technology industrial systems. Under these circumstances the successful economic operation of an automated industrial plant requires 'failure-free' operation. Thus 'failure prevention' becomes the prime requirement of maintenance activities, superseding both the 'time-based servicing' and 'repair on failure' actions currently used by many organizations. Movement in this direction will become inevitable, despite the cost involved, as increased demands are made on maintenance to provide seven-day, twenty-four hour use of industrial plant and equipment.

Currently, this trend is most pronounced in the process, utility and offshore industries, where the rapid development of automatic computer-based condition monitoring, diagnostics, and maintenance management has taken place. Thus, the utilization of available information technology, in terms of both the hardware and the software, has been shown by proven applications, to improve the effectiveness of traditional condition monitoring techniques. Hence, the majority of current research, development and application efforts are all focused in the information technology area. To give a flavour of the work being done, some of the major aspects will now be outlined in the sections which follow, together with a few speculative comments on the future of condition monitoring.

23.2 DATA ACQUISITION SYSTEMS

The development of new sensors and sensing arrangements is a very active area of engineering research. In many respects it is the key to effective condition monitoring, for without the ability to acquire accurate information, the quantitative monitoring of process machinery would be an impossible task. As the economics of industrial production are increasingly moving equipment in the direction of full automation, the efficient gathering and processing of plant information becomes an essential requirement of any control system. Accurate real-time knowledge of what is happening to both the equipment and the process, is necessary to optimize the overall performance of the machine and hence maximize plant utilization, productivity and operating life.

It is essential to remember, therefore, that these two basic information channels, one of which controls the automated process and the other monitoring the health of the system, need to interact continuously in the machine controller, so that suitable modifying action may be taken when non-optimal or fault conditions are detected. Data acquisition can thus be seen to be a key requirement in this type of technology, and the sensor systems necessary to obtain useful data are self-evidently critical elements in the makeup of this type of automated plant. In recent years great strides have been made in this area, and much of the sensing aspect of this technology is already available. With ongoing research being conducted into reducing sensor size and cost, while at the same time improving capability, resilience and reliability (Weck *et al.*, 1991).

Recent development in integrated circuit technology and optoelec-tronics have now progressed to the point where the assimilation of various sensor functions in a single unit has been achieved. In addition, information from several sources may also be combined within the package to form what might be termed an 'intelligent' sensor. Such sensors can contain some data evaluation capability, including that required to improve the signal to noise ratio, filter and refine the data acquired to essential information, or to combine several input signals. Currently, the technology exists to manufacture prototypes of these sensors, and via the use of microelectronics and microminiaturization, it is also possible to incorporate them in an integrated sensing system. However, it is important to note that further research is being conducted into the potential and characteristics of 'intelligent' sensors, with the main areas of effort being as follows.

- The determination of the precise circumstances under which the operation to transform the generated signal should be integrated within the sensor. This may only be justified if an output feature such as the autocorrelation function (AF), or Fast Fourier Transform (FFT) of a signal, is found to be useful, in for example a sensor-based machine tool management system, and where there is a cost or efficiency benefit to be achieved from 'local' processing (Towill, 1985).
- The method by which the identification, selection, and combination of input signals is organized into a meaningful quantity which consti-tutes the output of such an 'intelligent' sensor system. Typical multi-dimensional output features could be pictorial data, surface information or health index values depending on the application. Very Large Scale Integration (VLSI) miniaturization techniques now permit the flexible coupling of input information, so that various combina-tions of signal may be obtained with the goal of achieving, for example, indirect measurement (Ohba, 1992).

The increasing use of computer-controlled equipment in virtually every sector of manufacturing industry, and the associated demand for its high-productive availability, is now serving to emphasize the need for, and importance of, real-time actual condition knowledge. This information, the effective provision of which, can only be undertaken by 'intelligent' sensing systems, is now being correctly recognized as an absolutely essential requirement in the design of automated production systems (Weck *et al.*, 1991). Because the operating environment is frequently difficult, these sensing systems need to have the following characteristics:

They should be fast, accurate and self correcting in operation. Simple in design if possible and rugged in construction. They should also be non-intrusive, preferably non-contacting, highly reliable and induce no increase in system complexity (Grosvenor and Martin, 1985).

Optoelectronic measuring methods meet these requirements fairly well, and although they do have some drawbacks, currently, they seem to offer the most promising research and development avenue for industrial applications. Current design specifications in respect of these sensors and their use on automated systems call for them to be able to:

- reduce product defect frequency by continuous surveillance to less than one in a million;
- eliminate the possibility of random process error induced by uncontrollable environmental effects;
- supply information concerning the product and process for preventive fault avoidance, quality assurance, control and diagnostics.

It is important to note, that in the case of automated systems, the final point above implies the use of a computer-based adaptive control system, which in all probability would utilize some form of 'intelligent' software. Optoelectronic sensors if used in this context have the advantage of being fast, flexible and non-destructive during the measurement process. However, few have so far been applied, although many different prototypes have been developed and tested successfully. A lack of knowledge in respect of their characteristics is mostly to blame for this, along with only a few documented applications. Where they have been used, they appear to provide good performance being impervious to heat, dirt and vibration (Weck *et al.*, 1991).

Figure 23.1 illustrates the main groupings of optoelectronic measuring techniques, with some, such as the light barrier and reflex sensors, being well-known in industry. These have in the past been used extensively for positioning, controlling and counting tasks, which make full use of their speed, rugged construction and low cost. The further development of lasers, optics, electronic signal processing and filtering have now allowed some of the other techniques shown to be implemented via robust, and in some cases miniaturized sensors, although cost, and thus the number of applications, are limited by sensor complexity. This in turn is dependent upon the number of dimensions to be measured in any application, and as we saw in the case of visual inspection, conspires to hold back implementation.

Triangulation is one of the most mature of the optoelectronic sensing methods, with the basic principle being relatively easy to apply in one, two, or three dimensions. Thus straightforward length, space and contour measurements, which form the bulk of industrial sensing applica-

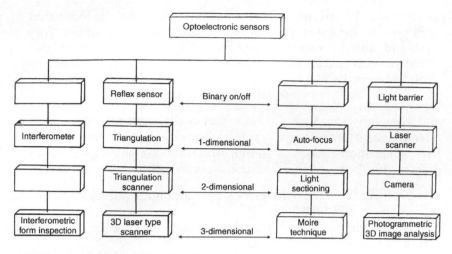

Fig. 23.1 Optoelectronic sensing techniques.

tions, can all be carried out by using optoelectronic sensors applying this technique. In addition, speed can be evaluated via laser-based sensors, which interpret the interference phenomena in the illumination of a moving object, resulting from the doppler shift of the light wave length. A disadvantage of optoelectronic sensors, particularly where they are used for tight tolerance measurement, is that the item concerned may have to be cleaned prior to gauging. Typical applications in which these sensors have been used are automatic tube forming, co-ordinate measuring machines and form inspection for car bodywork shells (Weck *et al.*, 1991).

Although there are considerable developments taking place in other aspects of sensor technology, optoelectronics appears to hold out the most promise for further advances towards the 'ideal' RTACK sensor. A reduction of disturbance in the measuring technique, pre- and post-processing of the measurement data, plus information transfer, are all current areas of optoelectronic sensor research. Colour and surface reflectance are additional problems under investigation, as the surface structure of the item to be inspected has a major influence on the measurement process. The use of fibreoptics with these sensors has improved their position in this respect via high-speed, disturbance-free and accurate light guidance. Application examples of this technology include, the inspection of machined threads and surfaces for accuracy, plus wear in the case of cutting tools.

A further advantage which optoelectronic sensors have over other sensing systems is their flexibility to help automate industrial processes. They can, for example, be used with robotic systems to provide fully automated handling and measuring arrangements in manufacturing assembly cells. In addition, the development of high-resolution digital cameras allows the inspection and measurement of workpart features and geometry. In some cases high-speed inspection is possible, making these systems compatible with production-line technology. The flexibility and integration capability of camera systems is currently being investigated and hopefully further developed to provide three-dimensional contour measurement of components (Weck *et al.*, 1991).

23.3 CONDITION EVALUATION TECHNIQUES

The accurate interpretation of the data obtained via sensor systems such as those outlined above is obviously a key requirement in the advancement of condition monitoring. Accordingly, considerable research is being undertaken in the area of evaluation techniques to provide both current, and precise knowledge of plant or system condition. This knowledge is essential for the following reasons:

1. to identify trends which lead to the failure of plant or equipment;
2. to diagnose prior to failure the cause of degenerative trends in the plant or equipment;
3. to permit corrective action to be taken prior to failure such that plant or equipment breakdown does not take place.

Thus, for condition-based maintenance to operate successfully, it may be seen that accurate data interpretation is vital. One possible solution to this problem is to identify, store and recall all known failure trends, along with their causes and proven solutions. This information would then need to be compared with current surveillance data to see if any failure trend was present. The practical implementation of this approach is restricted to equipment with a fairly well-defined failure domain, in terms of both the cause of a breakdown and the trend of its symptoms. In addition, the 'matching' of data to known failures in such a system, would also require an element of 'learning' to overcome 'fuzzy' situations, where there are two or more possible interpretations of the data.
  Such a system can be defined as:

  A tool which has the capability to understand problem specific knowledge and to use this domain information intelligently to suggest alternate paths of action (Kumara *et al.*, 1988).

Coding the knowledge within a computerized expert system is an obvious and cost-effective solution. However, there are problems to this approach which include:

- ensuring that all possible failure/repair modes are covered by the expert system, or failing that, giving it the ability to be updated on new failure/repair modes either manually, or by some form of automatic learning;

- ensuring that false diagnosis, as a consequence of poor data matching, is eliminated by virtue of the fault inference technique employed.

Both these areas are the subject of ongoing research in the use of Artificial Intelligence (AI) and expert systems as applied within condition monitoring and fault diagnosis. The use of simulation modelling, as shown in Figure 23.2, together with differing expert system shells, to design or customize a knowledge base/inference engine for particular applications, is one avenue in this research area which is currently under investigation (Davies, 1992). The great value of simulation in this context is that:

- It can exercise the plant, equipment or System Under Test (SUT) model through a range of practical operating conditions and thereby drive the AI/expert fault-finding system via typical problem output symptoms.
- By utilizing the AI/expert system output, via a feedback loop and adaptive control, the simulation model can assess the accuracy of the knowledge-based system's answers to questions involving individual SUT element failure, or poor SUT overall performance.
- Through recursive use of the simulation, the AI/expert system's knowledge base can be improved, SUT stability regions mapped, inference engine techniques optimized and SUT parameter interaction understood.

The assessment and development of existing and new inference methods for speed and efficiency in use are, as indicated above, particularly important for the future application of AI/expert condition monitoring systems. As an example of a typical inference technique, Figure 23.3 illustrates the 'nearest neighbour' method. This employs a pattern recognition methodology to assess changes in key features for real-time applications.

In operation the technique designates a directional vector of unit length to the recorded symptom data. This vector is then compared to previously known and stored failure vectors in a fault dictionary. The failure vector with the smallest metric value is then identified as the nearest recorded fault to the observed symptoms. It should also be noted that by using this approach, the fault dictionary can be updated or amended by the user in the light of experience. Thus, if a new fault with new symptoms is detected, its details can be recorded along with the repair action required without alteration to the inference logic.

The nearest neighbour technique is one of a family of cluster analysis methods which like most algorithmic inference arrangements makes use of the fact that the majority of performance data is quantitative, and

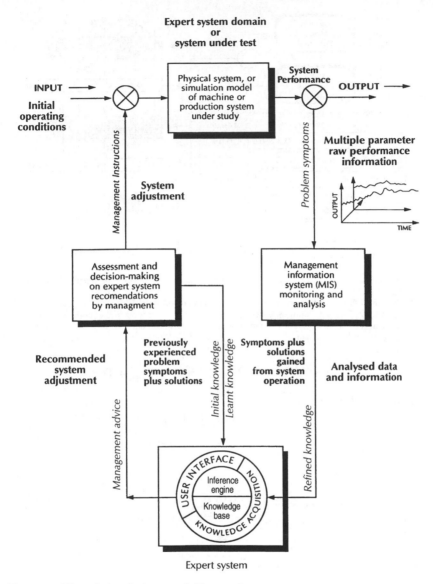

Fig. 23.2 Use of simulation modelling and expert systems.

carries information regarding change, its magnitude and reason for occurrence. Detection is conducted by scanning for abnormal values and numeric clues in respect of cause. One obvious advantage of the method is that there is no dependence on the repeatability of data relating to a particular problem for correct pattern recognition to occur, although the correct coding of information relating to specific problems is necessary.

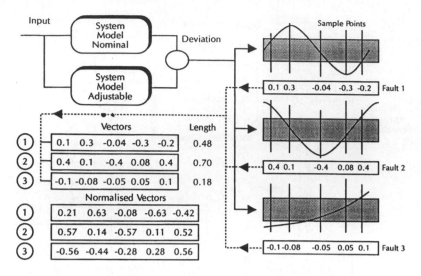

If s = [s_1, s_2, s_3, ...s_n]^T, where 'T' indicates a column vector then the length of s is:- |s| = $\sqrt{(s_1^2 + s_2^2 + s_2^3 + \ldots\ldots + s_n^2)}$ = L say. So for the fault vectors above L_1 = 0.48, L_2 = 0.70 and L_3 = 0.18. Then the signature is modified to its normalised form using s* = [s_1/L, s_2/L, s_3/L, ...s_n/L]^T. The procedure of matching the fault vector with one of the stored signatures is different from a quantised scheme. The method used is to use a 'nearness' measure [8]. Nearness can be measured by the distance between the vectors. The distance between two vectors a and b is given by: -

$$d\,a,b) = \sqrt{(a_1 - b_1)^2 + (a_2 - b_2)^2 + (a_3 - b_3)^2 + \ldots\ldots + (a_n - b_n)^2]}$$

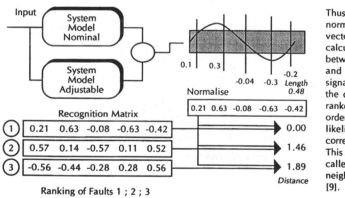

Ranking of Faults 1 ; 2 ; 3

Thus the scheme is to normalise the fault vector and then calculate the distance between this vector and the stored signatures. Finally the distances can be ranked in ascending order which gives the likelihood of the corresponding faults. This scheme is often called the nearest neighbour method [9].

Fig. 23.3 Nearest neighbour inference method (reproduced with permission from Williams *et al.*, 1994).

In other inference methods, tolerance banding of the data patterns may be necessary to ease recognition difficulties provided the base fault signatures are sufficiently diverse (Williams, 1985; Williams *et al.*, 1994).

Research into alternative inference methods for automatic fault diagnosis in machinery can involve predicate logic and neural networks. An example of the latter technique applied to signal identification being set out in Chapter 8. These techniques are, however, limited to specific applications in well-defined problem domains. A more generic paradigm, which can mimic accurately the knowledge and ability of a human expert, to fault-find in complex machinery, is simply not yet available. Part of the problem is that human knowledge is complex, unstructured and ill-formulated, with a percentage of expertise operating at the unconscious or intuitive level. This it is impossible to identify and code satisfactorily, with the result that current AI/expert systems accept task specific domain data from the user and select/use specific inference logic to come up with an appropriate answer.

23.4 KNOWLEDGE-BASED CONTROL METHODS

In the discussion so far, little account has been taken of how fault prediction, diagnostic reasoning and problem resolution actually proceeds in practice. Indeed it is important to note that this process can be sub-divided into the following stages:

- **detection:** the awareness that something has gone wrong with the operation process or machine under surveillance;
- **direction:** effectively the orientation and navigation of the search process to the fault area;
- **location:** the isolation and identification of the specific fault within the area in question;
- **solution:** the provision of a procedure by which the fault may be rectified and the operation, process or machine restored to the nominal functioning state.

Reasoning or inference in many current expert systems applied to condition monitoring spans 'direction' and 'location' only, with 'detection' being part of a human or sensor-based monitoring scheme, and 'solution' being left to the operator or perhaps forming part of the knowledge base. The nearest neighbour method for example, can detect a fault in the recognition matrix but is not initially directed to any specific area within it. Where the recognition matrix is small, as in Figure 23.3, its complete evaluation to determine the fault, poses little problem to a modern high-speed computer control system. In a complex real-time situation, however, where the recognition matrix for the plant may be extensive, a significant time delay may occur in fault identification and consequentially in any corrective control action.

Thus, research into inference techniques is quite clearly an important topic for integrated real-time automatic control and monitoring systems, along with that being undertaken on sensor development and knowledge-based solutions. In the case of the nearest neighbour method, for example, an addition to the technique which involves the use of Pareto analysis can identify significant elements in a large matrix very rapidly, and subsequently direct the search to the most likely fault area within the matrix (Williams, 1985). Alternatively, where time is less critical, a multi-layered approach can be used. This incorporates a number of different directing mechanisms via which the fault is identified. An example of this second arrangement is shown in Figure 23.4 which outlines the use of the nearest neighbour method in conjunction with fault tree logic and a rule-based fault location method (Thomas, 1993).

The multi-layered reasoning/fault location mechanism shown in the diagram, was devised for use on a high-intensity hydraulic press, Figure 23.5 illustrating the associated software user interface specification. In addition to these techniques, many other reasoning and directing methods can be used for diagnostic fault location which operate in a similar manner to those outlined above. Hence, extensive research interest is being shown into the use of Influence Diagrams (Agogino *et al.*, 1988), Petri-Nets (Abu Bakar, 1993) and Fuzzy Logic (Zeng and Wang, 1991), for use in knowledge-based condition monitoring. The necessity of researching and developing these alternative methods is linked to the requirement for speed, accuracy and efficiency of diagnostic operation in real-time applications. Accordingly, and to illustrate some of the practical difficulties involved in the design of such software, an example of a scheme involving the Petri-Net approach to machine tool diagnostics will now be outlined.

The overall concept is presented in Figure 23.6, which illustrates the interaction envisaged between a simulator and an expert Machine Tool Breakdown Diagnostic (MTBD) system. In this approach, generic simulation software is used to model the operation of a machine-tool system, via the use of input data, which configures the simulator to represent the target machine. Note that this approach is not restricted to machine-tool systems and that, within reason, any item of industrial plant or equipment could be similarly modelled. In this case when in operation, the simulation software produces typical output failure symptoms for subsequent diagnostic analysis by the expert MTBD system. This allows the expert system software to be tested for accuracy and efficiency before application on the target machine.

As shown in the figure, in part, the simulator is configured via the use of actual failure distribution data which relates to the major functional subsystems in the target machine. Ultimately, for this research system, it is intended that each representational level in the modelling hierarchy will become more detailed as the simulation is refined. This is to provide the ability to more closely reproduce, in a generic sense, any machine-tool system. Ideally, the simulation software used ought to be capable of modelling any item of industrial plant. However, there are practical limitations to most simulation software and research into a fully generic

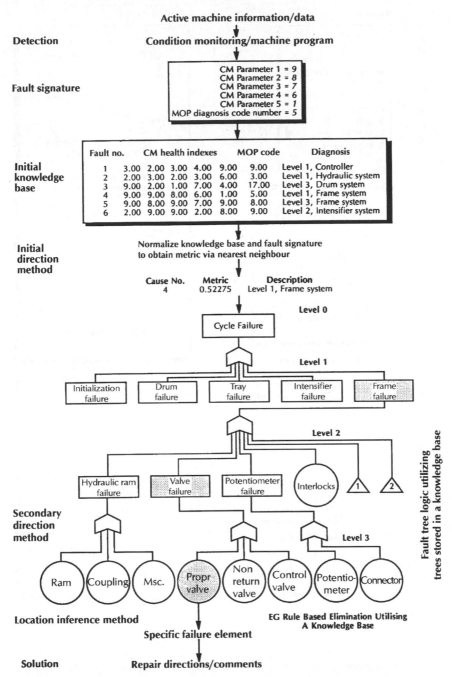

Fig. 23.4 Multi-layered fault location reasoning.

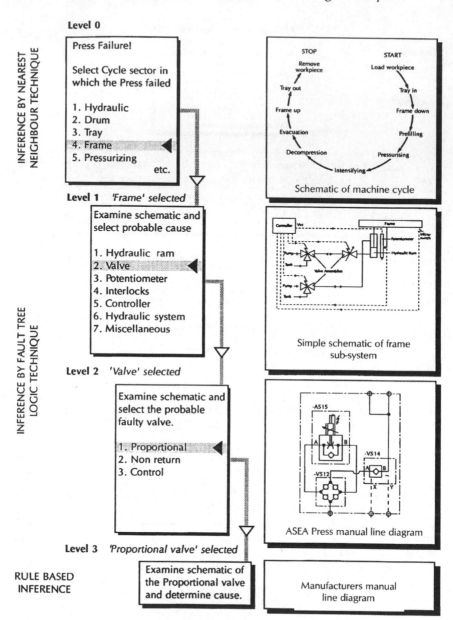

Fig. 23.5 The user interface for a multi-layer diagnostic system.

system is ongoing. The current simulator outlined here employs both plant and context knowledge to mimic the operation of the target machine, and is 'driven' in the same way as a machining system, i.e. via the use of Computer Numerical Control (CNC) part programs.

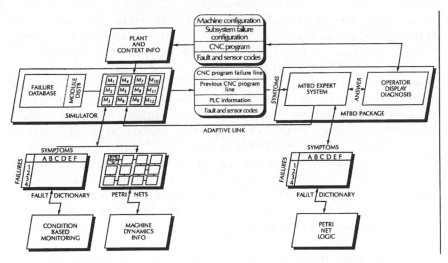

Fig. 23.6 Simplified linked layer representation of MTBD and simulator.

Each identified subsystem is represented by a module in the simulator, and these are based on a Petri-Net representation of the systems which make up the target machine. At present, the current development software permits a module to be failed according to the Normal, Exponential or Weibull distributions. However, there is no restriction on adding additional discrete or continuous distributions to the simulator if this is required. In operation, each module is assigned a failure time based on the distributions held in a database. This documents failures over a range of similar machines, so that each module can interact in the simulator to mimic the observed failures of the target machine. Note that the modular subdivision outlined here can apply equally to any item of plant or equipment and is not restricted to machine-tool systems.

Thus as the simulation proceeds, a set of characteristics similar to those of the target machine is provided via this mechanism, for the expert system to analyse. The simulator is driven in time by use of a CNC part program, in exactly the same way as a real machine tool would machine a component, or other item of plant perform its duty cycle, by, in this case executing the program line by line. At present, this is effected by assigning an execution time to each line of the CNC program. In future, however, when generic Petri-Net, process and machine dynamics are incorporated, the simulator will be capable of mimicking any machine tool exactly, thereby producing a range of output data relating to the machine system and process operation as machining proceeds. This data may be used on failure as symptoms for fault location, or in adaptive mode to detect and adjust to degradation failures (Boey, 1993).

The expert system software is designed to link with the simulator as shown in Figure 23.6 and fitted to the target machine after customization. In operation the expert system continuously interrogates either the simulation during customization, or the machine tool in use, thereby providing system monitoring and failure diagnostics. The inputs to the software were carefully identified during its design with the objective of using all the available information for failure diagnostics, while holding to a minimum the number of additional sensors required for fault location. In the current software a rule-based approach is used to identify the failures and, at present, the CNC part program plus the Programmable Logic Controller (PLC) signals are used to identify the faulty subsystem. Petri-Net logic and sensor information is then used to identify the exact failure within the subsystem (Abu Bakar, 1993).

The determination of failure is undertaken in three stages. Initially, all modules are suspect, and as the CNC program is executed, the PLC signals are monitored to identify, on failure, the subsystem concerned. PLC signals are generated within the simulator, in exactly the same way as they would be by the machine control system in actual operation; it being assumed that the PLC has been programmed using the Petri-Net approach. In the second stage of fault location, the token position within the failed system is used to narrow the search area and to prompt as the third stage 'chaining' within the expert system for precise fault location. For each state of the subsystem described by the Petri-Nets, diagnostic knowledge is stored in the software as a fault dictionary. The Petri-Net approach enables the implementation of systematic, fast-search diagnostic logic through an ordered knowledge base (Gopal, 1992).

In future work on this scheme, it is envisaged that 'hard' faults will be identified via the use of Petri-Net logic. This would be extracted from generalized net formulation software, capable of analysing both machine and process states. To detect 'soft' faults, the monitoring of existing machine signals is required. These would be identified via the Petri-Nets, and used whenever possible. This reduces the need for extra system sensors, lowers complexity and minimizes cost. The current software, which implements this scheme for a hypothetical Flexible Manufacturing Cell (FMC), utilizes the KAPPA shell, with a simulator written in C (Abu Bakar (1993) and Boey (1993)). It should be noted that although this example is based on a machine tool system, the approach is equally applicable to most items of industrial plant.

23.5   AUTOMATED SYSTEM IMPLEMENTATION

The final product of the research outlined above, into sensor systems, condition evaluation techniques and knowledge-based control has to be its practical implementation in a flexible, generically applicable and inexpensive data acquisition, monitoring and control system for industrial plant and equipment. Such a system is currently under development in the School of Engineering at the University of Wales Cardiff (UWC), as part of a European Union Project (BREU-CT91-0463). The project is funded under the BRITE initiative and entitled 'Machine Management for Increasing Reliability Availability and Maintainability – MIRAM'. Other partners involved in the project include the German machine-tool manufacturer Heckler & Koch GmbH whose machine tools, located in various factories in Germany, are being used as test beds (Drake *et al.*, 1995).

The aim of the project is to produce a machine surveillance and management system that will provide rapid fault diagnosis, and enable the implementation of a predictive maintenance strategy via the provision of condition monitoring. The design of the system has been undertaken by adopting a 'top-down' approach, wherein the design methodology commenced with a technical audit of the machine to be monitored. This was necessary to establish the presence and functionality of all the subsystems and components therein. Each subsystem was then subjected to a Failure Modes and Effects Analysis (FMEA); from this, the tests and corresponding signals within the machine which permit failure detection were determined (Jennings and Whittleton, 1993). The machine's subsystems were identified and examined in turn during this research (Whittleton and Jennings, 1992), as was the available field failure data (Sandner and Berges, 1992).

In design, the monitoring and control system has been arranged to be as flexible and as generally applicable as possible. The overall hardware architecture as shown in Figure 23.7 consisting of an IBM PC AT computer and a signal conditioning rack (Drake, *et al.*, 1995). This rack houses the interface arrangement between the monitored machine signals and the PC system, which in turn is linked with a central maintenance management computer via a radio modem. The complete cost of the system, excluding labour, amounts to less than £5.5k, and thus it meets one of the main objectives listed above, that is to be inexpensive. A relational database management system is used to control the monitoring arrangement, with the user specifying the surveillance to be performed. Full details of the system, including the database and software, which uses existing and specially mounted sensors for machine monitoring, are given in (Drake *et al.*, 1995).

The field of automatic machine tool monitoring and control, in common with many other industrial process applications, is an extremely complex research area, and the system described above represents at best a working testbed for data acquisition and analysis (Prickett and Grosvenor, 1995). As an example of the complexity involved, a brief outline will now be given indicating the problem areas which remain, and where possible, parallels will be drawn with other industrial

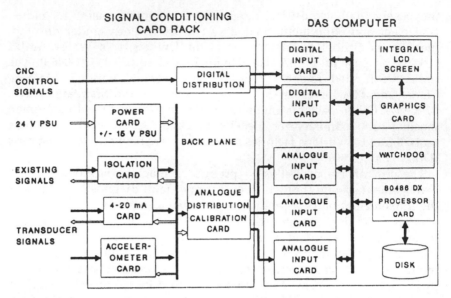

Fig. 23.7 Data acquisition hardware (reproduced with permission from Drake *et al.*, 1995).

machine monitoring and control applications. One major difficulty is the integration, control and diagnostics of several different items of equipment which are linked to form a manufacturing cell or system. As these equipment configurations may well be unique, and the individual items supplied by different manufacturers, the integration of their control systems can be quite difficult (Weck *et al.*, 1991).

Similar situations to that above can arise in complex process plants, where the adaptation and testing of individual equipment control programmes may in particular be an obstacle to startup, along with possibly a high PLC failure rate. The restarting of individual machines in an integrated system after failure can also be a time-consuming problem. This requires research to be conducted into software which assists in the automatic and synchronized startup of all machines, from the point of programme stoppage, without error introduction to cause another failure. In a machining contest, strategies for recognizing tool wear and breakage also need to be further developed, along with those for adaptive control to optimize production. The development of process stability monitoring and product quality surveillance is also an area of research activity which requires further work, to permit the practical implementation of indirect industrial system control.

Establishing a correct monitoring scheme for items of industrial plant requires, as we have seen, the examination, evaluation and selection of sensor signals plus the appropriate analysis procedures. In addition, knowledge of process or machine characteristics is needed, along with suitable decision strategies, for in many applications the link between the available monitoring signals and process performance can be extremely complex or even unknown. Additional research is therefore required in this area to develop tools for the identification of correct process monitoring strategies which are suitable for particular items of equipment. The large number of potential failures associated with complex manufacturing or process plants also give rise to concern in respect of diagnostic procedures. Techniques which can handle straightforward situations may not be able to cope with large number of different fault causes in real time. Accordingly, further research is required in this area on rapid inference and knowledge-based fault location techniques.

23.6 CONCLUSIONS

The above discussion has centred around the development and usage of modern information technology systems, applied in a condition monitoring context. Because of the limitations of space, it has only touched upon what the author considers to be the essential and exciting research areas in the 'enabling' aspect of condition monitoring technology. It should be noted, however, that many research projects and interesting advances are taking place in a multitude of different aspects of condition monitoring. Some of these have been touched upon in the individual chapters of this book, and their complete documentation would require a completely separate volume. It is hoped, however, that the reader now has a better understanding of the diversity of techniques available in the area of condition monitoring, and appreciates that this field of study is a dynamic research area, which provides a direct contribution to the well-being of industry.

23.7 REFERENCES

Abu Bakar, S.B. (1993) The Development of an Expert System for Machine Tool Breakdown Diagnostics, MSc Dissertation, UWCC, Cardiff, UK.

Agogino, A.M., Srinvas, S. and Schneider, K.M. (1988) Multiple sensor expert system for diagnostic reasoning monitoring and control of mechanical systems, *Mechanical Systems Signal Processing*, 2(2), 165–85.

Boey, G.K.H. (1993) The Development of an Expert Simulator for Machining Processes, MSc Dissertation, UWCC, Cardiff, UK.

Davies, A. (1992) *Initial Progress Report on the Design of a Software Simulator for Reseach into Machine Tool Breakdown Diagnostics*, Technical Note EP 185, UWCC, ELSYM, Cardiff, UK.

Drake, P.R., Jennings, A.D., Grosvenor, R.I. and Whittleton, D. (1995) A data acquisition system for machine tool condition monitoring, *Quality and Reliability Engineering International*, **11**, 15–26.

Gopal, P.G. (1992) An Expert System for Machine Tool Breakdown Diagnostics using the KAPPA Shell, MSc Dissertation, UWCC, Cardiff, UK.

Grosvenor, R. and Martin, K.F. (1985) *Review of Sensors for Machine Tool Application*, UWIST Technote EP 141, Cardiff, UK.

Jennings, A.D. and Whittleton, D. (1993) *Identification of BA25 Machine Tool Signals*, MIRAM project report MIRAM-UWC-2.2-EXT-ADJ&DW-3-B, available via UWCC, Cardiff, UK.

Kumara, A., Kashyap, R.L. and Soyster, A.L. (1988) An introduction to artificial intelligence in expert systems, *Industrial Engineering and Management Press*, 3–9.

Ohba, R. (1992) *Intelligent Sensor Technology*, John Wiley & Sons, Chichester, UK.

Prickett, P.W. and Grosvenor, R.I. (1995) A Petri-Net based machine tool failure diagnosis system, *Journal of Quality in Maintenance Engineering*, **1**(3), 47–57.

Sandner, R. and Berges, C. (1992) *Reliability Analysis of Existing Machine Tools*, MIRAM project report MIRAM-IAO-2.3-INT-SA&BE-3-A, available via UWCC, Cardiff, UK.

Thomas, P.V. (1993) The Application of an IT Tool to a Business Problem, MSc Dissertation, UWCC, Cardiff, UK.

Towill, D.R. (1985) *Sensor Based Machine Tool Management Systems*, UWIST Technote EP 110, Cardiff, UK.

Tuenschel, L. (1987) The key role of maintenance in future manufacturing strategies, *Proceedings of the International Conference on Modern Concepts and Methods in Maintenance*, London, UK, 19–20 May.

Weck, M. *et al.* (1991) *Production Engineering – The Competitive Edge*, Butterworth-Heinemann, Oxford, UK.

Whittleton, D. and Jennings, A.D. (1992) *Data Acquisition Requirements*, MIRAM project report MIRAM-UWC-1.1-EXT-DW&ADJ-1-C, available via UWCC, Cardiff, UK.

Williams, J.H. (1985) *Transfer Function Techniques and Fault Location*, Research Studies Press, Letchworth, UK.

Williams, J.H., Davies, A. and Drake, P.R. (1994) *Condition Based Maintenance and Machine Diagnostics*, Chapman & Hall, London, UK.

Zeng, L. and Wang, H.P. (1991) Machine-fault classification: a fuzzy-set approach, *International Journal of Advanced Manufacturing Technology*, **6**, 83–94.

Zorkoczy, P. (1982) *Information Technology: An Introduction*, Pitman, Bath, UK.

Index